DARWIN'S UNFINISHED SYMPHONY

DARWIN'S UNFINISHED SYMPHONY

HOW CULTURE MADE THE HUMAN MIND

KEVIN N. LALAND

PRINCETON UNIVERSITY PRESS
Princeton & Oxford

Published by Princeton University Press,
41 William Street, Princeton, New Jersey 08540

In the United Kingdom: Princeton University Press,
6 Oxford Street, Woodstock, Oxfordshire OX20 1TR

press.princeton.edu

Jacket design by Michael Boland for thebolanddesignco.com.
Images courtesy of iStock

ISBN 978-0-691-15118-2

Library of Congress Control Number: 2016944396

British Library Cataloging-in-Publication Data is available

This book has been composed in Adobe Text Pro and Trade Gothic LT Std

Printed on acid-free paper. ∞

Printed in the United States of America

10 9 8 7 6 5 4 3 2 1

This book is dedicated to Henry Plotkin,
who started me off on this journey.

CONTENTS

FOREWORD

This book is the product of a collective endeavor. Although I am the sole author, I set out to portray the efforts of a team of researchers—the members of my research laboratory and other collaborators—who, over a period of 30 years, have shared the scientific challenge of trying to understand the evolution of culture. I hope to provide a compelling scientific account for the evolutionary origins of the human mind, our intelligence, language, and culture; and for our species' extraordinary technological and artistic achievements. More than that, however, this book sets out to capture something of the scientific process—to lay bare, in an honest way, our struggles, false starts, moments of insight and inspiration, and our triumphs and failures in a scientific journey of discovery. I present our story; that is, I introduce the members of the Laland lab, past and present, and depict our efforts to understand the tremendously exciting puzzle that comprises the evolutionary origins of human culture. I am no novelist and, although this book is written in a style designed to be accessible, it inevitably cannot possess the pace, thrills, or drama of fiction. I hope, nonetheless, that a little something of a detective story comes across, and that the reader experiences a modicum of excitement as they read how our experimental and theoretical findings provided the clues that fueled our investigation.

My first note of thanks must, of course, go to the researchers whose work is described in these pages. I have been privileged to work with some extraordinarily gifted individuals, and have constantly benefitted from the hard work, good ideas, clever experimentation, and ingenious theoretical work of countless undergraduates, Master's students, PhD students and postdoctoral researchers, as well as numerous collaborators both in my own and other institutions. These include Nicola Atton, Patrick Bateson, Neeltje Boogert, Robert Boyd, Culum Brown, Gillian Brown, Hannah Capon, Laura Chouinard-Thuly, Nicky Clayton, Becky Coe, Isabelle Coolen, Alice Cowie, Daniel Cownden, Lucy Crooks, Catharine Cross, Lewis Dean, Magnus Enquist, Kimmo

Eriksson, Cara Evans, Marcus Feldman, Laurel Fogarty, Jeff Galef, Stephano Ghirlanda, Paul Hart, Will Hoppitt, Ronan Kearney, Jeremy Kendal, Rachel Kendal, Jochen Kumm, Rob Lachlan, Hannah Lewis, Tim Lillicrap, Tom MacDonald, Anna Markula, Alex Mesoudi, Tom Morgan, Sean Myles, Ana Navarrete, Mike O'Brien, John Odling-Smee, Tom Pike, Henry Plotkin, Simon Reader, Luke Rendell, Steven Shapiro, Jonas Sjostrand, Ed Stanley, Sally Street, Pontus Strimling, Will Swaney, Bernard Thierry, Alex Thornton, Ignacio de la Torre, Natalie Uomini, Yfke van Bergen, Jack van Horn, Ashley Ward, Mike Webster, Andrew Whalen, Andrew Whiten, Clive Wilkins, and Kerry Williams. To the extent that we have contributed to a scientific understanding of the topics discussed, this book is their achievement every bit as much as mine.

Many people too have helped with the writing of the book. I would like to thank those who read the entire manuscript, one or more chapters, and/or provided helpful feedback or insights: Rob Boyd, Charlotte Brand, Alexis Breen, Gillian Brown, Nicky Clayton, Michael Corr, Daniel Cownden, Rachel Dale, Lewis Dean, Nathan Emery, Tecumseh Fitch, Ellen Garland, Tim Hubbard, Hilton Japyassú, Nicholas Jones, Murillo Pagnotta, Simon Kirby, Claire Laland, Sheina Lew-Levy, Elena Miu, Keelin Murray, Ana Navarrete, John Odling-Smee, James Ounsley, Luke Rendell, Peter Richerson, Christopher Ritter, Christian Rutz, Joseph Stubbersfield, Wataru Toyokawa, Camille Troisi, Stuart Watson, Andrew Whalen, and two anonymous external referees. Through their help, this book has been greatly improved, becoming both more scientifically accurate and more accessible to the general reader. Katherine Meacham also merits a special note of thanks for administrative support in numerous guises, from formatting, to editing notes, to compiling references, all of which were always conducted with extraordinary efficiency and attention to detail.

The idea of my writing this book was first devised as a graduate student at University College London, nearly thirty years ago. I was inspired on reading John Bonner's wonderful monograph *The Evolution of Culture in Animals* (1980, Princeton University Press). I loved the grand sweep and vision of Bonner's book, and was enraptured by the sheer scale of the question it addressed. However, an equally inspirational conversation with University of McMaster psychologist Jeff

Galef, doyenne of the field of animal social learning, helped me to set Bonner's contribution within the broader framework of the field that had Galef led so impressively for decades. With Jeff's help, I was able to recognize that, for all its merits, Bonner's book did not provide a thorough explanatory account of how human culture could have evolved from the social learning and tradition observed in other animals. That conversation with Jeff also brought home how a great deal of scientific work would be required before the mysteries underlying the evolution of culture could be unraveled. Bonner's visionary conception and Galef's demand for explanatory rigor combined to hatch the idea in my mind that perhaps one day I might rise to this particular challenge.

I would also like to thank Alison Kalett at Princeton University Press for commissioning this book, and pushing me to write it at least ten years before I felt I was ready, and also Betsy Blumenthal, Jenny Wolkowicki and Sheila Dean for help with the production. I am grateful to all at PUP for support, encouragement, and patience throughout a writing process that proved extremely protracted.

Much of this book was written while I was on sabbatical, based in Nicky Clayton's laboratory in the Department of Experimental Psychology, at the University of Cambridge in the United Kingdom. I am indebted to Nicky and the members of her Comparative Cognition Laboratory for making me feel at home and providing an environment, both tranquil and stimulating, that was conducive to productive writing. The final chapters of the book particularly benefitted from these exchanges. I am also very grateful to Gillian Brown, Sean Earnshaw, Julia Kunz, Ros Odling-Smee, Susan Perry, Irena Schulz, Caroline Schuppli, and Carel van Schaik for kindly providing images.

I would like to thank the BBSRC, NERC, The Royal Society, EU Framework 6 and 7 programs, Human Frontier Science Programme, European Research Council, and John Templeton Foundation for financial support for my research. I am particularly indebted to Paul Wason, Kevin Arnold and Heather Micklewright at the John Templeton Foundation who have supported my investigations over many years.

Finally, and most of all, I would like to thank my thesis advisor, Henry Plotkin, to whom I owe so much. Henry taught me the ropes of the academic business with unfailing patience, generosity, and enthusiasm. He trained me in how to design experiments, how to think critically, how to balance theory and empirical work, and where attention to detail is important. Our regular, Friday morning discussions were a highlight of my PhD years, and I consider myself hugely privileged to have shared so much of his time.

KEVIN LALAND
March 2016
St Andrews, United Kingdom

PART I

FOUNDATIONS OF CULTURE

CHAPTER 1

DARWIN'S UNFINISHED SYMPHONY

It is interesting to contemplate an entangled bank, clothed with many plants of many kinds, with birds singing on the bushes, with various insects flitting about, and with worms crawling through the damp earth, and to reflect that these elaborately constructed forms, so different from each other, and so dependent upon each other in so complex a manner, have all been produced by laws acting around us. . . . Thus from the war of nature, from famine and death, the most exalted object which we are capable of conceiving, namely, the production of the higher animals, directly follows.

—CHARLES DARWIN, *ON THE ORIGIN OF SPECIES*

As he looked out on the English countryside from his study at Down House, Charles Darwin could reflect with satisfaction that he had gained a compelling understanding of the processes through which the complex fabric of the natural world had come into existence. In the final, perhaps the most famous, and certainly the most evocative, passage of *The Origin of Species*, Darwin contemplated an entangled bank, replete with plants, birds, insects, and worms, all functioning with intricate coherence. The tremendous legacy of Darwin is that so much of that interwoven majesty can now be explained through the process of evolution by natural selection.

I look out of my window and see the skyline of St Andrews, a small town in southeastern Scotland. I see bushes, trees, and birds too, but the view is dominated by stone buildings, roofs, chimneys, and a church steeple. I see telegraph poles and electricity pylons. I look south, and in the distance is a school, and just to the west, a hospital fed by roads dotted with busy commuters. I wonder, can evolutionary biology explain the existence of chimneys, cars, and electricity in as convincing a fashion as it does the natural world? Can it describe the origin of prayer books and church choirs, as it does the origin of species? Is there an evolutionary explanation for the computer on which I type, for the satellites in the sky, or for the scientific concept of gravity?

At first sight, such questions may not appear particularly troubling. Clearly human beings have evolved, and we happen to be unusually intelligent primates that are good at science and technology. Darwin claimed, "the most exalted higher animals" had emerged "from the war of nature,"[1] and our own species is surely as high and exalted as species come. Isn't it apparent that our intelligence, our culture, and our language are what has allowed us to dominate and transform the planet so dramatically?

With a little more thought, however, this type of explanation unravels with disturbing rapidity, in the process generating a barrage of even more challenging questions. If intelligence, language, or the ability to construct elaborate artifacts evolved in humans because they enhance the ability to survive and reproduce, then why didn't other species acquire these capabilities? Why haven't other apes, our closest relatives, who are genetically similar to us, built rockets and space stations and put themselves on the moon? Animals have traditions for eating specific foods, or singing the local song, which researchers call "animal cultures," but these possess no laws, morals, or institutions, and are not imbued with symbolism, like human culture. Nor do animal tool-using traditions constantly ratchet up in complexity and diversity over time as our technology does. There seems a world of difference between a male chaffinch's song and Giacomo Puccini's arias, between fishing for ants by chimpanzees and *haute cuisine* restaurants, or between the ability of animals to count to three and Isaac Newton's derivation of calculus. A gap, an ostensibly unbridgeable gap, exists between the cognitive capabilities and achievements of humanity and those of other animals.

This book explores the origins of the entangled bank of human culture, and the animal roots of the human mind. It presents an account of the most challenging and mysterious aspect of the human story, an explanation for how evolutionary processes resulted in a species so entirely different from all others. It relates how our ancestors made the journey from apes scavenging a living on ants, tubers, and nuts, to modern humans able compose symphonies, recite poetry, perform ballet, and design particle accelerators. Yet Rachmaninoff's piano concertos did not evolve by the laws of natural selection, and space stations didn't emerge through the "famine and death" of the Darwinian struggle. The men and women who design and build computers and iPhones have no more children than those in other professions.

So, what laws account for the relentless progress and diversification of technology, or the changing fashions of the arts? Explanations based on cultural evolution,[2] whereby competition between cultural traits generates changes in behavior and technology,[3] can only begin to be considered satisfactory with clarification of how minds capable of generating complex culture evolved in the first place. Yet, as later chapters in this book reveal, our species' most cherished intellectual faculties were themselves fashioned in a whirlpool of coevolutionary feedbacks in which culture played a vital role. Indeed, my central argument is that no single prime mover is responsible for the evolution of the human mind. Instead, I highlight the significance of accelerating cycles of evolutionary feedback, whereby an interwoven complex of cultural processes to reinforce each other in an irresistible runaway dynamic that engineered the mind's breathtaking computational power.

Comprehending the distinguishing features of humanity through comparison with similar characteristics in other animals is another central theme in this book, and a distinctive feature of my research group's approach to investigating human cognition and culture. Such comparisons not only help to put our species' achievements in perspective, but help us to reconstruct the evolutionary pathways to humanity's spectacular achievements. We not only seek a scientific explanation for the origins of technology, science, language, and the arts, but endeavor to trace the roots of these phenomena right back to the realm of animal behavior.

Consider, for illustration, the school that I see from my window. How could it have come into existence? To most people the answer to this question is trivial; that is, workers from a building company contracted by the Fife Council built it. Yet to an evolutionary biologist the construction represents an enormous challenge. The immediate mechanical explanation is not the problem; rather, the dilemma is to understand how humans are even capable of such undertakings. With a little training, the same people could build a shopping mall, bridge, canal, or dock, but no bird ever built anything other than a nest or bower, and no termite worker deviated from constructing a mound.

When one starts to reflect, the scale of cooperation necessary to build a school is astounding. Imagine all of the workers who had to co-ordinate their actions in the right place at the right time to ensure that foundations are safely laid, windows and doors are put in place, piping and electricity wires are suitably positioned, and woodwork is painted. Imagine the companies with whom the contractor had to engineer transactions, buy the building materials, arrange for delivery, purchase or loan the tools, subcontract jobs, and organize finances. Think of the businesses that had to make the tools, nuts, bolts, screws, washers, paint, and windowpanes. Imagine the people who designed the tools; smelted the iron; logged the trees; and made the paper, ink, and plastic. So it goes on, endlessly, in a voracious multidimensional expansion. All of those interactions, that endless web of exchanges, transactions, and cooperative endeavors—the vast majority carried out by unrelated individuals on the basis of promises of future remuneration—had to function for the school to be built. Not only did these cooperative trans-actions work, but they repeatedly operate with seamless efficiency day in and day out, as new schools, hospitals, shopping malls, and leisure centers are put together all across the country and around the world. Such procedures are so commonplace that we now entirely take it for granted that the school will be built, and even complain if completion is a little late.

I earn my living in part by studying animals, and I am captivated with the complexity of their social behavior. Chimpanzees, dolphins, elephants, crows, and countless other animals, exhibit rich and sophisti-cated cognition that reveals an often impressive level of intelligence that through the process of natural selection has become suited to the worlds they each inhabit. Yet if we ever wanted a lesson in what an achievement of creativity, cooperation, and communication the construction of a building is, we only have to give a group of animals the materials, tools, and equipment to build such a structure, and then see what happens. I would imagine the chimpanzees might grasp pipes or stones to throw or wave about in dominance displays. The dolphins might plausibly play with materials that floated. Corvids or parrots would perhaps pick out some novel items with which to decorate their nests. I do not wish to disparage the abilities of other animals, whose achievements are striking in their own domains. Yet science has accrued a strong understanding of the evolution of animal behavior, while the origins of human cognition and the complexities of our society, technology, and culture remain poorly understood. For most of us in the industrialized world, every aspect of our lives is utterly reliant on thousands of cooperative interactions with millions of individuals from hundreds of countries, the vast majority of whom we never see, don't know, and indeed never knew existed. Just how exceptional such intricate coordination is remains hard to appreciate; nothing remotely like it is found in any of the other 5–40 million species on the planet.[4]

The inner workings of the school and the activities of children and staff are just as astonishing to an evolutionary biologist like myself. There is no compelling evidence that other apes will go out of their way to teach their friends or relatives anything at all, let alone build elaborate institutions that dispense vast amounts of knowledge, skills, and values to hordes of children with factory-like efficiency. Teaching, by which I mean actively setting out to educate another individual, is rare in nature.[5] Nonhuman animals assist one another in alternative ways, such as provisioning with food or collaborating in an alliance, but they mostly aid their offspring or close relatives, who share their genes and hence also possess their tendency to help.[6] Yet in our species, dedicated teachers devote vast amounts of time and effort with children entirely unrelated to them, helping them to acquire knowledge, in spite of the fact that this does not inherently increase a teacher's evolutionary fitness. Pointing out that teachers are paid, which might be regarded as a form of trade (i.e., goods for work), only trivializes this mystery. The pound coin or dollar bill have no intrinsic value, the money in our bank account has a largely virtual existence, and the banking system is an unfathomably complex institution. Explaining how money or financial markets came into existence is no easier than explaining why schoolteachers will coach unrelated pupils.

As I gaze at the school, I imagine the children sitting at their desks, all dressed in the same uniform, and all (or, at least, many) sitting calmly and listening to their teacher's instruction. But why do they listen? Why bother absorb facts about events in antiquity, or labor to compute the angle of an abstract shape? Other animals only learn what is of immediate use to them. Capuchin monkeys don't instruct juveniles in how their ancestors cracked nuts hundreds of years ago, and no songbird educates the young about what is sung in the wood across the road.

Just as curious to a biologist is the fact that the pupils all dress the same. Some of these children will come from less fortunate backgrounds. Their parents cannot easily afford to spend money on special clothes for school. When they finish their education many of these young people will exchange school attire for another uniform (probably equally uncomfortable), perhaps comprising a suit, or the white and blue attire of doctors and nurses in the hospital down the road. Even the students at my university, replete with liberal, radical, and freethinking values often dress the same, in jeans, T-shirts, sweatshirts, and sneakers. Where did these proclivities come from? Other animals don't have fashions or norms.

Darwin provided a compelling explanation for the protracted history of the biological world, but only hinted about origins of the cultural realm. When discussing evolution of the "intellectual faculties," he confessed: "Undoubtedly it would have been very interesting to have traced the development of each separate faculty from the state in which it exists in the lower animals to that in which it exists in man; but neither my ability nor knowledge permit the attempt."[7] With the benefit of hindsight, we should not be surprised if Darwin struggled to understand the origins of humanity's intellectual achievements; it is a monumental challenge. A satisfactory explanation demands insight into the evolutionary origins of some of our most striking attributes— our intelligence, language, cooperation, teaching, and morality—yet most of these features are not just distinctive, they are unique to our species. That makes it harder to glean clues to the distant history of our minds through comparison with other species.

At the heart of this challenge lies the undeniable fact that we humans are an amazingly successful species. Our range is unprecedented; we have colonized virtually every terrestrial habitat on Earth, from steaming rainforests to frozen tundra, in numbers that far exceed what would be typical for another mammal of our size.[8] We exhibit behavioral diversity that is unparalleled in the animal kingdom,[9] but (unlike most other animals) this variation is not explained by underlying genetic diversity, which is in fact atypically low.[10] We have resolved countless ecological, social, and technological challenges, from splitting the atom, to irrigating the deserts, to sequencing genomes. Humanity so dominates the planet that, through a combination of habitat destruction and competition, we are driving countless other species to extinction. With rare exceptions, the species comparably prosperous to humans are solely our domesticates, such as cattle or dogs; our commensals, such as mice, rats, and house flies; and our parasites, such as lice, ticks, and worms, which thrive at our expense. When one considers that the life history, social life, sexual behavior, and foraging patterns of humans have also diverged sharply from those of other apes,[11] there are grounds for claiming that human evolution exhibits unusual and striking features that go beyond our self-obsession and demand explanation.[12]

As the pages of this book demonstrate, our species' extraordinary accomplishments can be attributed to our uniquely potent capability for culture. By "culture" I mean the extensive accumulation of shared, learned knowledge, and iterative improvements in technology over time.[13] Humanity's success is sometimes accredited to our cleverness,[14] but culture is actually what makes us smart.[15] Intelligence is not irrelevant of course, but what singles out our species is an ability to pool our insights and knowledge, and build on each other's solutions. New technology has little to do with a lone inventor figuring out a problem on their own; virtually all innovation is a reworking or refinement of existing technology.[16] The simplest artifacts provide the test cases with which to evaluate this claim, because clearly no single person could invent, say, a space station.

Consider the example of the paper clip. You might be forgiven for assuming that what is, in essence, just a bent piece of wire was devised in its current form by a single imaginative individual. Yet that could not be further from the truth.[17] Paper was originally developed in first-century China, but only by the Middle Ages was sufficient paper produced and used in Europe to create the demand for a means to bind sheets of paper together temporarily. The initial solution was to use pins as fasteners, but these rusted and left unsightly holes, such that the pinned corners of documents sometimes became ragged. By the middle of the nineteenth century, bulky spring devices (resembling those on clipboards today) and small metal clasps were in use, and in the decades that followed a great variety of fasteners came into existence, with fierce competition governing their use. The first patent for a bent wire paper clip was awarded in 1867.[18] However, the mass production of cheap paper fasteners had to wait for the invention of a wire with the appropriate malleability, and a machine capable of bending it, both of which were developed in the late nineteenth century. Even then, the earliest paper clips were suboptimal in form—for instance, these included a rectangular-shaped wire with one overlapping side, rather than the circular "loop within a loop" design dominant today. A variety of shapes were experimented with for several decades of the twentieth century before manufacturers finally converged on the now standard paper clip design, known as the "Gem." What appears at first sight to be the simplest of artifacts was in fact fashioned through centuries of reworking and refinement.[19] Even today, in spite of the Gem's success, novel paper clip designs continue to emerge, with a wide range of cheaper plastic forms manufactured over the last few decades.

The history of the paper clip is broadly representative of how technology changes and complexifies, and such transformations occur in other areas too. Humanity's rich and diverse culture is manifest in extraordinarily complex knowledge, artifacts, and institutions. These multifaceted, composite aspects of culture are rarely produced in a single step, but are generated by repeated, incremental refinements of existing forms in a process known as "cumulative culture."[20] Our language, cooperativeness, and ultrasociality, just like our intelligence, are frequently lauded as setting us apart from other animals. But, as we shall see, these features are themselves more likely products of our exceptional cultural capabilities.[21]

I have dedicated my scientific career to investigating the evolutionary origins of human culture. In my research laboratory we do this both through experimental investigations of animal behavior, and through the use of mathematical evolutionary models that allow us to answer questions not amenable to experimentation. We are part of a wider community of researchers who have established that many animals, including mammals, birds, fishes, and even insects, acquire knowledge and skills from others of their species.[22] Through copying,[23] animals learn what to eat, where to find it, how to process it, what a predator looks like, how to escape that predator, and more. There are thousands of reports of novel behaviors spreading through natural populations in this way, in animals ranging from fruit flies and bumblebees, to rhesus macaques and killer whales. These behavioral diffusions occur too rapidly to be attributed to the spread of favorable genes through natural selection, and are unquestionably underpinned by learning. The behavioral repertoires of some species vary between and within regions, in a manner that is not easily explained by ecological or genetic variation, and is often described as "cultural."[24] Some animals appear to have an unusually broad cultural repertoire, with multiple and diverse traditions, and distinctive behavioral profiles in each community.[25] Rich repertoires are observed in some whales and birds,[26] but outside of humans, animal traditions reach their zenith in the primates, where various socially transmitted behavior patterns, including tool use and social conventions, have been recorded for several species, notably chimpanzees, orangutans, and capuchin monkeys.[27] Experimental studies of other apes in captivity provide strong evidence for imitation,[28] tool use, and other aspects of complex cognition;[29] at least these are complex relative to other animals. Yet, in spite of this, the traditions of even apes or dolphins just don't seem to ratchet up in complexity like human technology does, and the very notion of cumulative culture in animals remains controversial.[30] Perhaps the most credible candidate was proposed by the Swiss primatologist Christophe Boesch, who has argued that the use of hammerstones to crack open nuts by chimpanzees has been refined and improved over time.[31] Some chimpanzees have begun to deploy a second stone as an anvil on which to place the nuts that they smash, and a couple of individuals have even been seen to insert another stabilizing stone to wedge the anvil securely. While Boesch's claim is plausible, and would meet some definitions of cumulative culture if confirmed, it remains uncorroborated. Even the most complex variant of nut cracking could plausibly have been invented by a single individual, which means this tool use need not imply any building on the shoulders of chimpanzee predecessors.[32] The same issue arises for all chimpanzee behaviors that have excited claims of cumulative culture;[33] there is no direct evidence that any of the more elaborate variants have developed from simpler ones. Circumstantial evidence for cumulative culture in other species is equally contentious—notably in New Caledonian crows,[34] a bird renowned for manufacturing complex foraging tools from twigs and leaves.[35] Novel learned behavior frequently spreads through animal populations, but is rarely, if ever, refined to generate a superior solution.

In striking contrast, the invention, refinement, and propagation of innovations by humans is extremely well documented.[36] The most obvious illustration comes from the archaeological record;[37] this can be traced back 3.4 million years to the use of flake tools by a group of African hominins known as australopithecines, who may have been early human ancestors.[38] The technology, known as Oldowan because it was first discovered at the Olduvai Gorge in Tanzania, consisted of basic stone flakes struck off a core with a hammerstone that were used to carve up carcasses and extract meat and bone marrow.[39] By 1.8 million years ago, a new stone tool technology arose, known as Acheulian, and associated with other hominins, *Homo erectus* and *H. ergaster*. Acheulian technology consisted of hand axes that were more systematically designed and particularly well suited to the butchery of large animals.[40] Acheulian technologies, together with the appearance of hominins outside Africa and evidence for systematic hunting and the use of fire, leave no doubt that by at least this juncture in our history, our ancestors benefitted from cumulative cultural knowledge.[41] By around 300,000 years ago, hominins were combining wooden spears with flint flakes,[42] building dwellings with fire hearths,[43] and producing fire-hardened spears for big game hunting.[44] By 200,000 years ago, Neanderthals and early *Homo sapiens* were manufacturing an entire tool kit from the same stone.[45] African sites dated to 65–90 thousand years ago provide evidence of abstract art, blade tools, barbed bone harpoon points,[46] and composite tools, such as hafting implements and awls used to sew clothing.[47] Between 35 and 45 thousand years ago, perhaps earlier,[48] a plethora of new tools appear, comprising blades, chisels, scrapers, points, knives, drills, borers, throwing sticks, and needles.[49] This period also introduced tools made from antler, ivory, and bone; raw materials transported over long distances; construction of elaborate shelters; creation of art and ornaments; and ritualized burials.[50] Technological complexity escalated further with the advent of agriculture, which was swiftly followed by the wheel, the plow, irrigation systems, domesticated animals, city-states, and countless other innovations.[51] With the industrial revolution, the pace of change accelerated again.[52] Human culture continues relentlessly to grow in intricacy and diversity, culminating in the mind-boggling technological complexity of today's innovation society.

Whether or not chimpanzees, orangutans, or New Caledonian crows have managed some crude advancements over their basic tool-using habits, the scale of difference when compared with the monumental advances of humanity is breathtaking. In some limited respects, animal traditions resemble aspects of human culture and cognition,[53] yet the fact remains that humans alone have devised vaccines, written novels, danced in *Swan Lake*, and composed moonlight sonatas, while the most culturally accomplished nonhuman animals remain in the rain forest cracking nuts and fishing for ants and honey.

Tempting though it may be to view "culture" as the faculty that sets humans apart from the rest of nature, the human cultural capability obviously must itself have evolved. Herein lies a major challenge facing the sciences and humanities; namely, to work out how the extraordinary and unique human capacity for culture evolved from ancient roots in animal behavior and cognition. Understanding the rise of culture has proven a remarkably stubborn puzzle,[54] largely because many other evolutionary conundrums must be addressed in the process. We must first understand why animals copy each other at all, and we must isolate the rules that guide their use of social information. We then need to identify the critical conditions that favored cumulative culture, and the cognitive prerequisites for its expression. The circumstances leading to the evolution of the abilities to innovate, teach, cooperate, and conform must all be established. Also critical is knowing how and why humans invented language, and how that led to complex forms of cooperation. Finally, and crucially, we need to comprehend how all of these processes and capabilities fed back on each other to shape our bodies and minds. Only then can researchers begin to understand how human beings uniquely came to possess the remarkable suite of cognitive skills that has allowed our species to flourish. These are the issues with which my research group has wrestled for many years, and our studies and those of others in our field, are beginning to provide answers.

Some readers might be surprised by the suggestion that understanding the evolution of the human mind and culture has proven a major challenge. After all, Darwin wrote at great length about human evolution, and that was 150 years ago; unquestionably, extensive progress has been made in the intervening period.[55] In fact, in *The Origin of Species* Darwin did not mention human evolution at all, except to say in the final pages that "light will be thrown on the origin of man and his history."[56] Darwin took a long time, well over a decade, to elaborate on this enigmatic statement, but he eventually brought forth two huge books on the topic: *The Descent of Man and Selection in Relation to Sex* (1871) and *The Expression of the Emotions in Man and Animals* (1872). Strikingly, in these books, Darwin says rather little about human anatomy, but instead concentrates on the question of the evolution of "the mental powers of Man." This focus is highly significant. To Victorian readers, as to us, there seemed to be a far greater divide between the mental abilities of human beings and other animals than between their bodies. Darwin recognized that understanding the evolution of cognition was the greater challenge if he was to convince his readers that humans had evolved. The origin of mind was the key terrain over which the battle regarding human evolution was to be fought.

The account given in *The Descent of Man* is typical of Darwinian reasoning. Darwin maintained that there was variation in mental capacity and that being intellectually gifted was advantageous in the struggle to survive and reproduce:

> To avoid enemies, or to attack them with success, to capture wild animals, and to invent and fashion weapons, requires the aid of the higher mental faculties, namely, observation, reason, invention, or imagination.[57]

Darwin attempted to counter the widespread belief, brought to prominence through the writings of French philosopher René Descartes, that animals were merely machines driven by instinct, while humanity alone was capable of reason and advanced mental processing.[58] Instead, Darwin sought to demonstrate both that animals possessed more elevated cognition than hitherto conceived and that human beings possessed instinctive tendencies. Through extensive use of examples, such as rats learning to avoid traps and apes using tools, Darwin documented how many animals exhibit signs of intelligence, and how even simple animals are capable of learning and memory. Much of his analysis reads a little anthropomorphically today; he claimed that the songs of birds demonstrate an appreciation of beauty, that their behavior near a nest revealed some concept of personal property, and even that his dog showed the rudiments of spirituality. Yet the data Darwin presented were a serious challenge to the established, stark, Cartesian human-versus-animal mental divide.

Darwin also documented the evidence that human beings possess behavioral characteristics in common with other animals, cataloguing an amazing array of shared facial expressions.[59] For instance, he noted that monkeys, like human beings, have "an instinctive dread of serpents" and will respond to snakes with the same screams and the same fearful faces as many of us do. Through these efforts, Darwin established a scientific tradition that perpetuates to this day and that seeks to demonstrate that the differences in mental ability between human beings and other animals were not as great as formerly believed.

What is of relevance here is that Darwin's approach to explaining the evolution of the human mind is, in essence, identical to his strategy for accounting for the evolution of the human body. He sought to shrink the apparently chasmic gap between the intellectual abilities of human beings and other animals by showing that for any given character, humans are sufficiently animallike, or animals sufficiently humanlike, that it is possible a chain of intermediary forms could have been forged by natural selection. The data he presented did not demonstrate such chains; nor were they intended to. Darwin merely set out to illustrate that the construction of such a case for continuity of mind was, in principle, highly plausible.

Darwin's stance contrasted decidedly with that of his contemporary Alfred Wallace, who had struck upon the idea of evolution by natural selection around the same time. Wallace concluded that the complex language, intellect, and the music, art, and morals of human beings could not be explained solely by natural selection and must have resulted from the intervention of a divine creator.[60] History has perhaps judged Wallace harshly, with the fact that he despaired of a scientific explanation for the origins of mind leading some to interpret his position as indicative of some weakness of character, in comparison to Darwin's courageous stance.[61] Any such conclusion would be unjust. Wallace's evaluation of the evidence was primarily an honest reflection of the state of knowledge at the time. The explanations that Darwin offered to account for the evolution of mind were, as he conceded, "imperfect and fragmentary."[62] Darwin's position was based on the firm belief that in the future science would provide more concrete evidence to bridge the mental divide; a stance now being vindicated.

Comprehending the evolution of the human mind is Darwin's unfinished symphony. Unlike the unfinished compositions of Beethoven or Schubert, which had to be assembled into popular masterpieces using solely those fragmentary sketches left by the original composers, Darwin's intellectual descendants have taken up the challenge of completing his work. In the intervening decades great progress has been made, and rudimentary answers to the conundrum of the evolution of our mental abilities have started to emerge. However, it is only in the last few years that a truly compelling account has begun to crystallize. Darwin thought that competition, for food or mates, drove the evolution of intelligence and, in its broad thrust, this assertion is supported.[63] However, what was not recognized until recently was the central role played by culture in the origins of mind.

Darwin and his intellectual descendants have unearthed findings that have substantially shrunk the recognized differences between human and animal cognition relative to the strict dichotomy that was accepted in the Victorian era. We now know that humans share many cognitive skills with their nearest primate relatives.[64] A long list of strong claims of human uniqueness—humans are the only species to use tools, to teach, to imitate, to use signals to communicate meanings, to possess memories of past events and anticipate the future—have been eroded by science as careful research into animal cognition has revealed unanticipated richness and complexity in the animal kingdom.[65] Yet the distinctiveness of human mental ability relative to that of other animals remains striking, and the research field of comparative cognition has matured to the point where we can now be confident that this gap is unlikely to be eroded away completely.[66] A hundred years of intensive research has established beyond reasonable doubt what most human beings have intuited all along; the gap is real. In a number of key dimensions, particularly the social realm, human cognition vastly outstrips that of even the cleverest nonhuman primates.

I suspect that in the past, many animal behaviorists have been loath to admit this for fear that it would reinforce the position of those who denied human evolution altogether. A "good evolutionist" emphasized continuity in the intellectual attainments of humans and other primates. Dwelling on our mental superiority was portrayed as anthropocentric, and was often tainted with a suspicion that those who would set humans apart from the rest of nature must have some personal agenda. Humans might be unique, but then, it was argued, so are all species. At the same time the media has been rife with "talking" apes and Machiavellian monkeys, giving the impression that other primates were as cunning and manipulative as the most devious and sinister humans, with untapped potential for sophisticated communication, and possessing rich intellectual and even moral lives.[67] Political and conservationist agendas fed into this doctrine, leading to the assertion that other apes were so similar to us that they merit special protection or human rights, and it has even been suggested they actually are people.[68] Reinforcing this perspective is a long-standing and highly successful genre of popular science books that challenged readers to contemplate their animal selves. We have been vividly portrayed as "naked apes" adapted to a small-group forest existence, and then thrust suddenly into a modern world with which we are ill equipped to cope.[69] We (at least, the males among us) have been designated "man the hunter," shaped by natural selection for a life of brutal aggression.[70] Other tomes depict us as so laden with baggage from our animal heritage that we will be driven to destruction.[71] The authors of such books were often authoritative scientists, who explicitly drew on knowledge of animal behavior and evolutionary biology to justify their assertions.

In my view, too much has been made of superficial similarities between the behavior of humans and other animals, whether by inflating the intellectual credentials of other animals or by exaggerating humanity's bestial nature. Humans may be closely related to chimpanzees, but we are not chimpanzees, and nor are chimpanzees people. Any agenda to "prove" human evolution by demonstrating continuity of our mental abilities with those of other living animals is no longer required; it has become anachronistic. We now know for certain what Darwin could only suspect: several extinct hominin species existed over the intervening five to seven million years since humans and chimpanzees shared a common ancestor. Archaeological remains leave little doubt that these hominins possessed intellectual abilities intermediate to that of humans and chimpanzees.[72] The gap between apes and humans is real, but this is not a problem for Darwinism, because our extinct ancestors bridge the cognitive divide.

Nonetheless, demonstrating the authenticity of the mental ability gap between humans and other living primates is a necessary platform for this book. That is because, ostensibly, we humans live in complex societies organized around linguistically coded rules, morals, norms, and social institutions, with a massive reliance on technology, while our closest primate relatives do not. Were these differences illusory, either because human cognition is dominated by bestial tendencies that can be explained in the same manner as that of other animals, or because other animals possess hidden powers of reasoning and social complexity, the problem of explaining the origins of mind would melt away in the manner that evolutionists have anticipated, and perhaps hoped, for a century. However, the differences, as we shall see, are not illusory, and the challenge does not melt away.

Consider the genetic evidence. Perhaps the most misunderstood statistic in science is that humans and chimpanzees are 98.5% similar genetically. To many people, this statistic implies that chimpanzees are 98.5% human, or that 98.5% of chimpanzee genes work in the same way as ours, or that the differences between humans and chimpanzees are attributable to the 1.5% of genetic differences. All such inferences are wildly inaccurate. The 98.5% figure relates to similarity in the DNA sequence level across the entire genomes. Human and chimpanzee genomes comprise a long series of DNA base pairs, with tens of thousands, even millions, of base pairs in each protein-coding gene. Humans have something in the region of 20,000 protein-coding genes, although these make up only a small portion of our genome. The 1.5% represents about 35 million nucleotide differences between the two species. Most of these do not affect the gene's function at all, but some have big effects. Even a single change can affect how a gene operates, which means that a human and chimpanzee gene could be virtually identical and yet function differently. Many of the affected genes code for transcription factors (proteins that bind to DNA sequences and thereby regulate the transcription of other genes), thereby allowing the small sequence differences between the species to be amplified.[73]

Further genetic differences between humans and chimpanzees result from insertions and deletions of genetic material,[74] differences in the promoters and enhancers that switch genes on and off,[75] and between-species variation in the number of copies of each gene. Copy number variation has arisen through both gene loss and the duplication of genes (typically in the hominin lineage); the latter can be adaptive in cases where more gene product is required.[76] One study found that 6.4% of all human genes do not have a matching copy number in chimpanzees.[77] In addition, genes can be read in a variety of different ways to produce multiple diverse products, as different regions of the gene (exons) are spliced together. This "alternative splicing" is not a rare phenomenon. More than 90% of human genes exhibit alternative splicing, and 6–8% of genes shared by humans and chimpanzees show pronounced differences in how they are spliced.[78]

More important than differences between genes, however, are between-species differences in how the genes are used. Genes might be thought of as children's building bricks—broadly similar blocks that are assembled in different species in dissimilar ways. Human and chimpanzee genes could be exactly identical and still work differently because they can be turned on and off to different degrees, in different places, or at different times. Allan Wilson and Mary-Claire King, the pioneering Berkeley scientists who first drew attention to the striking genetic similarity between humans and chimpanzees, speculated that the differences between the two species have less to do with genetic sequence differences and much more to do with when and how those genes are switched on and off.[79] The intervening years have confirmed this supposition.[80] The Encyclopedia of DNA Elements (ENCODE), a massive research project launched by the US National Human Genome Research Institute in 2003 to identify all functional elements in the human genome, recently found around eight million binding sites, and variation in these largely regulatory elements is thought to be responsible for many species differences.[81]

An instructive comparison here is between the English and German languages. In terms of their written symbolic form (i.e., the letters used), these two Indo-European languages are identical, although only German speakers make use of the umlaut, recognizable as two dots over a vowel, which changes its pronunciation.[82] Yet it would clearly be ridiculous to claim that all differences between the two languages are attributable to the umlaut, or that to master German, an English speaker merely has to master the rules of umlaut usage. The differences between the two languages relate far more to how the letters are used, to how they are combined into words and sentences, than to differences in the phonological elements. So it is with genes. Among the key empirical insights to emerge recently from the field of evolutionary developmental biology (or "evo-devo") is the finding that evolution typically proceeds through changes in the gene regulatory machinery—through "teaching old genes new tricks."[83] Such changes include the timing of protein production, the region of the body in which the gene is expressed, the amount of protein produced, and the form of the gene product. The differences between human and chimpanzees relate far more to how *all* our genes are switched on and off than they do to the small differences in the sequences.

Among the sample of genes that do differ between humans and chimpanzees, a disproportionately high number are expressed in the brain and nervous system.[84] Genes expressed in the brain have been subject to strong positive selection in the hominin lineage, with over 90% of such genes upregulating their activity relative to chimpanzees.[85] Such differences are likely to have a big impact on brain function. Unlike many other tissues, gene expression patterns in the brains of chimpanzees have been found to be far more similar to those of macaques than to humans.[86] In terms of their anatomy and physiology, chimpanzee brains resemble those of monkeys far more than those of humans.[87] Human brains are more than three times the size of chimpanzee brains and have been structurally reorganized in comparison; for instance, the former have proportionally larger neocortices and more direct connections from the neocortex to other brain regions.[88]

What this means is that humans and chimpanzees are not so biologically similar that we should assume they ought to be behaviorally or cognitively alike. Chimpanzees might be our closest relatives, but this is only because all other members of our genus—*Homo habilis, Homo erectus, Homo neanderthalensis*, and more[89]—as well as all the *Australopithecines*, and all other hominins (*Paranthropus, Ardipithecus, Sahelanthropus, Kenyanthropus*) are extinct. Had they endured, chimpanzees would surely have a lower status in the minds of humans, and less might have been expected of them.

Let us put aside any preconceived notions and consider what exactly *is* special about the mental capabilities of humans. Careful experimental analyses of the cognitive capabilities of humans and other animals over the last hundred years have allowed researchers to characterize the truly unique aspects of our cognition. This is no trivial matter, because history is littered with claims along the lines of "humans uniquely do X, or possess Y" that have subsequently fallen by the wayside when established in another species. Comparisons of humans with other apes have also isolated features that the former share with other animals. Indeed, examining shared traits has proven as insightful as investigating human uniqueness, because such comparisons help us to reconstruct the past; this allows inferences to be made about the attributes of species ancestral to humans so that the evolutionary history of traits seen in modern humans can be understood. Nonetheless, some striking differences remain.

Consider, for example, research into human cooperation, which in recent years has been subject to intense investigation through the use of economic games. One is called the "ultimatum game," where two players must decide how to split a sum of money. The first player proposes how to divide the sum between them, and the second can either accept or reject this proposal. If the second player accepts, the money is split according to the proposal, but if the second player rejects, neither player receives anything. The most interesting feature of the ultimatum game is that it is never really rational for the second player to reject, since any offer is better than nothing. Hence, we might expect the first player to offer the absolute minimum and then keep the bulk of the sum. However, that is not what humans typically do. Humans frequently make far more generous offers (the most common offer is 50%, a "fair" division), and are much more prone to reject offers (those less than 20% are typically rejected) than would be expected if behaving entirely rationally. Moreover, the magnitude of offers and rates of rejection vary from one society to the next in a manner consistent with a society's cultural norms. For instance, particularly generous offers may be observed in a culture of extensive gift giving.[90] Humans seem predisposed to cooperate, and expect the same of others. Our behavior is often motivated by a sense of fairness and consideration of others' perspectives, and frequently adheres to the conventions of society. We even feel a compulsion to be fair to absolute strangers, irrespective of whether they are likely to be seen again. These conclusions are echoed in literally thousands of experimental findings, set across a very wide range of contexts and spanning broad scales of interaction.[91]

What happens when chimpanzees are asked to partake in such games? Psychologists Keith Jensen, Josep Call, and Michael Tomasello presented a simplified version of the ultimatum game to chimpanzees. The clever experimental setup allowed the "proposer" chimpanzee to choose between two options, one that shared a food reward equally with another chimpanzee, and another that gave the proposer a greater proportion. They found that chimpanzees tended to select the option that maximized their own returns with little regard to whether or not this was fair to others.[92] Compared to humans, the chimpanzees might appear to have behaved in a selfish manner, but their behavior, rather than ours, is the rational response. Studies like these, and there are many, support the argument that hominins may have been subject to selection promoting both consideration of others and sensitivity to local norms of fairness.[93] This is not to suggest that other apes never cooperate; chimpanzees, much like most other primates, cooperate in restricted domains.[94] However, extensive experimental data has established that other apes do not cooperate as extensively as humans do.

Many prominent primatologists believe that cooperation is at least partly constrained in other primates by a lack of understanding of the perspective of other individuals with whom they are required to cooperate.[95] Research into this topic was initiated in a classic study by comparative psychologists David Premack and Guy Woodruff, who asked, "Does the chimpanzee have a theory of mind?" They questioned whether chimpanzees, like adult humans, understand that other individuals may have false beliefs, intentions, and goals.[96] Their study triggered a spate of experimental investigations comparing the performance of chimpanzees and young children. In the main, the data led many researchers to answer Premack and Woodruff's question in the negative. More recent studies, however, suggest that chimpanzees may have some precursors of a theory of mind.[97] For instance, there is evidence that chimpanzees can infer a human experimenter's intentions; they react very differently when a person refrains from giving food because they are unwilling to do so compared with when they are unable to do so, or when doing something on purpose rather than by accident.[98] Other studies suggest that chimpanzees can understand the goals, perception, and knowledge of others to a limited degree. However, these conclusions remain contested,[99] and crucially, such studies provide no evidence that chimpanzees understand that others may possess false beliefs.[100] In contrast, children typically understand that others can have false beliefs by the age of four years, and possibly much earlier,[101] which implies that this capability evolved in the hominin lineage. Moreover, humans readily comprehend many orders of belief and understanding; for instance, you could *understand* that I could *claim* my wife *believes* that her daughter *thinks* her mother's hair looks best short, whereas in fact my daughter is only saying that to make her mother happy. Such beliefs about beliefs about beliefs are a natural and common aspect of human cognition, and our species can comprehend up to six orders. Other apes struggle with first-order intentionality.[102]

A reader unfamiliar with research in comparative psychology might reasonably wonder why the field should contrast the performance of chimpanzees of all ages with that of human children in laboratory tests of cognition.[103] Ostensibly, the fairer comparison would be of the two species at the same age. The general rationale for comparing chimpanzees to children (often at nursery school age) rather than to adult humans is that adults have been greatly enculturated by human society; the use of children thus represents an attempt to tease out the inherent differences between the two species prior to culture becoming too great a confounding factor. However, whether this argument holds water is contentious; after all, even four- or five-year-old children will have been hugely enculturated. A more pragmatic rationale for the comparison may be closer to the truth; that is, with most cognitive tasks, there would be little point in comparing adult humans with adult chimpanzees, because the former would far outstrip the latter. Even human toddlers outperform the adults of other ape species in tests of mental ability. For instance, developmental psychologist Esther Herrmann and her colleagues gave a battery of cognitive tests to two-and-a-half-year-old children, as well as to chimpanzees and orangutans ranging from 3 to 21 years of age. These researchers found that, even at such a young age, the children already had comparable cognitive skills to adult chimpanzees and orangutans for dealing with the physical world (e.g., spatial memory, object rotation, tool use), and had far more sophisticated cognitive skills than both adult chimpanzees and orangutans for dealing with the social realm (e.g., social learning, producing communicative gestures, understanding intentions); they typically performed twice as well as (nonhuman) apes in the tasks.[104] While other experiments have established that chimpanzees do show impressive proficiency in social learning and social cognition,[105] those studies that directly compare species nonetheless consistently reveal strong differences between humans and other apes.[106] The hypothesis that social intelligence, in particular, blossomed among our hominin ancestors is now widely accepted.[107]

Communication is perhaps the most obvious respect in which there appears to be a major, qualitative difference between the mental abilities of humans and other primates. Animal communication comprises various classes of signals concerning survival (e.g., predator alarm calls), courtship and mating (such as the red sexual swellings of some monkeys), and other social signals (for instance, dominance displays).[108] Such signals each have very specific meanings, and typically relate to the animal's immediate circumstances. In contrast, language allows us to exchange ideas about matters distant in space and time (I could tell you about my upbringing in the English Midlands, or you could inform me of the new coffee shop in the next town). With rare exceptions, such as the honeybee waggle dance through which bees transmit abstract information about the location of nectar-rich flowers, animals do not communicate about phenomena that are not immediately present. Chimpanzees do not tell each other about the termite mound they found yesterday, and gorillas do not discuss the nettle patch on the other side of the forest. Some primate vocalizations do appear to symbolize objects in the world: famously, vervet monkeys, which range throughout southern Africa, are thought to possess three distinct calls that are labels for avian, mammalian, and snake predators,[109] and similar claims have been made for several other primates. However, primate vocalizations largely consist of single, unrelated signals that are rarely put together to transmit more complex messages, and any atypical composite messages are highly restricted. For instance, some monkeys simultaneously inform others of both the existence of a predator and of its location.[110] In contrast, human language is entirely open-ended, allowing humans to produce an infinite set of utterances and to create entirely new sentences through their mastery of symbols.

A romance exists around the notion that animals, such as chimpanzees or dolphins, might covertly harbor complex natural communication systems as yet unfathomed by humans. Many of us quite like the idea that "arrogant" scientists have prematurely assumed that other animals don't talk to each other when they failed to decode the cryptic complex of calls and whistles. Sadly, all the evidence suggests that this is just fantasy. Animal communication has been subject to intense scientific investigation for over a century, and few hints of any such complexity have arisen. To the contrary, it has proven remarkably difficult to provide compelling evidence that the signals of chimpanzees or dolphins possess a referential quality.[111] Chimpanzees are unquestionably smart in many respects, but their communication is not unambiguously richer, and may even be less language-like, than that of many other animals.[112] This means that communication systems cannot be arrayed on a continuum of similar forms, with human language at one end of the spectrum, closely aligned to some highly complex animal protolanguage, and passing through less and less sophisticated animal communication systems to end up with, say, simple olfactory messages at the other end. Rather, language appears qualitatively different. Even if the gulf between human language and the others were ignored, and animal communication systems were aligned on a continuum from simple to complex, current evidence implies that those species most closely related to humans are not the ones with the most complex natural communication systems.[113]

Perhaps apes are capable of more complex communication than they exhibit in their natural environments. A simple continuity argument might yet be resurrected if apes could be trained to talk, and several high-profile studies have pursued this dream.[114] Other apes, of course, are not anatomically suited to complex vocalization; their vocal control and physiology aren't capable of speech production. This much was established in the 1940s by American psychologists Keith and Cathy Hayes, who raised a young female chimpanzee called Viki from birth in their own home, endeavoring to treat her identically to their own children. Viki learned to produce just four words—"mama," "papa," "cup," and "up"—and by all accounts, the pronunciation was not compelling. If that sounds like a disappointment, it was at least more successful than the only previous attempt. This was made by Winthrop and Lu-ella Kellogg, another husband and wife team of psychologists, who reared a female chimpanzee called Gua with their son Donald; Gua was seven months old when they started and Donald was close in age. The Kelloggs were forced to abandon the exercise after a couple of years, when Gua hadn't learned a single word, but Donald had started to imitate chimpanzee sounds! Real progress had to wait until the 1960s, when a third couple, Allen and Beatrice Gardner, tried again, but this time with the ingenious idea of teaching American Sign Language to Washoe, their young chimpanzee. Washoe is reported to have learned over 300 signed words, many through imitation, and to even to have passed on some of these to a younger chimpanzee called Loulis. Washoe also spontaneously combined signs; for instance, on seeing a swan, Washoe signed "water" and "bird," to much acclaim. The investigation generated considerable excitement and triggered a series of studies of "talking apes," including Nim Chimpsky, Koko the gorilla, and Kanzi the bonobo who were all taught signs or to use a symbolic lexicon.

Yet the vaulted claims that apes had produced language do not stand up to close scrutiny, a point on which virtually all linguists concur.[115] The animals had successfully learned the meanings of signs, and were able to produce simple two- or three-word combinations, but they showed no hint of having mastered grammatical structure or syntax. Human languages differ from animal communication systems in the use of grammatical and semantic categories, such as nouns, adjectives, and conjunctions, combined with verbs in present, past, and future tenses, in order to express exceedingly complex meanings. Washoe, Koko, and Kanzi may have comprehended the meaning of a large numbers of words and symbols (although none was able to learn as many different words as a typical three-year-old child) but more to the point, none of them acquired anything resembling the complex grammar of human language. Even enthusiastic devotees of the complexity of ape commu-nication have acknowledged the contrast.[116] A world of difference sepa-rates a chimpanzee communication and a Shakespearean comedy.

Equally romantic is the notion that science has not yet gauged the full depth of the moral lives of animals, a premise that sells an awful lot of popular science books and flushes the coffers of Hollywood mov-iemakers. Television shows and storybooks are full of animals, from

Lassie, to Flipper, to Champion the Wonder Horse, who can grasp complex situations, often more effectively than humans, and who exhibit humanlike moral emotions such as sympathy or guilt. Once again, the scientific evidence is disappointingly dull; many popular books claim that animals understand the difference between right and wrong, but precious few scientific papers demonstrate this. Instead, claims of animal morality are heavily reliant on anecdotal reports, including stories of apes (but also dolphins, elephants, and monkeys) behaving as if they possess sympathy or compassion for another animal; for instance, these animals appear to console sick or dying individuals or "reconcile" after a fight.[117] However, such reports require careful interpretation.

Animals unquestionably lead rich emotional lives; strong scientific evidence demonstrates that many form attachments, experience distress, and respond to the emotional state of others.[118] Yet, that is not the same as possessing morals. Animals sometimes behave as if they can tell right from wrong, but there are usually alternative ways of interpreting such examples. The animals might be following simple rules without much reflection or care for others. For instance, grooming the victims of aggression might be beneficial if this provides a prime opportunity to forge new alliances. Primates may reconcile to obtain short-term objectives, such as access to desirable resources or to preserve valuable relationships damaged by conflict.[119] Rather than feeling guilt after being reprimanded, your dog may simply have learned that giving you "the eyes" will lead to more rapid forgiveness on your part. Instead of feeling sympathy for another individual that screams, an observing animal may respond emotionally out of fear for itself, a phenomenon known as emotional contagion.[120] Some writers have interpreted reconciliation after fights in monkeys as indicating that the protagonists feel "guilt" or "forgiveness," arguing on evolutionary grounds that it is parsimonious to assume that our close relatives experience the same emotions and cognition as ourselves.[121] However, this reasoning appears more questionable when we learn that fish behave in the same way.[122] Are we to assume that they also have a sense of forgiveness? Another concern is that for every anecdote suggesting particular animals possess moral tendencies, there are typically many more from the same species showing selfish and exploitative behavior.[123] The scientific literature is rife with reports of animals behaving indifferently to the distress of others, or taking advantage of the weak. Expressions of "moral" tendencies are, at best, rare events in other species.

Human beings are very much a part of the animal kingdom, and well over a century of careful research by scientists in several fields has established many continuities between our behavior and that of other animals. Yet despite this, important differences between the cognitive capabilities and achievements of humans and those of our closest animal relatives have been experimentally ratified. This divergence demands an evolutionary explanation. One-hundred-and-fifty years ago, Charles Darwin penned the first credible accounts of human evolution but inevitably, with fossil data scarce, the arguments brought to bear were designed more to illustrate the kinds of processes through which humans might have evolved, rather than to relate the actual story of our origin. In the intervening time, the unearthing of literally thousands of hominin fossils by paleontologists has allowed a detailed history of our evolutionary ancestry to be scripted.[124] Yet that history is largely written of teeth and bones, supplemented by clever inferences about diet and life history, together with stone tools and archaeological remains. Knowledge of the history of the human mind remains rare, speculative, and circumstantial.

Darwin recognized that a truly compelling account of human evolution would have to account for human mental abilities, including our culture, language, and morality, and in spite of extensive and productive scientific research for over a century, this remains a monumental challenge. The sheer magnitude of this task has not always been universally recognized. In the struggle to establish, and then to not undermine, the case for human evolution, the scientific community has perhaps been reticent to acknowledge that humans are cognitively very different from other apes. I confess that this is the mindset with which I began my scientific career. As data from comparative cognition experiments accumulated, however, and the striking differences between the mental abilities of humans and other apes began to crystallize, evolutionary biologists like myself have been forced to accept that something unusual must have happened in the hominin lineage to humanity. That supposition is reinforced by anatomical data, showing a near quadrupling in hominin brain size in the last three million years,[125] by genetic data showing massive upregulation of gene expression in the human brain,[126] and by archaeological data showing hyperexponential increases in the complexity and diversity of our technology and knowledge base.[127] Not all of the respects in which human beings excel are so flattering; we also exhibit unprecedented capabilities for war, crime, destruction, and habitat degradation. Yet these negative attributes also serve to highlight the distinctiveness of our evolutionary journey. How is it all to be understood?

This book sets out to explain the evolution of the extraordinary human capacity for culture, and in the process aims to provide answers to the conundrum of the human mind's emergence. An account is given of how the most singular and definitively human capabilities intermingled to forge a collective existence in our species. The explanation given for the origins of mind and culture cannot be the whole story—far from it, since indubitably many diverse and complex selection pressures must have acted on an organ as complex as the human brain and a cognitive capability that is so multidimensional. The story told is far from conjecture, however; it is supported all the way by scientific findings.

Yet this book is not just about the evolution of culture; it is a description of the scientific program of research dedicated to its unraveling. It synthesizes my work, and that of my students, assistants, and collaborators, who as a team have pursued this topic for over 25 years. It depicts how modern research proceeds, including how scientific questions are addressed, how serendipitous findings are capitalized on, how researchers can be led in new directions by data, and how different scientific methodologies (experiments, observations, statistical analyses, and mathematical models) are interwoven to construct a deeper understanding of a problem. I set out to depict, in an honest way, our struggles, false starts, and moments of insight and despair. In a very real sense, this book is a detective story, describing how one puzzle led to the next, how we followed the trail of clues, and how gradually our efforts were rewarded with a climax as rich and convoluted as in any whodunit mystery. The "answer" that gradually becomes clear as the book progresses, may perhaps be regarded as a new theory of the evolution of mind and intelligence.

Our story begins with the seemingly prosaic observation that countless animals, from tiny fruit flies to gigantic whales, learn life skills and acquire valuable knowledge by copying other individuals. Perhaps surprisingly, an understanding of why they should do so—that is, why copying should be so widespread in nature—had eluded science until quite recently. Indeed, the puzzle was sufficiently challenging that we were forced to organize a scientific competition to address it. The competition solved the conundrum by conclusively demonstrating that copying pays because other individuals prefilter behavior, thereby making adaptive solutions available for others to copy. Running the competition taught us a vital lesson: natural selection will relentlessly favor more and more efficient and accurate means of copying.

Once we understood why animals copy each other, we began to appreciate the clever manner in which they did so. Animal copying was far from mindlessly or universally applied; social learning is highly strategic. Animals follow clever rules, such as "copy only when learning through trial and error would be costly," or "copy the behavior of the majority," which have proven to be highly efficient methods of exploiting the available information. What is more, we began to find that we could predict patterns of copying behavior using evolutionary principles. Subsequently, our experimental and theoretical analyses started to reveal how selection for more efficient and accurate copying had seemingly led some primates to rely more on socially transmitted information. This process supported traditions and cultures comprising databanks of valuable knowledge that conferred on populations the adaptive plasticity to respond flexibly to challenges and create new opportunities for themselves. This heavy reliance on social learning had other, less obvious, consequences as well, including a transformation in how natural selection acted on the evolving primate brain, and its consequent impact on primate cognition. In certain primate lineages, social learning capabilities coevolved with enhanced innovativeness and complex tool use to promote survival. The same feedback mechanisms may have operated in other lineages too, including some birds and whales, but with constraints that did not apply in the primates. The result was a runaway process, in which different components of cognition fed back to reinforce and promote each other, leading to extraordinary growth in brain size in some primate lineages, and to the evolution of high intelligence.

One key insight was that, under stringent conditions identified by mathematical models, this runaway process favored teaching, which is defined here as costly behavior designed to enhance learning in others.

This high-fidelity information transmission allowed hominin culture to diversify and accumulate complexity. Experimental studies and other data suggested that selection for more efficient teaching may have been the critical factor that accounts for why our ancestors evolved language. In turn, the appearance of widespread teaching combined with language was key to the appearance to extensive large-scale human cooperation. As our investigation proceeded, further lines of evidence supported our account, and a picture of what had happened in our lineage began to emerge. Human genetic data, for instance, testified to an unprecedented interaction between cultural and genetic processes in human evolution, fueling a relentless acceleration in the computational power of our brains. The data suggested that the same autocatalytic process has continued right up to the present, with accelerating cultural change driving technological progress and diversification in the arts, leading directly to today's human population explosion and the resultant planetary-scale changes.

What surprised us most about our investigations, however, was that only when we finally felt that we were closing in on a reasonable understanding of the evolutionary origins of the human capability for culture, did it dawn on us that we had stumbled upon so much more. We had inadvertently assembled insights into the birth of intelligence, cooperation, and technology. We had a novel account of the origins of complex society, and a new theory of why humans, and humans alone, possess language. We could explain why our species practices 10,000 or so different religions,[128] and could account for a technological explosion that has generated tens of millions of patents.[129] We could also elucidate how humans can paint sunsets, play football, dance the jitterbug, and solve differential equations.

Something remarkable happened in the lineage leading to humanity. Such a dramatic and distinctive enhancement in mental ability cannot be observed in the ancestry of any other living animal. Humans are more than just souped-up apes; our history embraces a different kind of evolutionary dynamic. All species are unique, but we are uniquely unique. To account for the rise of our species, we must recognize what is genuinely special about us, and explain it using evolutionary principles. Doing so requires analysis of the evolution of culture, because it turns out that culture is far more than just another component, or an outgrowth, of human mental abilities. Human culture is not just a magnificent end product of the evolutionary process, an entity that, like the peacock's tail or the orchid's bloom, is a spectacular outcome of Darwinian laws. For humans, culture is a big part of the explanatory process too. The evolution of the truly extraordinary characteristics of our species—our intelligence, language, cooperation, and technology—have proven difficult to comprehend because, unlike most other evolved characters, they are not adaptive responses to extrinsic conditions. Rather, humans are creatures of their own making. The learned and socially transmitted activities of our ancestors, far more than climate, predators, or disease, created the conditions under which our intelligence evolved. Human minds are not just built *for* culture; they are built *by* culture. In order to understand the evolution of cognition, we must first comprehend the evolution of culture, because for our ancestors and perhaps our ancestors alone, culture transformed the evolutionary process.

CHAPTER 2

UBIQUITOUS COPYING

It is impossible to catch many [animals] in the same place and in the same kind of trap, or to destroy them by the same kind of poison; yet it is improbable that all should have partaken of the poison, and impossible that all should have been caught in the trap. They must learn caution by seeing their brethren caught or poisoned.

—DARWIN, *DESCENT OF MAN*

The brown rat does not, as its Latin name (*Rattus norvegicus*) misleadingly implies, originate in Norway, but rather in China, from which it has spread to all continents apart from Antarctica over the last few hundred years. It has been described as one of "the most successful nonhuman mammals on the planet."[1] Its range and versatility are remarkable; colonies of rats scavenge a living on human garbage in Alaska, subsist on beetles and ground-nesting birds in South Georgia, and flourish in almost all farms and cities in between.[2]

The rats' success in part reflects a long history of dependence on humanity, a relationship in which we have proven an unwelcoming and brutal partner. Yet, in spite of centuries of traps, poisons and fumigations, no pied piper has ever managed to eradicate this most perseverant of pests. The reason, as Darwin intuited, is that rats cunningly avoid all agents of extermination; and they do so through copying.

In Darwin's day, the presiding belief was that children and monkeys imitated, but that the behavior of most animals was controlled by instincts.[3] The adage "monkey see, monkey do" and the phrase "to ape" betray the widespread belief that primates, and perhaps primates alone, copy each other's behavior. As with so many scientific issues, Darwin was ahead of his time in recognizing that copying is ubiquitous in nature. Today, extensive and incontrovertible experimental evidence for social learning exists in a very wide variety of animals.[4]

31

Darwin suspected that a long history of trapping mammalian pests would select for their "sagacity, caution and cunning,"[5] and certainly rats possess these qualities. Decades of control attempts failed in part because rats react to any change in their habitat with extreme apprehension.[6] For several years I studied rat behavior. I observed how any novel food or new object is slowly and stealthily stalked, the body crouched so low that the belly is almost on the floor, with the rat ready to turn tail at the slightest provocation. If nothing bad happens the curious rat will eventually take some food, but feeding will be highly sporadic at first, with only very small amounts of any new food taken.

Up until the middle of last century, the poisons that humans used required rats to eat substantial amounts to be lethal, and the modest amounts of bait ingested frequently just left the rats ill; this would inadvertently train them to avoid the new food source. Despite the occasional initial success in reducing pest numbers, after a short period of trying a new poison, rates of bait acceptance would become increasingly poor, and colonies would rapidly return to their initial sizes.

In the 1950s, the advent of Warfarin, a slow acting poison, proved a successful innovation in the battle to control rats, because the pests felt unwell sufficiently long after consuming the food to not develop bait shyness. Warfarin-type poisons were used against rats and other rodents all over the world, but always with only partial success, eventually giving the population of survivors time to evolve a genetic resistance.

Frustration that rats should remain so stubbornly difficult to eradicate eventually became the impetus for detailed research into rat behavior in the middle of the last century. Fritz Steininger, a German applied ecologist who spent many years studying ways to improve methods of rodent control, was the first scientist to provide data that supported Darwin's belief that rats learn socially to avoid poisons.[7] Decades of observation and experiment led Steiniger to the view that inexperienced rats were dissuaded by experienced individuals from ingesting potential foods by individuals that had learned the bait was toxic. This was an important insight, although Steiniger's interpretation was not correct in the details. In fact, the information transmission mechanisms turn out to be multiple, diverse, and subtle. Decades later, a Canadian psychologist called Jeff Galef—the world's foremost authority on animal social learning—finally got to the bottom of this puzzle.

With a beautifully designed series of experiments conducted over more than 30 years, Galef and his students painstakingly revealed the multiplicity of means by which the feeding patterns of adult rats influence the food choices of other rats, particularly the young. Galef discovered that rats do not actively avoid consuming foods that make others sick, but do acquire strong preferences for eating foods that healthy rats have eaten. These mechanisms are so effective that they support colony-wide dietary traditions that efficiently exploit safe, palatable, and nutritious foods, while leaving toxic foods largely untouched.

Remarkably, the transmission mechanisms begin to operate even before birth. A rat fetus exposed to a flavor while still in its mother's womb will, after birth, exhibit a preference for food with that flavor. Feeding garlic to a pregnant rat enhances the postnatal preference of her young for the odor of garlic in food.[8] The flavors of eaten foods also find their way into the milk of lactating mothers, and suckling rat pups' exposure to such flavors is sufficient to culture a subsequent preference for the same food.[9] Later, when rat pups take their first solid meals, they eat exclusively at food sites where an adult is present,[10] primarily because they follow the adults to these sites and thereby learn cues associated with food.[11] Even when removed from the social group and presented with foods in isolation, youngsters will eat only those foods that they have seen adults eat.[12]

Rats do not even need to be physically present to shape the dietary decisions of the young. When leaving a feeding site, they deposit scent trails that direct young rats seeking food to locations where food was ingested.[13] Moreover, feeding adults deposit residual cues in the form of urine marks and feces, both in the vicinity of a food source and on foods they are eating.[14] As a graduate student at University College London, I investigated the role that these cues played in transmitting dietary preferences. I found that rats leave a rich concentration of marks and feces in the vicinity of food sites,[15] cues that effectively contain the message that "this food is safe to eat." If I disrupted the cues in any way, either by cleaning off the urine marks but leaving the feces, or by removing the feces but not the urine marks, or even by replacing the food with a different food, the "message" immediately lost its potency, and other rats no longer preferred that site. Rats seemed attuned to copy each other faithfully—unless they encountered anything suspicious, in which instance they would rapidly switch into a cautious mode.

I also found that I could establish experimental traditions for feeding on particular foods among groups of rats that never met.[16] I would place a bowl containing a flavored food on one side of a clean enclosure and allow rats to feed there for a few days. Over this period, the rats would mark the food site. Then I would remove the rats, and place an identical bowl containing a differently flavored but equally nutritious food on the other side of the enclosure. Thereafter, every day I would place a new rat in the enclosure, monitor its feeding and marking behavior, and then remove it. I found that the rats would maintain traditions, lasting several days, for eating the foods at the original, marked food bowls—traditions that were upheld over several iterations of replacing the inhabitants. The olfactory cues laid by the original rats lost their potency within 48 hours, which means that for the traditions to be maintained for days, rats must not only choose to feed at marked sites but also reinforce the markings of other rats.

Yet none of the aforementioned processes are thought to be the primary means by which rats transmit dietary preferences. After a rat feeds, other rats will attend to food-related odor cues on its breath, as well as the scent of food on its fur and whiskers, allowing them to identify the foods that others have eaten.[17] The effects of exposure to a recently fed rat on the food choices of its fellows can be surprisingly powerful, and sufficient to override prior preferences and aversions completely.[18] In combination with the other mechanisms for the transmission of dietary preferences, such as scent marks that stabilize transmission,[19] these cues can generate colony-specific traditions for eating particular foods.[20] In this manner, colonies of rats are able to track changes in the palatability and toxicity of diverse and changing foodstuffs efficiently, a critical adaptation for an opportunistic, scavenging omnivore that must subsist on a diverse and constantly changing diet in a dangerous and unpredictable world.

This chapter provides a brief overview of the evidence for social learning in animals. My objective is to demonstrate the ubiquity of copying in nature. Learning from others is an extremely prevalent trick that animals rely on to acquire the skills and knowledge necessary to earn a living in a tough and unforgiving world. All kinds of creatures, from elephants and whales to ants and wood crickets, exploit the wisdom others have accrued. That wisdom, whether it relates to foods, predators, or mates, is absolutely vital to the animal's survival. Later in this book I will show that the diverse roles that social learning plays in the lives of many social animals provides the foundations from which complex cognition evolved.

The ability shown by rats to exploit diet cues on the breath of others is found in several rodent species, as well as dogs and bats.[21] Other animals possess analogous mechanisms. For instance, fish are famously slimy because they produce a mucus secretion that coats their body; it helps them to swim efficiently by reducing drag and protects them from external parasites, which get washed off. My postgraduate student Nicola Atton found that the slime of some fish has evolved an additional quality. The fish secrete food cues in their mucus, as well as in their urine, to which other fish attend. If a recently fed fish emits chemical cues of stress at the same time as these food cues, other fish seemingly draw the inference that the new food is one to be avoided. Conversely, when there are no such stress chemicals in the water, the mucus cues are acted upon and observing fish rapidly develop a preference for the newly consumed diet.[22] Bumblebees possess a similar mechanism; when successful foragers bring home nectar to the nest, they deposit the scented solution in honeypots, where other colony members sample it and thereby acquire a preference for the floral scent.[23] Eating what others eat is a highly adaptive strategy, provided effective mechanisms are in place to prevent "bad" information from spreading.

The pervasiveness of animal social learning is a recent revelation that has surprised the scientific community.[24] Thirty years ago, when I first started studying animal social learning and tradition, there was a strong belief among researchers that social learning was predominantly found in large-brained animals. We were, of course, all aware of cases such as the spread of milk-bottle opening in birds, where a dozen or so species, including great tits and blue tits, starting pecking open the foil caps of milk bottles delivered by milkmen to European doorsteps, to drink the cream.[25] Also well established was the finding that many songbirds learn their songs from adult tutors, and that such learning could generate vocal dialects in different geographical locations.[26] Regional variation in the songs of several birds had been documented, notably in white-crowned sparrows and chaffinches, and this was often referred to as "cultural" variation.[27] However, milk-bottle opening and bird song were widely regarded as specialized mechanisms that did not imply the species concerned were capable of learning additional behavioral habits from others. Researchers tended to assume that natural selection had fashioned dedicated mechanisms in these animals that allowed them to acquire particular kinds of information socially, rather than resulting in a general copying competence. Likewise, the famous waggle dance of the honeybees,[28] which transmitted information about the location of food sources, was regarded as a specialized adaptation, tailored to a narrow species-specific context; it was thought to be a trait analogous, rather than homologous, to human culture.

If there was a paradigmatic exemplar of animal social learning it was sweet-potato washing in Japanese macaques. In 1953 a young female Japanese macaque called Imo, whose troop lived on the small Islet of Koshima in Japan, began washing sweet potatoes in a freshwater stream before eating them.[29] Imo's troop had been provisioned with this novel food on the beach by Japanese primatologists. Seemingly, the food washing functioned to remove dirt and sand grains prior to eating, and that a monkey should exhibit such hygienic behavior appeared remarkably civilized and humanlike, and excited considerable attention.

The habit spread, and soon other monkeys in the troop were washing the provisioned food, either in the stream or in the sea. When, three years after her first invention, Imo devised a second novel foraging behavior, that of separating wheat from sand by throwing mixed handfuls into water and scooping out the floating grains,[30] she was destined to become something of a celebrity. Renowned Harvard biologist Edward Wilson characterized Imo as "a monkey genius,"[31] while Jane Goodall, an eminent authority on chimpanzee behavior, described her as "gifted."[32] Whether such plaudits are justified is an issue taken up in a later chapter. What is not in doubt, however, is that Imo's inventions spread through the troop. What is more, this was no fluke; macaques exhibit many behavioral traditions.[33]

In the 1970s and 1980s, primatologist Bill McGrew compiled evidence for diverse behavioral traditions among chimpanzee populations in Africa.[34] Evidence began to emerge for traditional behavior in several other apes and monkeys too, and the impression that social learning was a distinctive characteristic of primates became highly prevalent.[35] As we humans are both cerebral and highly reliant on social learning, researchers, perhaps naturally, linked these attributes and began to assume that effective copying would be restricted to those species most closely related to ourselves. This intuition proved to be entirely fallacious.

Certainly, social learning is widespread in monkeys and apes. The most celebrated example concerns the distinctive tool-using traditions of chimpanzees throughout Africa, which were brought to prominence through a landmark article in the journal Nature by developmental psychologist Andrew Whiten and his colleagues.[36] Some chimpanzee populations use stalks to probe for termites, others fish for ants or honey in the same manner, and still others crack open nuts with stone hammers. Each region has chimpanzees with their own repertoire of habits,[37] and each repertoire extends far beyond the foraging domain. Less well known are learned traditions for grooming with particular postures, dancing in the rain, and using plants as medicines.[38] Developmental data provide evidence that these behavior patterns are acquired through social learning.[39] For instance, chimpanzees at Gombe National Park in Tanzania will insert stalks and other probes into termite mounds to extract the termites. Primatologist Elizabeth Lonsdorf found that the amount of time mothers spend termite fishing correlates strongly with the number of aspects of this fishing that young chimpanzees acquired.[40] Revealingly, young females spent lots of time watching their mothers, and thereby acquired the same technique, while sons spent far less time watching, and their foraging technique did not correlate with their mothers'.[41]

Orangutans,[42] another close relative of humans, also share distinctive group-specific traditions for feeding, nesting, and communicating.[43] Like chimpanzees, many orangutan cultural behaviors involve foraging with tools, such as using leaves to handle spiny fruits or scooping water out of a crevice in a tree. Others relate to building behavior, such as manufacturing an umbrellalike cover for protection from the elements, and communication signals such as the "kiss-squeak"; for the latter, orangutans use their hands as a sound box to make their calls sound deeper, thereby making themselves sound bigger in order to ward off predators. The function of some orangutan habits remains a puzzle. For instance, at least three populations have the curious habit of blowing raspberries as they go to sleep.[44] Other orangutan traditions are remarkably evocative of human behavior. That orangutans might make "cups" for drinking rainwater from leaves, or "beds" to sleep in, is perhaps not too much of a surprise. However, two populations of Borneo orangutans have been observed to make themselves a bundle of leaves which they cuddle at bedtime like a doll.[45]

Equally striking are the bizarre social conventions found in Costa Rica's capuchin monkeys, brought to prominence through many years of careful study by UCLA primatologist Susan Perry and her coworkers.[46] These researchers found that specific monkey populations possess some quite extraordinary regional habits, including sniffing each other's hands, sucking of each other's body parts, and placing fingers in the mouths and eyes of other monkeys.[47] For instance, in one group found in the Lomas Barbudal reserve, pairs of monkeys commonly insert their fingers in each other's nostrils simultaneously and remain in this pose for several minutes, sometimes swaying in a trance-like state. In two other groups (Cuajiniquil and Station Troop), hand sniffing is combined with finger-sucking behavior, while monkeys at Pelon engage in eyeball poking, where a finger is inserted between the other monkey's eyelid and eyeball up to the knuckle (figure 1). The monkeys are thought to use these group- or clique-specific social conventions to test the quality of their social relationships. Likewise, while the Japanese macaques' food-washing habits make functional sense, the tradition, observed in some populations of this species, to bang together rocks for hours on end remains a complete mystery.[48] Perhaps it is the precursor of some musical tendency, perhaps it is a social signal, or perhaps it is a dysfunctional byproduct of boredom, or excess time.

Yet while the prevalence and diversity of their traditions leave no doubt that social learning is vital to many primate species, they do not preclude the possibility that copying is equally central to other animals. As ever, Darwin was more perceptive than most. In an 1841 letter he wrote to a periodical called *The Gardeners' Chronicle*, Darwin noted that some honeybees had adopted the bumblebee's habit of cutting holes in flowers to rob them of nectar, and speculated that this trick had been acquired through interspecific copying. He wrote:

> Should this be verified, it will, I think, be a very instructive case of acquired knowledge in insects. We should be astonished did one

FIGURE 1. White-faced capuchins in Costa Rica possess extraordinary social conventions, which vary from one population to the next. Here two adult females (Rumor and Sedonia) from the Pelon group demonstrate the curious local traditions of hand sniffing and eyeball poking. Rumor, a serial innovator, is thought to have invented eyeball poking. By permission of Susan Perry.

genus of monkeys adopt from another a particular manner of opening hard-shelled fruit; how much more so ought we to be in a tribe of insects so pre-eminent for their instinctive faculties.[49]

Whether Darwin was right about honeybees copying bumblebees is difficult now to determine,[50] but we do now know that the bumblebee's habit of nectar robbing, no less than the monkey's use of tools to crack open nuts, is a socially transmitted tradition.

Not just knowledge of what to eat, but where to find food and how to process it, are often socially transmitted among animals. Countless species, from very diverse taxonomic groups, acquire relevant foraging knowledge through interaction with, or observation of, other animals. One of the most compelling studies was carried out by Norwegian Tore Slagsvold and Canadian Karen Wiebe, a team of animal behaviorists who studied social learning in the wild by moving eggs of blue tits to nests of great tits,[51] and vice versa (this experimental procedure is known as "cross-fostering").[52] These birds live close to one another and forage in mixed-species flocks, but have quite distinct feeding niches, which until recently had been assumed to be the result of evolved, unlearned preferences. Blue tits feed mainly from twigs high in trees, eating buds, grubs, and moths; whereas great tits feed mostly on the ground or on the trunks and thicker branches of trees, consuming larger invertebrate food items. Like many animals, these birds forage together in mixed-species groups because large numbers provide a more effective defense against predators compared to small aggregations, and gathering with these particular flock mates has the additional advantage of not having to compete for food.

Slagsvold and Wiebe were able to quantify the consequences of being reared by foster parents from a different species in an environment otherwise natural to the birds. The cross-fostering approach dramatically demonstrated an effect of early learning on a large number of behaviors.[53] Blue tits reared by great tits adopted great tit foraging habits, and vice versa. The height at which the birds foraged in trees, as well as their type and size of prey, shifted in the direction of the foster species as a result of this social learning experience. The great tits sometimes even tried to forage hanging upside down like their blue tit foster parents, even though they kept falling off! The birds' nest-site choices exhibited a similar shift toward the foster parents' inclinations,[54] as did mating preferences,[55] song variants,[56] and alarm calls.[57] The birds learned an enormous part of their species-typical behavioral repertoire socially.

Countless other studies provide evidence that diverse behavior patterns are learned socially. Dolphins possess traditions for foraging using sponges as probing tools to flush out fish hiding on the sea bottom.[58] Killer whales have seal-hunting traditions, including the method of knocking seals off ice floes by charging toward them in unison and creating a giant wave.[59] Archerfish, who dramatically shoot down flying insect prey by spitting droplets of water at them, can learn this habit through observing others.[60] Animals as distinct as meerkats and honeybees share population-specific bedtime habits, some groups being early and others being late risers—such traditions cannot be explained by ecological differences.[61] Even chickens can acquire bloodthirsty cannibalistic habits through social learning.[62] This experimental study found that watching other birds feed on blood sufficed to elicit canni-balistic tendencies. Cannibalism is widespread in the animal kingdom, both in wild populations and in factory-farmed poultry; it is a serious welfare problem in the latter, and understanding its causes has major economic ramifications.[63]

The ubiquitous influence of social learning in nature is beautifully illustrated by the example of mate-choice copying, where an animal's choice of partner is shaped by the mating decisions of other, same-sex individuals. This form of copying is extremely widespread, with exam-ples known among insects,[64] fishes,[65] birds,[66] and mammals,[67] including humans.[68] The fact that animals do not require a big brain to copy could not be more clearly demonstrated than by the tendency of tiny female fruit flies to select male flies that other females have chosen as mates.[69]

Nor is mate-choice copying restricted to cases where individuals di-rectly observe the courtship or mating of others; just like the little mes-sages left by rats with their excretory deposits, indirect cues of mating choices can have the same effect. In many fish species, males build nests and females select among these nests to decide where to lay their eggs. Usually this decision is based on the female's assessment of the male's quality, but in some species her choice depends more on the character-istics of the male's nest. In some species, females nest choice has been shown to be influenced by the number of eggs already within, with popular nests becoming increasingly successful.[70] Seemingly, female fish interpret the presence of a large haul of eggs as an indication that many females have chosen the nest's owner as a mate, and infer that he must be a high-caliber male. Getting a threshold number of eggs in one's nest is so vital to attracting females that in some species males have actu-ally been observed to steal eggs from other nests to increase their future success.[71] Evolutionary biologists tend to assume that male animals will do what they can to avoid being "cuckolded" and raising another male's offspring. However, here male fish embrace such cuckolding as a means to manipulate females and enhance their own reproductive success.

Perhaps the best-studied example of mate-choice copying is in the guppy,[72] a small South American tropical fish popular with aquarium enthusiasts. Biologist Lee Dugatkin at the University of Louisville con-ducted a series of experiments in which two male guppies were placed behind transparent partitions at either end of an aquarium, with a "demonstrator" female fish near one of the males, giving the impression that she had chosen him as a mate.[73] A focal female was then placed into the middle of the tank and allowed to observe the males. Subsequently, the demonstrator female was removed, and the focal female freed to swim throughout the aquarium, which allowed the two males to court her. The experiment found that focal females spent a significantly greater amount of time in the vicinity of the male that had been near the female; that is, her mate-choice decision had seemingly been influenced by the apparent choice of the demonstrator fish. As with the rats picking up cues on the breath of other rats, this mate-choice copying effect was strong enough to reverse prior preferences.[74] Males hitherto regarded as unappealing suddenly became of interest to a female, once other females appeared to choose them.

Another small tropical fish, the Atlantic molly,[75] also exhibits mate-choice copying.[76] However, here males also engage in copying behavior, preferring females that other males have selected as mates. Interestingly, this has led to natural selection favoring deceptive behavior in these fish as a male strategy to reduce the competition. When their courtship is being watched by rivals, male mollies switch to courting the lesser preferred of two females to mislead the observer into pursuing the less attractive quarry![77] Remarkably, humans aside, the male Atlantic molly's behavior is the only known example of deliberate deception to hinder social learning in the animal kingdom. In principle, one of the major problems with a copying strategy is that it may not be in the copied individual's interests to ensure that the copier receives accurate information. Why, in spite of this, animal social learning should remain largely honest is an issue to which we will return in later chapters.

Social learning proves important in domains other than foraging and mate choice. Previously mentioned is the extensive experimental evidence that many male songbirds learn their songs from their fathers, or more commonly, from neighboring adult males, with this learning frequently generating local song variant traditions known as dialects.[78] Recent studies demonstrate the existence of vocal traditions in many mammals as well, particularly in whales and dolphins.[79] Much of this research has focused on bottlenose dolphins,[80] killer whales,[81] and humpback whales.[82] For instance, all males in a humpback whale population share a song that changes gradually through the singing season, an alteration much too rapid to be explained by changes in genes.[83] Rather, humpback whales appear to acquire their songs through social learning, with continuously introduced changes then dispersed from whale to whale throughout the ocean. However, the songs that are sung by humpbacks in the Pacific, Atlantic, and Indian Oceans are quite distinct. Occasionally these tunes are seen to undergo a revolution. Strikingly, in 1996 in the Pacific Ocean just off the east coast of Australia, two humpback whales were first heard singing a novel song that differed substantially from the dominant song of the other eighty humpback whales in the vicinity. A year later, other whales were singing the new song, and by 1998, only two years after its introduction, all recorded whales in the Pacific were singing the new tune.[84] The novel variant resembled the song sung by Indian Ocean whales, on the other side of Australia, leading to the hypothesis that a small number of humpbacks had swum from one ocean to the other, bringing their catchy song along with them. More recent work suggests that such song revolutions may occur on a regular basis, and intriguingly always spread in the same direction, like cultural ripples extending eastward through populations in the western and central South Pacific.[85]

For many animals, important locations such as profitable food patches, areas safe from predation, resting sites, suitable areas to find mates and reproduce, as well as safe routes between these locations, must be learned. Fishes provide some of the best evidence for this form of social learning.[86] Many fish species exhibit learned traditions for reusing mating sites, schooling sites, resting sites, feeding sites, and pathways through their natural environments, repeatedly returning to the same locations for each activity on a regular daily, seasonal, or annual basis.[87] For instance, socially learned mating site traditions have been found to be present in bluehead wrasse,[88] whose mating-site locations in the Caribbean coral reefs remain in place over many generations. In theory, such traditions need not be indicative of social learning—genetic differences, or variation in the local ecology, could underlie any behavioral differences between populations. To investigate the role that learning played, evolutionary ecologist Robert Warner, from the University of California at Santa Barbara, removed entire populations of the wrasse and replaced them with other transplanted wrasse populations. Warner reasoned that if it was features of the environment or ecology that determined mating sites, then the new populations would adopt the same sites as had the old ones. Conversely, if these were learned traditions then there would be no reason to expect the new populations to adopt the same mating sites as the previous inhabitants.

Warner found that the wrasses established entirely new mating sites, which remained constant over the 12-year period of the study.[89] However, in a later study, when Warner replaced newly established populations after just one month, he found that the introduced fish used the same sites as their immediate predecessors.[90] Apparently, the fish initially choose mating sites and pathways based on their assessment of the optimal use of resources in the environment, and then these behavioral patterns become established as learned traditions. Subsequently, when aspects of the environment changed, the tradition was preserved, and the behavior of the fish was different than that expected from considerations of ecology alone. This phenomenon is known as "cultural inertia,"[91] named after cases such as the Viking settlement in Greenland, which collapsed because the settlers failed to adjust their culture to the new environmental conditions.[92] High levels of intermixing observed during the early life of the wrasse suggest that reef populations are not subject to significant genetic differentiation; combined with the observed traditionality, this research provides compelling evidence of cultural variation.

Field studies on learned migratory traditions like these in fish were the inspiration for some experiments that my students and I carried out in the laboratory. We wanted to evaluate the hypothesis that fish could acquire knowledge of the location of important resources simply by following knowledgeable individuals. Kerry Williams, an undergraduate student at the University of Cambridge, carried out a small-scale version of the fish migration studies to investigate the underlying mechanisms.[93] Over repeated trials, Kerry trained demonstrator guppies to take one of two alternative routes to a food source in laboratory aquaria. Then she introduced untrained fish into the populations, who tended to shoal with their demonstrators, and thereby take the same route to food. After five days of trials, the subjects were tested alone, and showed a significant preference for taking the same route as their demonstrators, despite the presence of an alternative route of equal distance and complexity. Kerry had shown that simply by shoaling with experienced individuals, fish could learn a route to food. Moreover, the more demonstrator fish that were swimming the route, the more effectively the experimental subjects learned. Multiple demonstrators reinforced each other's behavior to enhance their reliability and provide a very strong, clear indication of which route to take.[94]

We went on to conduct experiments using a transmission chain design, where small shoals were trained to take one of two routes, and these trained "founders" were then gradually replaced by naive individuals to see if the route preferences were retained in spite of the turnover in shoal composition.[95] Sure enough, several days after the original founders had been removed, the route preferences were still being maintained in the groups. Even when one route was substantially longer and more energetically costly than the alternative, it was still being widely used by individuals whose founders had been trained to swim that way.

Later, at the University of St Andrews, we demonstrated that not just routes, but also foraging techniques, could be maintained as traditions in laboratory populations.[96] We trained demonstrator fish to feed by swimming directly up into narrow vertical tubes that were closed at the top—a challenge that required them to swim in a manner not normally observed in these fish. In spite of its simplicity, this was a foraging task that the fish could not solve by themselves without training. While trained individuals reliably fed from these tubes, no naive fish presented with a vertical tube ever learned to feed from it on its own. However, when placed in groups with experienced demonstrators, untrained fish rapidly learned to feed from the vertical tubes, and traditions could be established that maintained this novel foraging behavior through social learning.

The laboratory traditions established by these experiments lasted days or weeks rather than years, but nonetheless suggest plausible mechanisms underlying the more stable traditions witnessed in natural populations.[97] Our experiments have established that fish prefer to join large shoals compared with small shoals,[98] and exhibit a tendency to adopt the majority behavior.[99] Simple processes like shoaling, copying the behavior of others when uncertain, and disproportionately attending to the behavior of groups collectively generate traditions that can become extremely stable, even to the point of preserving arbitrary and even maladaptive behavior.[100] It is these simple mechanisms that generate the cultural inertia observed in wild populations. Evolutionary biologists tend to expect that animals will match their behavior optimally to the environment, and often that appears to be the case. However, field experiments, like Robert Warner's wrasse study, show how the mating and schooling sites of natural populations cannot always be predicted from features of the environment, while controlled laboratory experiments help to unravel why.[101]

Similar processes may underpin the long-distance annual migrations exhibited by birds. A recent study by ecologist Thomas Mueller of the University of Maryland provides compelling evidence that, among migrating whooping cranes, more experienced birds transmit route knowledge to less experienced individuals.[102] Mueller and his colleagues devised an innovative training regime for a reintroduced population of migratory whooping cranes using ultralight aircraft. Captive-bred birds were trained to follow the aircraft on their first lifetime migration. For subsequent migrations, in which birds flew individually or in groups, the researchers found a dominant influence of social learning on migratory performance. The data strongly imply that younger birds typically learn aspects of the route by flying with more experienced birds. The same pattern is even observed in insects. For instance, when they are novice foragers, honeybees are more likely to follow the instructions encoded in dances for locating food sites rather than to search independently, while experienced foragers typically only follow dances if their previous trip was unsuccessful.[103]

This brief overview of the extent of social learning in nature would be incomplete without mention of one last domain in which it proves critical—recognizing and escaping predators. Avoiding being eaten is obviously a major priority for any animal, but gaining accurate knowledge of predators is not easy. While many species possess evolved antipredator mechanisms, overreliance on preestablished predator-evasion strategies would be disastrous if any novel predator with a new tactic appeared on the scene. A changing world requires animals to update their antipredator behavior continuously through learning. Yet this is a domain in which learning through trial and error is extremely difficult, because with the very first error an animal will likely end up inside the predator's stomach. No surprise, then, that the social transmission of fears and antipredator behavior should be one of the most prevalent forms of copying in nature.

Rhesus monkeys,[104] which live in the grasslands and forests of Asia, are vulnerable to a number of predators, including big cats, dogs, raptors, and particularly snakes. However, rhesus monkeys reared in captivity exhibit no fear of snakes, which shows the antipredator behavior found in natural populations is learned. In fact, youngsters only learn that snakes are a threat when they see more experienced monkeys responding fearfully to the snake with screams, facial expressions of terror, and desperate attempts to escape. Careful experiments have allowed researchers to establish that this observational experience allows young macaques to learn the identity of predators by developing an association between the snake stimulus and the fearful response of other monkeys, which triggers an emotional response in them.[105] The experimenters showed monkeys either live presentations, or video footage, of other monkeys reacting fearfully to snakes or, through clever experimental manipulations, to objects that do not normally induce fear—such as flowers—and subsequently tested their response to the same stimuli.[106] When ecologically relevant objects such as snakes were used, the resulting fear learning in the observer was rapid and strong.[107] A single social encounter with a fearful monkey combined with a snake produced a robust fear response in the observer that lasted several months.[108] However, no such conditioning occurred with fear-irrelevant stimuli. Findings such as these strongly imply that the observational fear-learning mechanism has been tailored by natural selection to be biased toward the recognition of genuine threats. An advantage of learning about predators in this manner is that it potentially allows monkeys to acquire a fear of any kind of snake, irrespective of its color or size, and to do so very rapidly, but not to acquire superstitious fears of safe objects in their environment, such as flowers.

The specificity of the monkey's fear learning stands in contrast with the findings of a similar study of predator learning in European blackbirds.[109] These birds often aggregate to drive off threats, swooping down to harass owls, hawks, and other predators. Young birds learn to recognize danger in part through witnessing this mobbing behavior. Through clever experimental manipulations, Ernst Curio and his colleagues from

Ruhr University Bochum in Germany, were able to trick young blackbirds into thinking the adults were mobbing a stuffed owl, a harmless friarbird, and even a plastic bottle; afterward, the young birds would mob all of these stimuli, seemingly convinced that they were dangerous.[110] Apparently, in these birds, unlike the monkeys, natural selection has not yet effectively fine-tuned the selectivity of their fear learning.

Given the clear adaptive value of acquiring fears through the comparative safety of observing others, it should come as no surprise that many animals, including insects, fishes, birds, mice, cats, cows, and primates, all do so.[111] Researchers are currently exploiting this observational learning capability to enhance conservation and restocking efforts.[112] For instance, Culum Brown, an Australian biologist who spent a period as a postdoctoral researcher in my laboratory at Cambridge, discovered that showing young salmon "video nasties" of other salmon being eaten by a pike was sufficient to train them to avoid large predators, a crucial life skill for a young fish. He was also able to "teach" salmon fry to consume appropriate novel foods by watching more experienced fish.[113] Subsequently, some Queensland hatcheries exploited our social learning protocols as part of their efforts to enhance the returns of salmon and other fish introduced into rivers for restocking. Hatchery fish are typically reared in huge vats in unnaturally high densities and fed pellet food, and when released in their millions must rapidly learn to recognize foods and predators, or die. Historically, survival rates have been only a few percent. Just a little prerelease training can make a big difference to both survival rates and hatchery returns.

This brief tour of the prevalence of copying in nature only scratches the surface of the myriad of different ways in which animals exploit information provided by others. The animal behavior research literature is replete with social learning experiments, reports of novel behavior spreading through animal populations, and traditional differences between populations, which number into the thousands. I have presented examples from some of the better-studied functional domains in some of the most intensively researched animal systems. However, social learning is so useful that it crops up in contexts that are far less intuitive, including some instances in which science has yet to understand the function of the transmitted behavior. Far from being restricted to clever, large-brained, or cognitively sophisticated animals, or to those closely related to ourselves, copying is everywhere in nature, at least among animals complex enough to be capable of associative learning. Animals regularly invent new solutions to problems, and these innovations often spread through the population, sometimes generating behavioral differences akin to "cultures." Darwin was correct in his belief that animal behavior is not completely controlled by "instincts" and "innate tendencies,"[114] but is also influenced by learned and socially transmitted wisdom. The prevalence of copying, and the success that it brings animals as different as bees, rats, and orangutans testifies to its utility.

We humans also exhibit extensive social learning. Like the monkeys, children can acquire a strong and persistent aversive response to a fear-relevant object—including a toy snake—after seeing it paired with their mothers' fear expressions.[115] Children with animal phobias or extreme fears toward certain situations, such as darkness, often report having observed parents fearful in the same or similar situations.[116] While such phobias may appear problematical, they are the outcome of a highly adaptive process. As a general strategy, it makes perfect sense for us to become fearful of anything that elicits fear in other humans. Copying others is a highly adaptive strategy and one in which, as further chapters will document, humanity has become particularly adept.

The research described in this chapter raises a rather obvious, but no less fascinating, question: What is so good about copying that it should be so widespread in nature? This seemingly innocent question is packed with hidden complexity. At first sight, the answer seems obvious— copying allows animals rapidly to acquire valuable knowledge and skills. However, evolutionary biologists have struggled with this answer for decades, because mathematical models show that intuition is not quite correct. The theory implied that copying was often as likely to lead to the transmission of inappropriate or outdated ideas as good ones, and hence would not guarantee success. Such analyses suggested that asocial learning was what allowed populations to track their changing environments. Why it should pay to copy others escalated into a major scientific conundrum known as "Rogers' paradox," after the University of Utah anthropologist Alan Rogers, who first drew attention to it.[117] Only in the last few years has the answer finally become clear. An international competition finally solved the problem, and that competition and the insights that it gleaned are the topic of the next chapter.

CHAPTER 3

WHY COPY?

The proudest moment of my academic career was when a photograph of my three-year-old son appeared in the pages of *Science*. The picture, which showed me mowing the lawn with a small boy joyfully pushing a toy mower in my wake, was featured in a commentary accompanying a scientific article of mine in the same edition (figure 2).[1] Our article was about copying—it presented findings that explained both why copying is so widespread in nature, and why we humans happen to be so good at it—leaving the picture wonderfully apt. Rarely does parental pride and academic achievement coincide so perfectly.

You might be forgiven for thinking that I had staged a photo shoot for the magazine's pages, but in fact the photograph had been taken years earlier, in the garden of our previous home, where that toy lawn mower had been pushed up and down on hundreds of occasions. Every time I mowed the lawn my son would rush out and grab his mower to accompany me. For years he did this, and didn't completely stop until he was around ten. On rational grounds, it is hard to understand why imitating a father in this way should have brought so much pleasure, but it did.

Readers who are parents will likely recall similar imitative tendencies among their own offspring. Young children very commonly go through a phase when they copy a person with whom they identify, or to whom they are emotionally attached. Between the ages of two and four my son seemed constantly to be copying everything I did. I remember a little toy shaving kit, with plastic razor and fake shaving cream, gleefully opened by a jubilant toddler each time I shaved. With the arrival of his little sister, the imitator became the imitated. Our young daughter latched on to her big brother, following him everywhere and copying what he said and did. On one occasion my son decided to try to turn off the light switch by hitting it, hurting his hand in the process. In spite of his yelp of pain and tears, his sister immediately tried the same trick.

FIGURE 2. Like father, like son. The author's lawn-mowing behavior was enthusiastically copied for many years. The tendency for imitation among infants is not only critical to child development, but may have played a pivotal role in the evolution of the human mind. By permission of Gillian Brown.

The phenomenon of children being so imitative has been the focus of intense scientific research by developmental psychologists for decades.[2] Classic experiments by Stanford psychologist Albert Bandura in the 1960s established that young children could acquire violent behavior after watching adults behave aggressively toward an inflatable "Bobo doll."[3] Bandura's experiments are widely credited with changing the face of modern psychology; they demonstrated how human beings frequently learn through observation, rather than through direct reward or punishment. Obviously, children do not just acquire aggressive tendencies through social learning, but also pick up useful skills and knowledge in this manner. However, the fact that imitation waxes and wanes at an early age and peaks around age four, but never completely disappears, suggests that childhood imitation also serves a social function—to cement relationships.[4]

Imitation is far from the only form of social learning in which humans engage. Much information is acquired through direct instruction, or through subtler motivational or attentional processes, but imitation is unquestionably an important form of human social learning. Even where, as in the above examples, social learning appears irrational and slavish, the copying is still discriminating. Children do not copy everything that they see and hear, but imitate strategically, according to a set of rules. Those rules might sometimes appear curious or even bizarre, but social learning researchers have made sense of them with the use of principles derived from evolutionary theory.

Most human beings—even those with no particular scholastic bent— exhibit an insatiable thirst for knowledge. From the moment we are born to the day we die, we drink up a virtual ocean of cultural information. So much knowledge is acquired from others that it is easy to forget we are actually highly selective about how and what we learn. Even leaving formal education to one side, our formative years entail a constant uptake of knowledge and skills as we learn from parents and other important people in our world how to walk, talk, and play, what is good behavior and what is bad, how to throw a ball, how to cook, clean, drive, shop, pray, and what to think about money, religion, politics, drugs, and countless other matters. Yet, even though human children may be especially prepared by their evolutionary past to absorb what others tell them, and despite the fact that we are more culturally dependent than any other species on earth,[5] we remain highly discriminating about what we copy.

If our social learning was genuinely mindless, then each time we went to a musical we would burst out singing. Were we really indiscriminating about our imitation, then every time we saw a violent movie we would turn into vicious fiends. Of course, the possibility that television, cinema, and computer-game violence might cause aggression is a serious and legitimate concern. Numerous studies have found a correlation between violent video-game play and aggressive behavior.[6] However, such findings are not straightforward to interpret because even if elevated levels of violence were detected in *Grand Theft Auto* game addicts, it is difficult to rule out the possibility that people with a violent disposition are drawn to such video games, rather than made violent by them. What is clear from such studies is that if causal effects of media violence on aggression do exist, their influence is relatively subtle. Media violence may have a terrifying influence on a tiny minority of susceptible viewers, or perhaps exert a weaker, or shorter-lasting, effect on a broader subset of users; however, clearly a majority of people are able to watch movies such as *Rambo* or *Natural Born Killers* without themselves becoming murderers. Copycat crimes do occur, but many of the people who mimic crimes seen in the media have mental-health problems or histories of violence.

Copycat suicides also occur, and the possibility is sufficiently real that as a means of prevention, it is customary in many countries for the police and media to discourage detail in reporting. On rare occasions, spates of suicide mimics have spread contagiously through a school or local community, or—following a celebrity suicide—created a blip in national statistics. Marilyn Monroe's death, for instance, which resulted from an overdose of barbiturates, was followed by an increase of 200 more suicides than average for that August month.[7] However individually tragic, such cases remain exceptional; many hundreds of millions of people learned of Marilyn Monroe's death without following suit.

Relevant here are experimental studies of childhood social learning that report a tendency for humans to engage in what has been dubbed "overimitation," whereby when learning to perform a task, children, but not chimpanzees, will copy "irrelevant" actions performed by the demonstrator.[8] A small industry of studies has built up around this basic finding. However, the characterization of such copying as "unselective,"[9] or "inefficient,"[10] is highly misleading, as this tendency almost certainly has a social function,[11] with any appearance of blanket copying a likely artifact of the impoverished experimental setup. More recent experimental investigations have established that where children see multiple demonstrations with some individuals, but not others, performing the irrelevant actions, the children rapidly infer that the irrelevant actions are unecessary; rates of overimitation then plummet.[12] Likewise, when children take part in transmission chain studies in which the solution to a puzzle box task is passed along a chain, any irrelevant actions initially introduced by the demonstrators are rapidly dropped and the transmitted knowledge converges on necessary actions.[13] Human beings copy; they copy a great deal. But they do not copy slavishly. Slavish copying would not be adaptive.

Copying, or *social learning* is, of course, not the only means through which humans acquire new knowledge—we, and other animals too, can learn through our own efforts, such as through trial and error, which is called *asocial learning*. Several theoretical analyses using evolutionary models have concluded that some mixture of social and asocial learning is usually necessary for animals to thrive in a variable and changing environment.[14] An intuitive way to see this is by analogy. Wherever some animals are able to find or produce food, other animals will typically come along and try to steal it from them. At least for larger or dominant individuals, scrounging food produced by others is easier than producing it for themselves. As a result, a group of animals—say, a flock of starlings or finches that forage together—typically comprises a balance of food producers and food scroungers.[15] In such groups, producers and scroungers typically receive roughly equivalent amounts of food.[16] This is no coincidence. Animals will switch strategy, from scrounging to producing, or vice versa, if the alternative strategy proves more productive. If there are many food producers, it is easy and cheap to scrounge, but when producers are rare, such that scrounging is not profitable, then individuals are forced to find their own food. The net result is a frequency-dependent balance comprising a mix of producers and scroungers.

The same reasoning applies to learning. Some individuals solve novel tasks through trial and error, interacting directly with the environment rather than observing others, and in the process they *produce* the knowledge of how to solve the problem. For instance, they might have to carry out a protracted search to find water or shelter, risk consuming potentially hazardous substances in order to identify novel foods, or learn the identity of predators by narrowly escaping being eaten. Such individuals, known as "asocial learners," thereby incur significant costs through their learning.

Asocial learning may be costly but, in contrast to the alternative strategy of social learning, it garners accurate, reliable, and up-to-date information. Social learning, on the other hand, is *information scrounging*. Through observation, individuals obtain information cheaply from others—concerning, for instance, where to find shelter or how to escape predators. However, social learners are vulnerable to acquiring out-dated information or knowledge that is more germane to the individual that they have copied than to themselves, particularly in a changing or spatially variable environment. To get reliable information, individuals need to copy those individuals who have directly interacted with the environment, including, for instance, asocial learners.[17] Consequently, theoretical studies predict a mixture of social and asocial learning in the population. In the same way that the foraging returns to producers and scroungers are expected to be equivalent, mathematical models predict that the population will reach an equilibrium at which the payoff to asocial and social learning strategies will be equal. The logic is identi-cal—if any strategy were more profitable, individuals would switch. In the language of evolutionary biology, at equilibrium the two strategies of asocial and social learning are expected to have equal fitness—that is, to have an equivalent effect on the chances of an individual surviving and reproducing.[18]

Anthropologist Alan Rogers first pointed out the "paradox" inherent in the observation that the fitness of social learners at this equilibrium would be no greater than that of asocial learners; he made this conclu-sion with the help of a mathematical analysis, as mentioned in the pre-ceding chapter.[19] At one level, the finding makes perfect sense. When social learning is rare, its payoff exceeds that of asocial learning, since reliable information generated by the prevalent asocial learners is com-mon in the population. As they possess higher fitness, the proportion of social learners initially increases through natural selection. However, as the frequency of social learners rises, there are fewer asocial learners producing reliable information, and the former become more likely to pick up misinformation; then the payoff to social learning starts to decline. At the extreme, if there were no asocial learners present, ev-eryone would be copying everyone else, but nobody would directly interact with the environment to determine the best behavior. Then, if the environment changed—for example, if a new predator appeared on the scene—the results could be disastrous, because no one would have learned to identify or evade the novel threat. Under such circum-stances, the fitness of asocial learning exceeds that of social learning, and asocial learners start to become more prevalent. Accordingly, the population is expected to evolve to reach a balance of social and asocial learning, which is known as a mixed evolutionarily stable strategy (ESS),[20] where by definition, the fitness of social learning equals that of asocial learning.[21]

As noted earlier, this finding is known as *"Rogers' paradox,"*[22] so called because it ostensibly conflicts with the commonly held assertion that culture enhances biological fitness. Ultimately, fitness in evolutionary terms comes down to how many descendants one leaves. Characters with high fitness are those that help organisms to survive and reproduce, and thereby leave lots of descendants. Human culture appears to confer high fitness since the spread of technological innovations has repeatedly led to increases in population size, which implies more individuals survive and reproduce. Indeed, the main reason why human culture is thought to be instrumental in our species' success is that it is associated with population growth. The world's population, which was around a million just 10 thousand years ago, now exceeds 7 billion.[23] With the agricultural and industrial revolutions, birth rates and life expectancy have increased dramatically.[24] These data imply that the spread of advances in technology can increase the average number of surviving offspring. Against this backdrop, Rogers' result seems paradoxical, as it appears to challenge the observation that social learning underlies our species' success.

Mathematical models are useful to scientists because they allow us to play out "what if?" scenarios. For instance, we can't re-run the tape of human evolution, but we can use mathematical models to explore how our ancestors would have evolved if they had certain properties, or were exposed to particular forms of natural selection. The models provide answers to such questions. When theory and data don't coincide it does not mean that the modeling exercise has failed; to the contrary, such instances can be highly informative. Rogers' model assumed that social learners copied indiscriminately. His findings clearly demonstrated that unselective copying does not increase absolute fitness over and above what can be achieved through asocial learning. This leads us to an important insight: if social learning does truly underlie the human success story, then our copying cannot be indiscriminate.[25]

In other words, it pays to copy strategically, but not mindlessly. The models, like the common sense observations with which this chapter began, imply that individuals must be selective with respect to when they rely on social learning and from whom they learn, if their learning is to be adaptive.[26] Through the operation of natural selection over time, a tendency on the part of humans and other animals to utilize specific decision-making rules should have evolved;[27] we call these rules *social learning strategies*,[28] and they specify the circumstances under which individuals should exploit information from others (and equally, when they should not).

One such rule is that animals should *copy when asocial learning is costly*. This rule specifies that when animals can solve problems easily and cheaply on their own through trial and error, they should do so. However, when individuals are confronted with a particularly challenging task that would require a lot of energy or risk to resolve—perhaps a complicated food-processing task that requires multiple steps—then they should look to what others are doing, and emulate that.

Another strategy is that animals should *copy when uncertain*. This is the suggestion that when individuals are in familiar territory, when they understand the problem and know ways to resolve it, they should rely on their own experience. Conversely, when they are thrust into a new situation—a new environment, for instance, or when confronted with a novel predator—and they are uncertain of the optimal way to behave, then they should copy what others are doing.

A third rule is to *copy if dissatisfied*; that is, when the current behavior reaps rich dividends, stick with it. But if the behavior leads to poor returns, imitate what others are doing in the hope of increasing payoffs. These are all examples of what are known as "when strategies," because they dictate when individuals should utilize social information.[29]

There are also "who strategies" that specify from whom individuals should acquire their knowledge.[30] For instance, individuals could copy the majority behavior, copy the most prestigious individual, or copy the individual exhibiting the most successful behavior. All of these rules have been subject to empirical and theoretical investigation, and all command some support.[31]

The trouble is, researchers can easily dream up a very large number of ostensibly plausible social learning strategies. Individuals could be biased toward copying kin, familiar individuals, or dominants; they could prioritize learning from older, more experienced, or more successful animals; they could watch trends, monitor payoffs to others, or seek out rapidly spreading variants; or they could copy in a state-dependent way—for instance, imitating others when pregnant, sick, or young. Moreover, they can combine these options into convoluted conditional strategies, such as *copy when uncertain and the demonstrators are all behaving in a consistent way*, or *copy the dominant when dissatisfied with current payoffs*.[32]

Such reflections immediately raise the question of which is the best social learning strategy—or perhaps, more realistically, which strategy is optimal in a given circumstance. The traditional means to address such questions is to build mathematical models using, for instance, the methods of evolutionary game theory or population genetics, which compute the strategy that has the highest fitness or is expected to be evolutionarily stable. The reasoning here is that natural selection, acting over millennia, will have resulted in animal minds that favor the use of optimal decision-making rules. Working out through mathematics what strategy is optimal thus leads to a clear prediction regarding what will be found in nature. This approach is widely used in evolutionary disciplines, such as evolutionary biology and behavioral ecology, and is generally very effective. However, it has enjoyed only limited success when applied to the problem of determining the optimal social learning strategy.[33] That is because such methods allow the relative merits of only a small number of strategies to be analyzed simultaneously. There are so many possible social learning strategies that the hypothetical strategy space is huge. Furthermore, the approach is obviously constrained to those strategies that the mathematically minded researcher chooses to analyze. In principle, far superior social learning strategies that nobody has yet considered could be implemented in the real world.

This problem troubled me for a long time. Members of my laboratory had carried out experiments that strongly implied animals were copying strategically. Our findings hinted at the strategies the animals might be using, although rarely in a truly definitive way. We had also developed mathematical models to investigate which strategy ought to be implemented, but we were always haunted by the possibility that what we thought was the best strategy could actually be superseded by any number of unconsidered options. How, when we had focused on just two or three of the most prominent strategies, could we have confidence that we had found the optimal one, when there were so many alternative possibilities?

There was another problem too, which also worried me. The data that we had generated seemed to imply that conditional social learning strategies—for instance, those that took account of the animal's state, the payoff to the copied individual, or the number of individuals performing each option—would yield higher payoffs than fixed, inflexible copying strategies. However, this suggested that if and when we ever found the "optimal" social learning strategy,[34] it might require individuals to engage in quite complex calculations to decide whether to utilize social information. Were animals really clever enough to make such computations? I could believe it of chimpanzees, or Japanese macaques, but studies had shown that fruit flies and wood crickets copy each other. Was it feasible that even invertebrates were computing payoffs to others and monitoring frequency dependence? We knew that if social learning was to be adaptive then it must be used selectively, and there was every reason to believe that natural selection would refine animal decision-making to be highly efficient. But that seemed to imply the copier ought to be smart, and social learning was being reported in animals that were not renowned for their intelligence. It was all a bit of a riddle.

What we really needed to make headway was a means to compare the relative merits of reliance on a very large number of social learning strategies, including strategies that we hadn't even dreamed of, all at the same time. I wrestled with this conundrum for a long time before a solution arose. Ironically, the answer had been in front of our noses all along—we just had to copy it.

It struck me one day that the challenge confronting researchers in the field of social learning was similar to that faced by another group of researchers in the 1970s investigating the evolution of cooperation. We wanted to know what was the best way to copy, whereas those researchers had wanted to know what behavioral strategies were most likely to lead to cooperation. An economist named Robert Axelrod, who was professor of political science and public policy at the University of Michigan, famously made great progress with the cooperation problem by organizing a tournament (in fact, two tournaments) based on a game known as the *"prisoner's dilemma."* The game is a useful model for many real-world situations that involve cooperation.

The prisoner's dilemma game can be described as follows. Imagine two criminals are captured by the police and held in solitary confinement on the same charge. The police don't have enough evidence to convict the criminals unless they incriminate each other by testifying that the other is guilty. The criminals could cooperate with each other and remain silent, in which case they would both get away with a minor sentence. Or they could defect, and testify that the other is guilty. However, if both defect they both get heavy sentences. If one of them defects, the defector gets off free but the other criminal gets a heavy sentence. The game is set up in such a way that betraying a partner offers a greater reward than cooperating. This means that purely rational, self-interested prisoners would betray their associates, leading to the two criminals incriminating each other. The game is called the prisoner's dilemma because if the two prisoners could both cooperate they would both be better off than if they both defected, yet each has an incentive to defect and blame the other for the crime.

Where two players play the prisoner's dilemma more than once in succession and they remember the previous actions of their opponent and adjust their strategy accordingly, the game is called the "*iterated prisoner's dilemma.*" Axelrod invited academic colleagues from all over the world to devise cooperative strategies and compete in an iterated prisoner's dilemma tournament.[35] The entered strategies, which varied widely in their complexity, initial cooperativeness, capacity to forgive past defection, and so forth, were played off against each other to determine their effectiveness. The winning strategy, called *TIT-FOR-TAT*, was entered by Anatol Rapoport, a psychologist at the University of Toronto in Canada. Individuals playing *TIT-FOR-TAT* cooperate on the first round of the game, and after that copy what the opponent did on the previous move. Axelrod's study is widely regarded as one of the most innovative pieces of behavioral research of the twentieth century, and proved a real boost for cooperation research, which grew into a major field of evolutionary biology and in no small part as a result of attention generated by the tournaments.

Thus inspired, I wondered whether we might be able to provide a similar impetus to our field of research by organizing a tournament to work out the best way to learn. We could arrange a competition based on a game of our own devising; it would be free to enter, open to everyone, and we would invite people to send in their ideas on how to copy optimally. We could then investigate how effective each of these ideas were by pitting them against each other in computer simulations and comparing their relative performance. If we attracted many entrants, then a rich vein of new ideas about how best to copy would be generated. We could even offer prize money to stimulate interest. Whether anything useful would come out of the exercise was hard to predict. We certainly hoped that the competition would lead to some truly general insights into why it paid to copy, and how best to do so, but this was far from guaranteed. Given the huge amount of work required, such a competition would be an enormous gamble. Fortunately, the tournament we organized was to prove a major success, not only solving the conundrum of why copying is widespread in nature, but also generating key insights into the mechanisms through which cultural processes drove the evolution of human cognition.

I managed to secure funding to carry out this project through a grant from the European Union to myself and colleagues from Sweden and Italy. The project was a component of a larger research program investigating cultural evolution called "*cultaptation*."[36] The wider program of research combined a variety of empirical and theoretical approaches to studying social learning and evolution; my role included overseeing the tournament. The funding allowed me to recruit a postdoctoral researcher, who would do the bulk of the work in organizing the competition and analyzing the entries. I took on Luke Rendell who had a rare background combining social learning research in whales with expertise in computational biology, and this proved to be an excellent decision because Luke was superb in the role.

The most challenging initial decision was to devise the tournament game. Here Axelrod had a major advantage over us, since the prisoner's dilemma was already a well-established vehicle for exploring cooperation; it was a familiar game that everyone knew. However, no equivalent, established social learning game existed. In making plans, it became rapidly clear to Luke and me that the whole exercise hung critically on us getting this game right. The more that we thought about it, the more it became apparent that it would be very easy for us to "screw up." That is, it would be all too easy to come up with a boring game that no one wanted to enter, or a potentially worthless game with no meaningful resemblance to any real life problem, or, perhaps most embarrassing of all, a trivial game for which a rush of entrants would find the solution.

To guard against these concerns, we decided to recruit a committee of expert advisers from the fields of social learning, cultural evolution, and game theory, who could help us to set up the tournament in the most sensible and productive manner. These advisors were Robert Boyd at UCLA, Magnus Enquist and Kimmo Eriksson at Stockholm University, and Marcus Feldman at Stanford. They are among the world's leading authorities on cultural evolution and game theory. We also benefitted from additional help and advice from Robert Axelrod, Laurel Fogarty at St Andrews, and Stefano Ghirlanda from Bologna University. We were thrilled to have recruited such an authoritative team.

Over the next 18 months we discussed the structure of the tournament intensively, trying out various options with computer simulations and competitions among ourselves. The game went through three separate iterations, with us twice forced to abandon a design after problems were recognized, even though we had poured a great deal of work into it. The second time this happened, when Kimmo and Magnus pointed out some deficiencies in our planned tournament structure, Luke and I were devastated. Fortunately, this led to us devising a new framework, with a neat simplicity to its design.

The framework on which we eventually settled is known as a "multi-armed bandit." You will probably be familiar with a "one-armed bandit," which is the slot or "fruit" machine found in gambling arcades that is operated by pulling a lever (or "arm") on the side. The gambler puts money in the slot, pulls the lever, and (with a certain probability that ensures the owner makes a healthy profit) may get a cash payoff. Now imagine a fruit machine with a hundred separate levers, each with a different probability of giving a payout. Given sufficient practice, a committed player could work out which levers give good or poor returns. That challenge of working out which levers to pull is analogous to our game.

We imagined a hypothetical population of organisms—let's call them agents—that had to survive in a novel, challenging, and changing world. For instance, the agents could be castaways on a tropical island, forced to survive and find food through their own devices. Agents could hunt rabbits, fish in rivers, dig for tubers, gather fruit, sow crops, and so forth.

We contrived a hundred such alternative behavior patterns that agents could perform, each with its own characteristic payoff. In our simulated world, a small number of these behavior patterns had very high payoffs, but a much larger number of behavior patterns gave poor returns.[37] Hence, like the gambler at the multi-armed slot machine, successful agents needed to identify which of the behavior patterns were really profitable, and to perform these extensively. In evolutionary terms, the greater the payoffs they accumulated throughout their lives, the fitter the agents became.

Whether it pays to, say, grow barley, or hunt buffalo, varies from one time to the next, depending on the weather, season, or fluctuations in prey availability. So it was in our game, with the simulated environment changing regularly and leading to changes in the payoffs associated with each behavior. This framework, known as a "restless" bandit, has the advantage of proving extremely difficult, perhaps impossible, to optimize analytically,[38] which meant we could be confident that our tournament would prove a tough challenge for the entrants. We also simulated evolution by choosing agents at random to die, and replacing them with the descendants of those other agents who had accrued elevated fitness through performing high-payoff behaviors. An agent's offspring inherited their parent's social learning strategy, and as a result, effective strategies would increase in frequency within the population through natural selection.

The tournament was structured into a number of rounds, and in each round each agent must perform one of three possible moves. These moves were INNOVATE, OBSERVE, or EXPLOIT. INNOVATE represented asocial learning. Playing INNOVATE led the agent to learn a new behavior,[39] as well as the payoff to that behavior, without error. Agents had to learn new behaviors because they were born with no behavior in their repertoire and therefore needed to build up a range of actions in order to find one with high returns. The second move, OBSERVE, represented any form of social learning. Playing OBSERVE allowed agents to copy the behavior performed by one or more agents who were selected at random from those performing a behavior in the previous round, and again learn the payoff or payoffs associated with those behavior patterns. However, learning through observation incurred two kinds of error; observing agents might misjudge the behavior being performed (i.e., get the wrong behavior), or they might incorrectly estimate the payoff to the demonstrator's behavior. Unlike INNOVATE, OBSERVE did not guarantee that a new behavior would be added to the agent's repertoire. If the observed agent was performing a behavior that the observer already knew, then nothing would be learned, and playing OBSERVE would be unproductive on that round. The probabilities of errors in social learning, the number of agents observed, the rate of environmental change, and a number of other factors, were parameters that we varied systematically in the tournament. Finally, the third move was called EXPLOIT, which represented the performance of a behavior from the agent's repertoire, equivalent to pulling one of the levers and getting the cash. Obviously agents could only EXPLOIT behavior patterns that they had previously learned. We also assumed that agents would remember behavior learned in previous rounds, and the payoff received.

The game would thus require entrants to achieve a good balance between exploring and exploiting.[40] Agents needed to learn through INNOVATE or OBSERVE to build up a repertoire of high-payoff behavior, but they could only actually obtain a payoff, and hence accrue any kind of fitness, by playing EXPLOIT. People entering our tournament were required to specify a set of rules detailing how the agents under their control—that is, deploying their strategy—would utilize the three possible moves.[41] The winning strategy would be the one that combined INNOVATE, OBSERVE, and EXPLOIT most effectively. By systematically varying the conditions—for instance, sometimes causing the environment to change rapidly and other times more slowly, or manipulating the error rates associated with OBSERVE—we would be able to determine when it was beneficial to copy others and when to learn for oneself.

The tournament was to be evaluated in two phases. The first phase would be a round robin, as in Axelrod's tournaments, with each strategy repeatedly played off against each of the others.[42] The top 10 best-performing strategies across all these pair-wise matches would then progress to a second phase, called the "melee," in which all 10 strategies compete simultaneously over a far broader range of simulation conditions than in the pair-wise contests. The strategy with the highest average frequency over all melee contests would be the winner.

Once we had settled on the rules, we advertised the tournament widely with posters, conference talks, e-mail lists, websites, and by targeting the research groups of potential participants. To maximize interest we offered a prize of €10,000 (about 13,650 USD) to the person or persons who entered the winning strategy.[43] Our biggest fear was that nobody would participate, and there were many sleepless nights when Luke and I worried that our efforts would be ignored. In the end, such fears proved unfounded—the response was fantastic.

Our tournament attracted an incredible 104 entries (significantly more than either of Axelrod's tournaments), stemming from 15 academic disciplines (including biology, computer science, engineering, mathematics, psychology, and statistics),[44] and with entries from no less than 16 different countries (Belgium, Canada, Czech Republic, Denmark, Finland, France, Japan, Netherlands, Portugal, Spain, Sweden, Switzerland, United Kingdom, and United States). The tournament had turned out to be a genuinely multidisciplinary, international competition.

Most, but not all, of the entries were from academic researchers, particularly university professors, postdoctoral researchers, and graduate students. However, we did get a small number of entries from interested members of the general public, and even had a few submissions from school children. Indeed, one of the best-performing strategies came from Ralph Barton and Joshua Brolin, two pupils from Winchester College, an independent, secondary school in the United Kingdom; these students' entry was placed ninth in the first phase, an incredible achievement. I found it tremendously gratifying to see how two bright young people were excited by this competition and, through their own good ideas, hard work, and initiative, had come up with a strategy that had outcompeted those submitted by professors of statistics and professional mathematicians. In recognition of this accomplishment, we awarded Ralph and Joshua a bonus prize of £1,000.

Judging from the caliber and complexity of the entries, people had taken the competition seriously. Frequently groups of individuals had entered in teams. Often entrants had written their own computer programs to try out ideas, sometimes conducting simulations that mimicked our multi-armed bandit game. People even ran preliminary minitournaments of their own devising to determine what strategies worked best.

Some of the entries were enormously complex, with all kinds of features, ranging from neural networks to genetic algorithms. Luke and I could barely believe the work that some participants had put into their entries. This must be among the most cost-effective forms of science ever conducted. For a fee of just €10,000, we had effectively employed hundreds of research assistants from around the world; these were tremendously bright and inventive people, who poured weeks, often months, of their time into solving the puzzle of how best to learn.

The next stage was to analyze the entries and try to understand which strategies had done well, and why. In principle, scores in the first (round robin) phase of the tournament could range from 0 (if a strategy lost every single pair-wise encounter) to 1 (if it won all of them). We found that the actual scores ranged from 0.02 to 0.89, indicating considerable variation in strategy effectiveness. This spread in performance was a considerable relief to us. It implied that we hadn't made the tournament so difficult that all entered strategies had performed poorly (known as a "floor effect"), nor had we made the task so easy that a multitude of strategies did equally well (a "ceiling effect"). The observed variation in performance was a modest indication that we had designed the tournament structure appropriately. More importantly, the variation would allow us to conduct a statistical analysis to determine which characteristics were associated with success. Strategies could be classified according to, for instance, whether they were deployed in a fixed or flexible manner, how much copying they did, whether they monitored the rate of environmental change and adjusted behavior to it, and so forth. We could then carry out statistical analyses to determine which of these properties made a strategy successful.

The first finding that jumped out at us was that it is possible to learn too much! In the tournament, investing lots of time in learning was not at all effective. In fact, we found a strong negative correlation between the proportion of a strategy's moves that were INNOVATE or OBSERVE, as opposed to EXPLOIT, and how well the strategy performed. Successful strategies spent only a small fraction of their time (5–10%) learning, and the bulk of their time caching in on what they had learned, through playing EXPLOIT. Only through playing EXPLOIT can a strategy directly accrue fitness. Hence, every time a strategy chooses to learn new behavior, be it through playing INNO-

VATE or OBSERVE, there is a cost corresponding to the payoff that would have been received had EXPLOIT been played instead. This implied that the way to get on in life was to do a very quick bit of learning and then EXPLOIT, EXPLOIT, EXPLOIT until you die. That is a sobering lesson for someone like myself who has spent his whole life in school or university.

On the other hand, we also established that when strategies did deploy a learning move, the best means to do so was through copying. We observed a strong positive correlation between the proportion of a strategy's moves that were OBSERVE, as opposed to INNOVATE, and how well that strategy performed in the tournament. The most successful strategies did not play learning moves often, but almost always played OBSERVE when they did. This seemingly straightforward relationship between copying and success, however, belied a degree of complexity that emerged only on closer inspection. Among the top-performing strategies that progressed to the melee, by and large, the more the strategy learned through OBSERVE rather than INNOVATE, the better it did. However, among the poorer performing strategies we actually witnessed the reverse relationship—the more they copied the worse they did. That told us something very interesting—copying was not universally beneficial. Copying only paid if it was done efficiently.

Poorly performing strategies had incurred a cost for copying, having missed the opportunity to cash in through playing EXPLOIT, but their playing OBSERVE had failed to bring new behavior into the agent's repertoire. Indeed, there turned out to be a huge cost to social learning, because playing OBSERVE failed to introduce new behavior into an agent's repertoire in 53% of OBSERVE moves in the first tournament phase. This was principally because agents observed behaviors that they already knew. In contrast, playing INNOVATE always returned a new behavior. The tournament confirms the intuitions with which this chapter began—copying badly is not a recipe for success. If copying is going to pay, if it is to increase individual fitness, it *must* be done efficiently.

Next, we set out to isolate the properties of the best-performing strategies that allowed them to excel. We discovered that the timing of their learning was a critical factor. Successful strategies timed bouts of learning to coincide with when the environment changed. Recall that successful strategies played EXPLOIT on most rounds, repeatedly performing the behavior in their repertoire with the highest payoff. However, when the environment changed, the payoff to that behavior would alter, typically for the worse. Behavior patterns that hitherto were yielding dividends would no longer do so. That was the time to play OBSERVE, because this would likely pick up a higher-payoff behavior. After all, agents with behavior in their repertoire that was suited to the new conditions would continue to EXPLOIT, and hence these high-payoff behaviors would be available to copy. Conversely, other agents whose returns had just plummeted would commonly switch to learning, and hence poorer-return behaviors would not be available to copy. In timing their learning in this manner, successful strategies were thus more likely to acquire behavior suited to the new conditions.

In contrast, poorly performing strategies not only chose to learn too much, but learned at the wrong times. Copying in the absence of environmental change would frequently return a behavior that was already in the agent's repertoire. If a strategy attempted to learn at the wrong time, it would be better off playing INNOVATE, as that guaranteed a return of new behavior. The result is a negative relationship between copying and fitness among less successful submissions.

The winning strategy, called *DISCOUNTMACHINE*, was submitted by two graduate students from Queens University in Ontario,[45] Dan Cownden and Tim Lillicrap. Dan is a mathematician and Tim a computational neuroscientist; together they were a formidable team. Dan and Tim had worked for months on their strategy, pouring huge amounts of effort into devising the optimal submission. Their triumph was both well deserving and convincing. *DISCOUNTMACHINE* was the top-performing strategy in both the round robin, where it won 89% of its contests, and the melee.[46] Tim and Dan called their strategy *DISCOUNTMACHINE* because it discounted what it learned according to its age, placing more weight on recently acquired rather than older information.[47]

The best performing strategies capped the amount of learning undertaken to ensure that high payoffs were maintained. *DISCOUNTMACHINE* stood out by spreading learning more evenly across the agent's life-span than any other strategies. Its success was partly due to the fact that it was able to spend less time learning and more time playing EXPLOIT than any other strategy, which in turn was because it was able to learn more efficiently than competitors. *DISCOUNTMACHINE* did this by estimating expected future payoffs from either learning, through playing OBSERVE, or by playing EXPLOIT.[48] In other words, the top strategy engaged in a form of mental time travel. *DISCOUNT-MACHINE* looked back into the past, and forward into the future, and used the information gathered to work out which move would be optimal on each round.

Strikingly, both *DISCOUNTMACHINE* and the runner-up, *INTER-GENERATION*, relied almost exclusively on OBSERVE as their means to learn, and at least 50% of the learning of all of the second-phase strategies was OBSERVE. We wondered to what extent *DISCOUNTMACHINE*'s success in the tournament could be attributed to its copying, and Luke came up with an ingenious idea to explore this. He edited *DISCOUNT-MACHINE*'s computer code to produce a mutant version of the strategy that was identical in every respect to the original, except that every time it would have played OBSERVE it played INNOVATE instead. We then re-ran the entire melee phase of the tournament using the other nine top strategies in their original form plus the mutant version of *DISCOUNT-MACHINE*. We reasoned that if *DISCOUNTMACHINE*'s success was at all attributable to its reliance on copying then it would do less well in the re-run melee than the original; conversely, if *DISCOUNTMA-CHINE*'s success was more to do with its other features, then it might still thrive. To our amazement, *DISCOUNTMACHINE*'s performance plummeted. The INNOVATE-only mutant of *DISCOUNTMACHINE* didn't just do badly, it came last! Clearly the success of the winning strategy was to no small degree attributable to its reliance on social learning.

Luke and I realized that we now had two versions of *DISCOUNT-MACHINE*, one that relied almost exclusively on social learning and the other dependent on asocial learning, and that we could play these against each other in a broad range of simulation conditions as a means of exploring the relative merits of the two forms of learning. Similar analyses along these lines had been carried out before, but without using such a smart algorithm, nor in such a rich simulation environment; this led us to believe that there would be more realism to our analyses than to previous studies. We were completely unprepared for the findings of this analysis, which astonished us: copying beat asocial learning hands down over virtually all plausible conditions. For instance, when we manipulated rates of environmental change, we found that the payoffs to each behavior would need to change with a greater than 50% probability in each round before the INNOVATE-only version of *DISCOUNTMACHINE* could gain a foothold. In other words, learning by oneself is only more effective than learning from others in extreme environments that change at extraordinarily high rates—rates so high that such conditions are probably rare in nature.

Our findings went against much of our previous knowledge and many of our intuitions. For instance, a widely held view among psychologists was that copying pays because it allows individuals to examine the behavior of a large number of individuals simultaneously.[49] Rapidly sampling multiple individuals' behavior allows the learner to implement strategies like conforming to the majority, which is thought to underlie a great deal of human learning.[50] Yet we found we could reduce the number of individuals copied when an individual plays OBSERVE to just one, and it still paid to copy; that is, the original *DISCOUNTMACHINE* dominated its asocial cousin, and more generally, "copy-happy" strategies still won out in the melee.

Within economics, social learning is widely thought to be advantageous because individuals can monitor payoffs to others and thereby adopt high-payoff behavior. Yet we found in simulations that we could ramp up the error rate associated with the estimate of the payoff for the observed behavior to the point where there was so much noise that the copying agent received no reliable information about payoffs at all, and yet the strategies most reliant on OBSERVE still won out.

In addition, many social learning researchers, myself included, had thought that a major disadvantage of social learning was that it inevitably generates errors in copying, such that individuals pick up the wrong behavior or fail to pick up any behavior at all. However, we found that the error rates could be extraordinarily high—50%, 60%, 70% of the time playing OBSERVE would fail to bring new, higher-payoff behaviors into the agent's repertoire. And yet, astonishingly, it still paid to copy.

Why is copying so robust? What is it that makes social learning so advantageous compared to asocial learning over such a broad range of conditions? Here the tournament yielded a major insight: copying pays because other individuals filter behavior, making adaptive information available to copy. Entrants to our tournament specified in their strategies that agents should first build up a repertoire of behavior, and then perform the learned behavior associated with the highest payoff. However, this meant that when individuals played EXPLOIT, they were not performing a randomly chosen behavior, but rather a select, tried-and-tested, high-payoff behavior. Accordingly, agents that played OBSERVE drew from this pool of high-payoff options, since that was what was performed. Playing OBSERVE was far more likely to pick up a behavior with a very high return than playing INNOVATE, because the latter acquires behavior at random and most of that behavior confers a low return. When we ran test simulations in which agents playing EXPLOIT drew from their repertoire at random, rather than performing the best behavior pattern that they possessed, the innovate-only version of *DISCOUNTMACHINE* dominated the original. The selective performance of behavior by the copied individual is what makes social learning so profitable to the copier.

That is why copying pays. That is why we see copying not just in animals with large brains, such as humans, chimpanzees, and Japanese macaques, but also in fruit flies and wood crickets. An animal does not need to be smart to benefit from copying, because a lot of the smart decision making has already been done for it by the copied individuals who have already prefiltered their behavior. We in the field of social learning had been so focused on what the observing individual would need to do to ensure it acquired adaptive information that we had neglected to consider how the observed individual makes the copier's job so much easier. Even relatively simple copying rules would be more likely to lead to high-payoff behavior than trial-and-error learning under many circumstances. That explains why copying is widespread in nature.

There were other ways in which our tournament challenged existing theory. Earlier analyses, such as Rogers' model, predicted that evolution would lead to a stable equilibrium where both social and asocial learning persisted in the population.[51] However, when we allowed the two versions of *DISCOUNTMACHINE* to compete, the original completely outcompeted the asocial learning variant under most conditions. In earlier analyses, social learners had been modeled as inflexible agents that continued to perform the same behavior even when the environment changed. This assumption had a double impact on the perceived fitness of social learning, because when those individuals were copied by fellow social learners, the latter also acquired the sub-optimal behavior. Conversely, agents in our tournament possessed a repertoire of behaviors that they exploited flexibly. Following environmental change, successful strategies like *DISCOUNTMACHINE* would not just stick with outdated behavior, but switch to the behavior in their repertoire with the next highest payoff. In turn, when agents played EXPLOIT, other copying agents playing OBSERVE also acquired a behavior with a reasonable return. Unlike in Rogers' model, social learners were not stuck in a frequency-dependent relationship with asocial learners, reliant on the latter to track environmental change. Provided there was a small amount of copy error, playing OBSERVE would generate enough behavioral diversity to allow social learners to respond adaptively to environmental change.

Earlier theoretical work had suggested that reliance on social learning would not necessarily raise the average fitness of individuals in a population,[52] and may even depress it,[53] thereby leading to Rogers' paradox. The tournament revealed how we could resolve the apparent conflict between such findings and our species' demographic success: simple, poorly implemented, and inflexible social learning does not increment biological fitness, but smart, sophisticated, and flexible social learning does.

Strategies that did well in competition with other strategies were not, however, those that maximized the returns to agents. Rather, we found a strong inverse relationship between the mean fitness of individuals in populations containing only one strategy, and that strategy's performance in the tournament.[54] This finding illustrates the parasitic effect of strategies that rely heavily on OBSERVE (some of the best performers, including *DISCOUNTMACHINE, INTERGENERATION, WEPREYCLAN,* and *DYNAMICASPIRATIONLEVEL,* ranked 1, 2, 4, and 6, all played OBSERVE on at least 95% of learning moves). Strategies using a mixture of social and asocial learning are vulnerable to being outcompeted by those using social learning alone, which may result in a population with lower average returns. These findings are evocative of an established rule in ecology; this specifies that, among competitors for a resource, the dominant competitor will be the species that can persist at the lowest resource level.[55] An equivalent rule may apply when alternative social learning strategies compete: the strategy that eventually dominates will be the one that can persist with the lowest frequency of asocial learning.[56]

Some population-level consequences of the strategic reliance on copying observed in the tournament are equally surprising. Consider the diversity of the population—that is, the number of behavior patterns known about across the entire population (i.e., those held in all agents' repertoires combined). If we align strategies according to the extent to which they rely on OBSERVE as opposed to INNOVATE when they learn, a positive correlation is found between the proportion of copying and the amount of behavioral diversity. This is an extraordinarily counterintuitive finding. After all, when an agent plays OBSERVE it does not increase the number of different behavior patterns in the population at all—the agent just adds a behavior that already existed in the population to its own repertoire. In contrast, playing INNOVATE guarantees a new behavior for the agent, and accordingly will frequently introduce new behavioral variants into the population. Why should increasing reliance on social learning lead to greater behavioral diversity?

Copying does not typically introduce new variants, except in those rare circumstances when copy error adds a new behavior. However, what copying does do is reduce the rate at which behavioral variants are lost. That is because social learning generates multiple copies of a piece of knowledge or behavior, retained across the repertoires of several individuals, so that when an individual dies its knowledge need not die with it. Here, the positive effect of copying on reducing the rate at which behavioral variants are lost typically exceeds the negative effect of reducing the rate at which new behavior is introduced, leading to a net increase in diversity across the population with an increasing reliance on OBSERVE over INNOVATE. Indeed, above certain levels of copying the population's knowledge base becomes completely saturated, and knowledge of all possible behaviors is retained.

It does not follow, however, that as the rate of copying goes up, the number of behaviors actually performed increases—to the contrary. Increasing reliance on social as opposed to asocial learning leads to fewer behavior patterns being exhibited, because the population starts to converge on a small number of high-payoff variants. At the extreme, when all agents learn through copying, then the population appears to exhibit behavior highly evocative of conformity. Everyone does what the majority are doing.

Now consider the longevity of knowledge in the population—that is, how long a behavior persists within the population once it has been introduced. High levels of reliance on social learning automatically generate extreme durability of cultural knowledge. Populations appear to pass through a threshold level of reliance on social learning, above which cultural knowledge becomes extremely stable and persists virtually indefinitely. The persistence of behavior patterns for many thousands of rounds witnessed in the tournament is equivalent to human beings retaining insights first gained by the ancient Greeks or Egyptians.[57] At the same time, with increasing social learning the behavior patterns actually performed change with greater rapidity, generating fads, fashions, and rapid turnover in cultural traditions.[58]

The social learning strategies tournament was a huge success.[59] It made sense of several conundrums concerning copying. The tournament established that there *are* genuine fitness benefits to copying provided it is done efficiently—that is, strategically and with high fidelity.[60] This finding, combined with the observation that the winning strategy was the one that learned with greatest efficiency, implies that natural selection should favor the implementation of optimally strategic copying rules, a central thesis of this book.

Copying, even "blind" copying, offers advantages over trial-and-error learning, because copiers benefit from the adaptive prefiltering of behavior by the individuals that are copied. This insight helps us to understand why social learning is so widespread in nature, even in animals that we do not think of as smart. Bumblebees, fruit flies, and wood crickets benefit from copying others because finding rich sources of pollen, fertile females, and means of escaping predators through trial and error when there are so many possible flower patches, female flies, and dangerous predators around is a hard problem. Social learning confers a rapid and effective shortcut to high-payoff behavior much of the time. At the other extreme of cognitive sophistication, the same insight could help explain the extreme reliance of children on imitation, leading them faithfully to copy even superfluous actions in a demonstrated task.[61] When children copy adults, they are inadvertently taking advantage of decades of information filtering. Trusting the adult is a highly efficient rule of thumb.

Yet, the tournament also teaches us that if individuals are able to copy in a "savvy" fashion—for instance, if they can be selective about when, and how frequently, they copy—there are real fitness dividends. Successful strategies were able to time copying for when payoffs drop, evaluate current information based on its age, judge how valuable information would be in the future, and use all this knowledge to maximize copying efficiency. Empirical evidence suggests that some animals are able to discount information based on the time since it was acquired.[62] The tournament's winning strategy also possessed the ability to project current conditions into the future, a rare attribute in nonhumans.[63] Few animals would be capable of implementing a strategy as sophisticated as *DISCOUNTMACHINE*, but there can be little doubt that humans can. This cognitive ability could be one factor behind the gulf between human culture and any nonhuman counterpart. The tournament hints that the adaptive use of social learning could be linked to the cognitive abilities underlying mental time travel, a theme picked up in later chapters.

Moreover, if efficient copying is adaptive, such that natural selection should favor greater and greater reliance on social as opposed to asocial learning, then the tournament also establishes that a number of characteristics strongly evocative of human culture will follow automatically. With increasing copying inevitably comes greater behavioral diversity; the retention of cultural knowledge for long periods of time; conformity; and rapid turnover in behavior such as fads, fashions, and changes in technology. Provided copying errors or innovation introduce new behavioral variants, copying can simultaneously increase the knowledge base of a population and reduce the range of exploited behavior to a core of high-performance variants. Similar reasoning accounts for the observation that copying can lead to knowledge being retained over long periods of time, yet trigger rapid turnover in behavior. Low-level performance of suboptimal behavior is sufficient to retain large amounts of cultural knowledge in social learning populations, over long periods. A high level of copying increases the retention of cultural knowledge by several orders of magnitude.

These observations suggest that social learning confers an adaptive plasticity on cultural populations; it allows them to respond to changing environments rapidly, drawing on a deep knowledge base. In biological evolution, the rate of change is positively related to genetic diversity,[64] and formal analyses suggest a similar relationship between the rate of cultural evolution and the amount of cultural variation.[65] Accordingly, we might envisage that populations heavily reliant on culture would rapidly diverge behaviorally, exploiting the rich levels of variation retained. The social learning strategies tournament teaches us that the ecological and demographic success of our species, our capacity for rapid change in behavior, our cultural diversity, our expansive knowledge base, and the sheer volume of cultural knowledge we exhibit, may all be direct products of the heavy, but smart, reliance of our species on social learning.

CHAPTER 4

A TALE OF TWO FISHES

For most researchers seeking to understand the evolutionary roots of culture, the obvious animals to compare to humans are monkeys and apes, but I myself have learned more about this topic through studying fishes. Such a claim, no doubt, would sound extraordinary to anyone who wrongly believed that fish are dim, instinct-driven creatures with three-second memories, a stereotype that Hollywood and the media ceaselessly perpetuate, indifferent to the scientific evidence. Yet, as described in chapter 2, extensive experimental data now show that social learning and tradition play important roles in the behavioral development of countless fishes, most of which are highly social animals. Fish behavior is far from rigidly controlled by a "genetic program,"[1] but rather is constantly and flexibly adjusted to exploit information and resources in the environment, including information provided by other fish.

Even given the knowledge that fishes are both competent at, and widely reliant on, copying, most anthropologists would surely balk at the suggestion that we can learn about culture through studying them. Yet, fishes have proven a terrific model system for investigating social learning processes because they offer major practical advantages over many other vertebrates, thereby leading to valuable insights. The key factor here is that animal traditions and the diffusion of new innovations are group-level phenomena, and if they are to be studied reliably, scientists require not just replicate experimental animals, but *replicate populations of experimental animals*. Leaving aside the nontrivial ethical considerations, it would be financially ruinous and a bureaucratic nightmare, for a researcher like myself to set up large numbers of populations of chimpanzees or Japanese macaques for behavioral experimentation. However, it is extremely straightforward and cheap to set up large numbers of populations of small fishes in the laboratory and investigate their behavior. Fish experimentalists enjoy the twin luxuries of the multiple conditions that good experimental design frequently demands and the healthy sample size that confers statistical power; both bring experimental rigor to any social learning investigation.[2] For researchers interested in animal culture, working with fishes makes a lot of sense.

Prior to the social learning strategies tournament, the idea that animal copying might be strategic had been intimated to me by a marvellously instructive series of experiments that we had conducted on sticklebacks. The sticklebacks are a family of sixteen species of fishes that are extremely common in the rivers, streams, and coastal ocean zones of the temperate Northern Hemisphere;[3] they are closely related to pipefish and seahorses. The distinctive features of sticklebacks are their spines and the absence of scales, which are replaced with bony armour plates. Chances are, if you live in Europe, North America, or Japan, there are sticklebacks in your local lakes, rivers, and streams. These fish are easily caught with a simple dip net, and thrive in the laboratory, where they make highly effective subjects in animal behavior studies. Partly for this reason, sticklebacks have long been a favorite experimental system for many ethologists and evolutionary biologists, including myself. For over two decades, my research group has investigated social learning and tradition in over thirty different species of animals, including rats, chickens, starlings, budgerigars, lemurs, capuchin monkeys, and chimpanzees,[4] yet the insights into these topics that we have gleaned through studying sticklebacks and other small fishes are among the most valuable.[5]

This chapter describes a protracted program of experimental research conducted over 20 years that set out to comprehend a fascinating difference in the social learning of two closely related species of sticklebacks. I present it in detail because it illustrates how a dedicated line of research using a flexible model system can provide valuable insights into more general issues related to the evolution of culture. The investigation also demonstrates how the science in this field is done.[6] Scientific questions in this domain are rarely answered with a single experiment, but often require an extensive series of studies, each chipping away at the problem. Beginning as a curious anomaly, prolonged experimentation on the question developed into a wonderful glimpse at the bigger picture of how social learning evolves. Later in the book, I will show how what was learned through studying fishes proved to be highly relevant to making sense of the evolution of primate cognition.

This particular line of investigation was initiated by Isabelle Coolen, a French behavioral ecologist who came to work with me at Cambridge in the late 1990s. Isabelle had just completed her PhD on the topic of "public-information use" in birds. The term *public-information use* has a more specific meaning than it *prima facie* implies. It is defined as the capability of an animal to assess the quality of a resource, such as the richness of a food patch, through vicariously monitoring the success and failure of others.[7] Public-information use is thus a form of social learning that allows individuals to collect information from a distance through observation, without incurring the costs associated with personal exploration and sampling, such as increased exposure to predators or the energy spent traveling between food patches to make comparisons. At the time, many researchers thought that public-information use required some high degree of intelligence, or sophisticated cognition. In questioning this suggestion, Isabelle began to wonder whether fish might be capable of public-information use. We resolved to explore the issue using threespine sticklebacks.[8]

Isabelle set out to collect threespines from local streams to be brought into the laboratory for testing. However, having been trained by working with birds, Isabelle was not particularly familiar with the subtle morphological differences between threespine sticklebacks and another fish that often shoals with them, the closely related ninespine stickleback.[9] Isabelle inadvertently collected both species, and as she had sufficient numbers, we decided we might as well test both—a beautiful illustration of how serendipity can play an important role in science. Had a trained fish researcher done the collecting we would probably have only tested the threespines, found little of interest, and dropped the investigation. As it was, Isabelle tested two species, and the differences in their behavior that her experiments revealed proved to be so absorbing that they initiated decades of fruitful research.

Isabelle's testing apparatus was quite simple—it was a standard 90 cm long aquarium tank, divided into three equal-sized (30 cm^2) sections by transparent partitions. At each end of the tank Isabelle positioned an artificial feeder, which simulated a natural food patch, and delivered food in the form of bloodworms to a tube that opened up at the bottom of the tank. She placed "observer" fish—our experimental subjects—one at a time in the central compartment from where, through the transparent partitions, they could observe two groups of three "demonstrator" fish being fed through the artificial feeders.

Isabelle delivered food to one feeder, which simulated a rich food patch, at three times the rate that it was delivered to the other feeder, called the poor patch. Suitably positioned transparent and opaque barriers meant that the demonstrators, but not the experimental subjects, could see the food as it fell down the feeder tubes to reach the base. The demonstrators would follow the worms as they sank, excitedly pecking at them, and eventually pull them out of the bottom of the tube to eat. Thus the observers could see two groups of three fish, both feeding at a food patch at an end of the tank, with one group feeding more rapidly than the other. After an observation period of 10 minutes, all the demonstrators and remaining food items were removed from the tank, and the observer was released.

Isabelle reasoned that if the sticklebacks were capable of public-information use, they would be able to distinguish between the rich and the poor patch based solely on the reactions of the demonstrators to the food. If so, on their release, the fish should tend to swim to the end of the tank formerly housing the rich patch, and would spend more time at that end of the tank than the alternative. Sure enough, Isabelle found that the ninespine sticklebacks predominantly swam to the rich-patch end of the tank, and spent more time at that end. The threespine sticklebacks, in contrast, exhibited no evident patch preferences, and seemingly swam to each end of the tank at random.

Isabelle's experiment hinted that the ninespine sticklebacks might be capable of public-information use since they were able to use the behavior of the demonstrators to establish which of the two food patches was the more profitable. The study also implied that threespines did not possess this capability. However, drawing these conclusions at that stage would have been premature, because there were a number of alternative explanations that first needed to be ruled out.

In her experiment, Isabelle had used fish of the same species as the subjects to act as demonstrators—that is, ninespine demonstrators for ninespines and threespine demonstrators for threespines. Perhaps the threespine demonstrators were not as effective at transmitting information about patch quality as the ninespine demonstrators, and a disparity in demonstration quality rather than in public-information use accounted for the species' difference. Isabelle repeated her first experiment, this time with heterospecific demonstrators: ninespine demonstrators for threespines and threespine demonstrators for ninespines. Yet this manipulation did not change the results; ninespine subjects swam disproportionately to the end that formerly housed the rich patch, while threespines swam to the former locations of rich and poor patches with apparent indifference.

We also wondered whether differences in the perceptual abilities of the two species might account for the findings. Maybe the threespine subjects could not see to the ends of the tank with sufficient acuity to discern that the demonstrators were feeding. Isabelle replicated the experiment a third time, on this occasion with food delivered at one end of the tank but not the other. If the threespines could not even discriminate between feeding and nonfeeding fish at that distance, then clearly they would not be able to make the subtler distinction between fish feeding at different rates. However, that was not the explanation either, since on this occasion the threespines, like the ninespines, swam to the end that formerly housed the feeding fish.

Another alternative explanation that we needed to rule out was that, even though we had removed all the food for the test, there might be residual olfactory cues left in the tank—a stronger smell of bloodworm in the region of the rich patch, for instance—to which the ninespines were more sensitive. This led Isabelle to conduct one of my favorite experiments, memorable for its oddness, and to my knowledge the only observational learning study ever conducted in which the observer could not actually see the demonstrators! Isabelle replicated the study a fourth time, this time with opaque partitions separating the demonstrators and observers; the partitions could not be seen through at all. Perhaps not surprisingly, the test revealed that neither species favored the rich patch.[10] Hence, there was no evidence for any perceptual differences between the species based on sight or smell, and clearly visual cues were critical to this form of learning. We began to believe that what we had discovered might genuinely be an adaptive specialization in social learning, with ninespines capable of exploiting public information, while their close relatives, the threespines, were not.

Yet before we could make authoritative claims about the cognitive abilities of these two species of fish we needed to test populations collected from different sites. We had to be confident that the observed disparity in the behavior of these fishes was found consistently across their entire range; or else we had to determine what factors accounted for the variation in performance. Gradually, over the following 15 years, we began to determine the robustness of this species difference. Through the work of Mike Webster, a postdoctoral researcher in my laboratory at St Andrews, who first tested ninespines and threespines collected from various locations around Britain and then went on to experiment with fish from around the world, we established that the species difference was extremely reliable and globally manifest.

Whether we tested sticklebacks from Cambridge, Scotland, the Baltic, Canada, or Japan, we invariably found the same pattern: ninespines were *always* capable of public-information use, but there was never a hint that threespines had the same capability. Mike tested freshwater populations, marine populations, armored fish, and spineless forms. He examined fish from high-predation sites, and fish from sites where predators were rare. None of this variation changed the results. Mike reared ninespines from eggs in captivity, but when he later tested them as adults he found no differences between lab-reared and wild-caught adults. Studies that manipulated the rearing conditions of the fishes also had little impact on their public-information use. This form of learning was seemingly little affected by manipulations of rearing density or environment complexity. Nor did any other factor, be it morphological, ecological, social, or developmental, explain variation in public-information use—the capability was simply always present in one species and not the other. All the evidence suggested that public-information use was indeed an unlearned, species-typical capability of ninespine sticklebacks.

I became intrigued with this case study and grew increasingly convinced that we had discovered an adaptive specialization in social learning. Such adaptive specializations were not unheard of (recall, for instance, the rhesus monkey's fear of snakes described in chapter 2), but such a specialization in stickleback social learning would be tremendously exciting because it would be so amenable to investigation. We could study the development of this capability in the laboratory, we could test other stickleback species to determine the trait's evolutionary history, we could carry out experiments to ascertain the function of public-information use, and we could investigate the underlying mechanisms at genetic, neural, endocrine, and behavioral levels. Isabelle's discovery seemed a "gift from the gods."

Before we could run with this ambitious program of research, however, we needed first to establish the specificity of our findings. Did ninespines and threespines really differ in a precise intellectual domain, or were our findings a manifestation of a more general difference in their cognition? At the extreme, perhaps ninespines were better than threespines at all forms of learning. Mike presented both stickleback species with a battery of learning tests. He tested their ability to learn the navigation of a T-maze, where food was found solely in one arm. He tested the fish in a color discrimination task, where they were required to learn that a particular color was associated with food. He also tested their ability to learn the location of resources through attending to the behavior of others (known as "local enhancement") as well their social foraging tendencies. In all of these tests, Mike found no significant differences in the two sticklebacks' performance. PhD student Nicola Atton tested the fishes' ability to utilize public information in a different context—rather than gleaning information about the quality of food patches, the observers potentially acquired knowledge about the worth of alternative rocky shelters, which would confer variable protection from predators. However, Nicola found no differences between the species in this case, with neither showing evidence of public-information use. Moreover, we had learned from published studies carried out by researchers in other laboratories that threespine sticklebacks were perfectly capable of other forms of social learning—to find food, identify kin, and learn about predators[11]—yet they were not capable of public-information use.

Collectively, these findings are genuinely intriguing. Here we have two very closely related species of fish, often collected from exactly the same rivers and streams, shoaling together throughout a broad region of their ranges, leading very similar lives, eating very similar foods, and comparable in their cognition in every other measured respect. Except the ninespine stickleback possesses a highly specific form of social learning—the capability for public-information use—that the other lacks. How can this be explained?

Strangely, the answer to this conundrum derives not from evolutionary biology, nor behavioral ecology, nor even comparative psychology, but from anthropology. The puzzle in this tale of two fishes is resolved by the work of biological anthropologists Robert Boyd and Peter Richerson, theoreticians and leading authorities in the field of cultural evolution. Boyd and Richerson,[12] following a theoretical analysis conducted with humans very much in mind, proposed a hypothesis that they called their "costly information hypothesis." This hypothesis is rich and multifaceted, but can be simplified here as the idea that humans should copy when asocial learning is costly. The hypothesis is relevant to our sticklebacks because the costs of learning asocially, through trial and error, differ between the two species due to differences in their morphology, specifically their physical defenses.

Threespine sticklebacks, as their name implies, typically have three large dorsal spines on their backs, as well some heavy-duty armor in the form of tough lateral plates that protect them against predators, usually birds or larger fishes (figure 3a). Remarkably, these morphological defenses are so effective that there are several reports of threespine sticklebacks actually surviving being eaten! Their spines get stuck in the throats of predators, who cough them up, and then the stickleback swims away, apparently unharmed. Such effective defenses mean that threespine sticklebacks can explore their environments in comparative safety, allowing them to sample alternative food patches directly, and to work out for themselves which is the richest food patch around. These fish don't need to copy in this case, because learning for themselves is not particularly costly.

Ninespine sticklebacks, on the other hand, have approximately nine spines on their backs,[13] but these are small and provide comparatively little protection. This species also typically has fewer, and thinner, lateral plates than the threespines (figure 3b). This leaves ninespines significantly more vulnerable to predators than their cousins. Indeed, studies have shown that predatory fishes display a preference for consuming ninespines over threespines.[14] Because they are more vulnerable to predation, ninespines typically respond by hiding when a threat appears. Isabelle noticed when she was collecting her fish that the ninespines

(a)

(b)

FIGURE 3. The threespine stickleback (*a*) has large spines and extensive protective plating, which the closely related ninespine stickleback (*b*) lacks. These morphological differences have evidently influenced how natural selection has fine-tuned the two species' social learning. By permission of Sean Earnshaw.

were much more likely than the threespines to be hiding in reeds and weeds. In contrast, the superior defenses of threespines mean that they are more likely to withstand the higher predation risk associated with foraging in open water, and therefore benefit more from maximizing their opportunities to feed.

For the ninespines, exploring the environment for themselves and sampling food patches through trial and error is sufficiently risky to incur real fitness costs, and the costly information hypothesis predicts that those costs should tip the balance in favor of reliance on social learning. Plausibly, natural selection has shaped the ninespines' ability to extract valuable foraging information through observation from a safe vantage point, leaving them able to swim to the richest patch when the coast is clear. If that hypothesis is correct, public-information use is indeed an adaptive specialization in social learning. Sure enough, when Isabelle repeated the experiment with cover in the enclosure, the ninespines, but not the threespines, spent a disproportionate amount of time hiding during the observation phase; but we saw their little heads poking out, as they carefully tracked the behavior of the demonstrators.

You might be wondering why the ninespines need to collect information on the payoffs to other fishes. Why do they have to monitor how often the other fish are feeding? Surely there is an easier way to solve this problem. The ninespines could simply swim to the patch visited by the most fish, which would surely be the most profitable one. There are at least two problems with such reasoning. First, a food patch may have had a lot of visitors in the recent past, but those animals will have eaten some of the food and reduced its profitability. Second, shoaling fishes, like any animal that aggregates for safety, do not move or make foraging patch decisions entirely independently of each other. A shoal of fishes could come across a food patch by chance and attract other fishes to that patch over an alternative location that might be more profitable. Basing decisions solely on the numbers of other individuals performing behavioral alternatives can lead animals to get locked into "information cascades,"[15] which at the extreme can be maladaptive.[16] For these reasons, raw numbers of fish can provide clues about the quality of a food patch, but such clues may be misleading.

Another set of experiments carried out by Isabelle shows this very neatly.[17] Isabelle replicated her original experiment, but this time manipulated the numbers of fish at each end of the tank, with six fish at one patch and two at the other. She found that if, during the observation phase of the experiment, the subjects did not see the demonstrators feeding (which meant the ninespine subjects were forced to base their patch-choice decision exclusively on the numbers of fish at the patches), then, after removal of the demonstrators, the subjects did indeed swim to the end that had contained the most fish. The presence of a companion shoal adjacent to the central compartment lent confidence to our conclusion that this was a foraging, rather than merely a shoaling decision, since the subjects effectively had to swim away from the safety of the shoal to choose a food patch.[18] However, when Isabelle traded off demonstrator numbers and patch quality, such that the ninespine subjects witnessed six fish feeding at a low rate at one patch, and at the same time two fish feeding at a high rate at an alternative, richer patch, the subjects later swam to the richer patch. This shows that ninespines will use the social cues provided by the numbers of fish at a patch when that is the only information to go on, but will preferentially utilize public information when that conflicts with the social cues. That is why extracting public information is beneficial; it is more reliable than social cues, and consequently helps animals guard against the acquisition and spread of misleading information.

The experimental procedure that Isabelle devised for investigating public-information use in sticklebacks has proven extremely flexible. It has allowed us to conduct many variations of the basic experiment, including manipulating the ratio of food delivered by each feeder in order to simulate rich and poor feeding patches; changing the number, characteristics, or species of the demonstrators; or providing the observer with different forms of prior experience about one or both of the patches. We have thus been able to explore the manner in which animals weigh different sources of information when they conflict. Such studies reveal that sticklebacks are capable of adaptive trade-offs in their reliance on social and asocial sources of information, mixing their prior knowledge of patch quality with the information gleaned from observation of others in a surprisingly sophisticated way.

Yfke van Bergen, a PhD student at the University of Cambridge, investigated these trade-offs in reliance on social and asocial information. In her experiment, Yfke first gave ninespines the opportunity to learn

88 CHAPTER 4

through direct foraging over repeated trials that one of the food patches, on average, yielded a larger number of prey items than the alternative. She was able to manipulate the reliability of this personal training regime by varying the number of training trials on which the patch that was richest over all trials was the richest on that particular trial. For instance, a training regime in which feeder *A* delivered more food than feeder *B* on 17 out of 18 trials, would lend more confidence that *A* was the richer patch than an alternative regime where *A* delivered more food than *B* on only 12 out of 18 trials.

Yfke then followed this personal training with the same procedures as Isabelle's original experiment, allowing the fish to observe three demonstrators feeding at a rich patch and three at a poor patch, followed by a test of patch preference. However, there was a twist; Yfke switched the patch qualities, such that what had been the lower-yielding patch according to the subjects' personal training became the richer patch in the public demonstration, and vice versa.[19] This design meant that asocial and social information were conflicting, and it allowed us to explore under what circumstances individuals will utilize information provided by others and when they will rely on their own prior knowledge.

We found that those fish that had received reliable and unambiguous private training almost completely ignored public information and when tested chose the patch they had previously experienced to be more productive, rather than the patch that was indicated to be the richer by the demonstrators. However, other fish that had experienced unreliable, noisy private training were more inclined to copy other fish, and to base their patch choice on what the demonstrators had indicated. The rate of copying increased with the degree of noisiness of the training; the more unreliable the ninespines' personal experience, the more they were inclined to copy.

In a second experiment, Yfke again subjected the fish to a personal training regime in which one patch was richer than the other, but then manipulated the time period before they received their conflicting public information and test, which in different conditions was one, three, five, or seven days. She found that the fish would base their patch choice on private information if only one day had elapsed since their training, but as their private information got more and more out of date, they increasingly copied the demonstrators. When seven days had elapsed since they last updated their private information, the fish switched completely to using public information, and copied at the same rate as individuals that had not previously sampled the patches.[20]

Our stickleback studies were teaching us about the strategic manner in which animals use social information. Isabelle's experiments had implied that the sticklebacks utilized a "copy when asocial learning would be costly" rule, according to the predictions of Boyd and Richerson's costly information hypothesis. Yfke's studies showed that the ninespines' copying was implemented even more subtly, with the fish restricting their use of social information to when past experience had left them uncertain as to the best option. What became manifestly clear was that these fish were not always, or unpredictably, using public information, but rather were switching between reliance on different sources of information in an extremely shrewd way.

Subsequently, we were to learn that these fish were not just copying efficiently, but optimally.[21] Jeremy Kendal, another postdoctoral researcher, now at Durham University, carried out an experiment that found ninespines apply an impressive "hill-climbing" strategy when they exploit public information.[22] We found that fish with prior experience of finding food at one patch would switch preferences when the prey capture rate of other fish suggested the yield of the alternative patch was greater than at their previously preferred site. Such a strategy allows individuals steadily to increase their foraging efficiency by gradually homing in on the most profitable foods or foraging locations exploited across the population, which is what lends this strategy its "hill-climbing" quality. Later, further experiments showed that the probability of a fish selecting a demonstrated prey patch depended solely upon the returns of the foraging demonstrators.[23] The degree of copying exhibited by the observing fish increased with the absolute rate of feeding by the fish they were watching. What is particularly interesting about this finding is the ninespines' behavior is precisely that predicted by a sophisticated evolutionary game theory analysis conducted by an economist in order to understand human behavior.[24] In other words, two species as different as humans and ninespine sticklebacks exhibit the same optimal payoff-based learning rule when they copy.[25] The use of this strategy by ninespines as they colonized new regions, for instance, would allow them gradually over time to increase the efficiency with which they exploit diverse prey items in their natural environments. We were excited by this finding. It suggested to us that through the use of a relatively simple rule (copying others in proportion to their payoff), our fish could achieve the surprisingly complex outcome of cumulative knowledge gain. While this obviously falls short of human cumulative culture,[26] the rule nonetheless possessed a "ratcheting" quality that, to my knowledge, had never previously been demonstrated in animals.

Ninespine and threespine sticklebacks are two closely related fishes that, because of small morphological differences, find themselves on opposite sides of a cost-benefit analysis specified by the costly information hypothesis. For ninespines, but not threespines, this specific form of social learning was sufficiently beneficial to evolve through natural selection. One might anticipate that the cost-to-benefit ratio of public-information use could also be changed by the personal circumstances of individual fish. Consider, for instance, the impact on female ninespines of being in reproductive condition. Pregnant females swollen with eggs would be more conspicuous and more attractive to predators than non-reproductive fish. They would also be slower to respond to predators than other females because their large bellies drag in the water. These changes in female condition effectively ratchet up even further the costs of learning through trial and error, and ought to make reproductive females even more reliant on social information than nonreproductive fish. Conversely, males in reproductive state are required to compete intensively with other males for females and territories, and need to invest heavily in paternal care; they are tied to the nest while they look after the eggs and fry, and during that time are usually unable to feed. Such circumstances favor males who take risks and seek out higher rewards through direct sampling of foraging sites, since replete energy reserves at the onset of courtship potentially yield a fitness bonanza. Being in reproductive condition ought to shift the balance toward male sticklebacks being less reliant on public information than other males.

These predictions were confirmed experimentally.[27] Mike Webster found that pregnant females relied almost exclusively on public information, copying to a significantly greater extent than nonreproductives, making fewer switches between patches, and spending more time in the cover of refuge than other fish. Their condition had led them to become more risk-averse, which favored heavy reliance on social learning. Reproductive males, on the other hand, exhibited no evidence of public-information use at all. These males exhibited the shortest patch selection times, the highest switching rates, and the least time in cover of all the fish that we tested. They also shoaled very little, and we suspected the same physiological changes that had led reproductive males to cease shoaling had also reduced their attention to the behavior of foraging conspecifics, and the loss of public-information use. A reduced shoaling tendency and a reduced tendency to spend time in or near cover are both highly risky, since individuals moving alone or in the open are known to be more vulnerable to predation.[28] Evidence from other animals suggests that heightened levels of circulating testosterone associated with the onset of the reproductive phase can reduce the sensitivity of males to risk.[29]

One finding that was of particular interest to us was that reproductive males were actually quicker than nonreproductive fish to solve a solitary foraging task. This meant that the transformation in the male sticklebacks' behavior with reproductive state could not simply be attributed to the effects of testosterone surging through their tissues and disrupting their learning. Rather, reproductive males appeared to be pursuing an alternative adaptive strategy that functions to maximize food intake prior to parental care through increased reliance on private sampling. Taking chances to gain a food windfall would be adaptive if it gives male sticklebacks a competitive edge in access to females, or in looking after the eggs and fry.[30] A transformation in the costs of asocial learning dependent on reproductive state explains this switch in foraging strategy.[31] In fact, ninespine reproductive males behave very much like threespines, raising the intriguing possibility that this difference between the two species might be mediated by a shift in hormone levels.

Ninespines and threespines are, of course, not the only species of sticklebacks—nor are they very closely related. The ninespine's sister species is the brook stickleback,[32] a species that very much resembles the ninespine but typically has five or six spines and lacks lateral bony plates. Given its physical similarity and close relatedness to the ninespine, we might expect that the brook stickleback would exploit public information too. Also more closely related to ninespines than threespines are the fourspine sticklebacks and the fifteenspine sticklebacks.[33] We were interested to know whether these species were capable of public-information use as well, since that would allow us to map the evolution of this social learning capability onto the stickleback family tree. For several years, Mike Webster has been traveling the globe collecting different species of sticklebacks, and testing them for public-information use. He has tested fish from over 50 populations of 5 separate genera and 8 species. Of these, it is only populations of ninespines and their closest relatives, the brook sticklebacks, that exhibit public-information use, while populations of three other genera seemingly do not. This implies that the public-information-use capability evolved among the ancestors of ninespine and brook sticklebacks after their divergence from the fourspine and fifteenspine sticklebacks, which would be around 10 million years ago.

This finding is a good illustration of a general pattern concerning the evolution of intelligence; that is, the mental abilities of animals are not best explained by how closely related animals are to humans. Different aspects of intelligence have evolved multiple times in diverse taxa through convergent selection.[34] Public-information use has evolved independently in animal groups that are not closely related, including humans, some birds, and a few fishes. These groups have in common little more than the cost-benefit balance that favors this form of learning. Later chapters will present further evidence that those cognitive capabilities that might be considered the rudimentary foundations of culture have similarly evolved through convergent selection in distinct primate lineages.

Let me sum up what we have learned through our studies of public-information use in sticklebacks. We have identified an adaptive specialization in social learning, with ninespine, but not threespine, sticklebacks capable of utilizing information about the richness of a food patch by monitoring the success or failure of other feeding fish. This species difference is found among sticklebacks from all around the world, and is unaffected by rearing conditions or any other tested experiential factor. The capability would appear to be highly specific; ninespines can extract public information about the quality of food patches, but not shelters. No other differences in the learning capabilities of the two species have been found. Ninespines have weak morphological defenses, as do the brook sticklebacks, which are the only other stickleback to use public information. This implies that public-information use is of benefit where it allows animals to acquire information concerning the quality of food patches safely, cheaply, and reliably. Among such species, our stickleback experiments suggest the use of public information has evolved to be highly strategic, and allows the fish to exploit resources in their environment with near optimal efficiency. The species' copying can nonetheless be predicted by evolutionary models, and is consistent with a number of distinct but complementary social learning strategies. Conversely, threespine sticklebacks, which are physically more robust, can sample patches directly at low cost and therefore have little need for public-information use. Indeed, to wait under cover while others feed would only mean that threespines miss out on feeding opportunities. This partly explains why these two fish are often found in the same stream and rivers together. Ninespines and threespines enjoy a mutualistic relationship, where each benefits from the others' presence. The opportunity to acquire public information from the threespines probably at least partly underlies the ninespines' preference for mixed-species shoaling. The threespines, in return, enjoy the safety that greater numbers bring, particularly since many predators preferentially target the ninespines as food.

Several valuable lessons about social learning emerged from this program of research. First and foremost, we learned that animals exploit information provided by others in a decidedly strategic manner. Ninespines do not copy at every available opportunity, but instead are highly selective. For instance, they tend to utilize social information when they have no relevant prior experience to rely on, or when the knowledge gained by that experience is unreliable, as when it is out of date. We also saw that ninespines were able to combine these two sources of information effectively, to maximize foraging returns and minimize risk. Prior to the findings of the tournament, this was an important and striking lesson. Once we had been alerted to the strategic nature of stickleback copying, we noticed that other animals copied very selectively too. My research group has studied the behavior of a lot of different animals, and in every species that we have worked with, without exception, the social learning observed is highly strategic. Social learning researchers around the world have overwhelmingly reached the same conclusion.

Subsequently, I coined the phrase "social learning strategy" in a deliberate attempt to equate the animal copying rules that were emerging from experimental studies with those strategies that were subject to analyses using evolutionary game theory.[35] The idea that humans, at least, might be copying strategically already existed in the anthropological literature,[36] and was supported by important theoretical findings.[37] However, there were clear opportunities to develop this theoretical foundation further, and because the concept of a social learning strategy had an intuitive appeal that resonated with the biological community, it became a growth area of social learning research.

One reason why the strategies approach proved productive is that it provided rich possibilities for integrating empirical and theoretical findings. Predictions from mathematical evolutionary models concerning the application of specific strategies could be tested with animal social learning experiments. These in turn provided data with which to ground theory and thereby ensure assumptions were sound. In our case, we were able to show that the pattern of copying exhibited by our sticklebacks fit with hypotheses derived from evolutionary theory, such as *copy when asocial learning is costly*,[38] *copy when uncertain*,[39] and *conform to the majority behavior*.[40] In this manner, the strategies approach helped to draw the field of social learning more closely into a general evolutionary framework.

The last decade has witnessed a rush of studies in this domain, leaving little doubt among researchers in the field that animal social learning is broadly, perhaps universally, strategic.[41] Honeybees were shown to follow the waggle dances of other bees more frequently when their own foraging had been unsuccessful.[42] The probability that minnows would copy the feeding sites of others was found to increase with predation risk.[43] Whether redwing blackbirds acquire a food preference through social learning hung critically on whether the demonstrator birds are sick or well.[44] Chimpanzees were more likely to copy a dominant over a subordinate animal.[45] And so forth. Strategic copying became the rule rather than the exception. It was experimentally demonstrated across a broad range of animals, including those in natural populations, where strategic copying was often found to increase biological fitness.[46]

Humans were no exception to this pattern. For instance, Tom Morgan, one of my PhD students at St Andrews, presented adult human subjects with a battery of experimental tasks. The experiments found conditional or strong support for the use of nine separate social learning strategies predicted by the cultural evolution literature, including conformity, payoff-based copying, copying when asocial learning is costly, and copying when uncertain.[47] These various influences operated simultaneously and interacted to produce behavior leading to effective decision making and higher payoffs.[48] In fact, the very term "copying" betrays the strategic quality of human social learning. Throughout this book I am using this term in a very general manner to refer to any form of social learning, but when "copying" is used in normal speech it often has a negative connotation. We think of the naughty schoolchild who copies during an exam. In fact, this imagery brings home the strategic nature of social learning beautifully. Nobody cheats in exams when they already know the answer! Cheating reprobates are violating norms because the examination is designed to establish what they alone know. However, the strategy to copy when uncertain is a smart rule that has served humans well throughout history.

Our success at confirming theoretical predictions brought with it fresh challenges. Over a few short years my laboratory generated experimental evidence that ninespine sticklebacks deployed no fewer than six separate social learning strategies. While not all tested strategies were confirmed,[49] the diversity of supported rules nonetheless suggested that strategic copying was far more complicated than first envisaged. For instance, any research agenda dedicated to working out *the* strategy implemented by a species of animal could no longer be tenable. Rather, animals typically use many social learning strategies, switching between them according to the circumstances, in order to exploit the available internal and external cues in a flexible and adaptive manner. This leaves the job of the social learning researcher even more challenging. It is not sufficient to work out which learning rule is being used; we must also work out the rules that specify which rule should be used. Currently, researchers are starting to envisage *meta*-strategies that dictate social learning strategy use in a context-specific manner,[50] or to think of strategies as biases that influence reliance on social information.[51]

Our stickleback experiments themselves provide clues as to how alternative strategies might be integrated. Fish rely on up-to-date and unambiguous personal information when available, but use social information when they lack relevant experience or where their knowledge is outdated or unclear.[52] Information about the payoffs to demonstrators is preferentially exploited in decision making,[53] but when such information is missing, the fish switch to the next most reliable source of information, which takes account of the numbers of individuals utilizing each option.[54] That information is, in turn, implemented through a conformist learning strategy,[55] which again has been shown to be highly adaptive.[56] Such observations suggest that animals may make judgments concerning learning strategies through mental processes that resemble hierarchically organized decision trees.[57]

A second challenge that arises with the identification of social learning strategies is to comprehend the mechanisms that allow animals to copy strategically. For instance, are the observed tendencies to conform to the majority or to copy the highest payoff behavior, biological adaptations that have evolved specifically to enhance the social learning performance of animals? Or have the animals learned through prior experience that attending to the majority, or to the payoffs to others, is a productive heuristic?[58] The strategies perspective has little to say on this matter, being inherently mechanism neutral,[59] and hence equally consistent with either possibility. However, studying the mechanisms that support social learning is no less important to a behavioral scientist than studying the functional rules that underlie the decision making of animals. Here again, our stickleback experiments prove instructive. In further experiments, Mike Webster analyzed the behavior of the demonstrator fish and identified their feeding strikes (where the fish suddenly dart toward and peck at the food) as the specific cue on which the observing ninespines focus while learning. The analysis suggests that that the capability for public-information use in ninespine sticklebacks is underpinned by a tendency to attend to, and quantify, the feeding strikes of other fishes. These findings, combined with the absence of evidence for general enhancements in the learning of ninespines compared to threespines, suggest that natural selection has enhanced the ninespines' public-information use by fine-tuning the perceptual, motivational, and information-processing capabilities associated with this form of social learning, rather than by directly enhancing their learning capabilities. This is consistent with the view that while *what* an animal learns may be ecologically specialized and may therefore vary among species, *how* animals learn (at least, at the level of the underlying associative processes) appears to be broadly similar across diverse taxa.[60]

Our stickleback experiments also reveal that animals often possess some impressively sophisticated social learning capabilities. Who would have imagined that a tiny freshwater fish would be found to share with humans the ability to utilize an optimally efficient hill-climbing learning rule, or to exhibit conformist transmission? However, much is made of the parallels between animal and human cognitive competences, and an honest comparative perspective demands that the differences be given equivalent attention. Our ninespine sticklebacks, proficient in acquiring knowledge about patch quality through observation, failed to learn that one shelter is better than another through use of public information. That inflexibility stands in contrast to humans who, no doubt, could assess the quality of a food patch by monitoring the returns to others, but equally could generalize across contexts to extract public information about mates, shelters, or any other resource. Nor could threespine sticklebacks solve the public-information use task, in spite of their being competent at other forms of social learning.

The following, I suggest, is a general pattern. Animals typically possess specific social learning competences, tailored by natural selection to address particular adaptive challenges relevant to the species in its natural environment, and which do not operate, or operate far less effectively, outside of the domains in which they have been selected to work. Macaques can acquire a fear of snakes, and any snakelike object, through associating the object's characteristics with the fear responses of other monkeys, but they seemingly can't acquire fears of other objects that way.[61] Young male songbirds appear predisposed to acquire the songs of their own species, but rarely those of others, implying evolved predispositions to pick up some sounds more readily than others.[62] More generally, most animals are social learning specialists; their capabilities are specialized solutions that have evolved in distinct lineages to fulfill specific functions, and which are operational in a comparatively narrow domain. Humans, in contrast, are social learning generalists; our copying is certainly applied strategically, but is seldom greatly constrained by our competences. We are not only capable of learning about foods, mates, and predators socially, but also about algebra, ballet steps, and car mechanics; these are phenomena that did not appear in our evolutionary past and were not part of the adaptive challenge that our minds were selected to overcome.

The same pattern applies to other aspects of cognition relevant to the evolution of culture. Honeybees can use their waggle dances to transmit information about food sources and nest sites, but unlike human language, the dances cannot communicate other forms of knowledge.[63] Meerkat helpers will actively teach pups how to process prey items but, in contrast to human teaching, show no signs of teaching youngsters other forms of wisdom.[64] The manufacture and use of tools by New Caledonian crows allows them to grub for food items hidden deep in crevices,[65] but, again unlike humans, the crows rarely use tools in other ways. In each instance, the cognitive capabilities of animals are found to be specific to particular taxa that share the same ecological challenges and, in marked contrast to humans, their functionality is largely restricted to domains in which they evolved.

The social learning strategies tournament taught us that there are benefits to copying strategically, and this chapter confirms that strategic copying is what animals do. The tournament suggested that natural selection will have favored learning rules that increased the efficiency of copying, and sure enough, we found that our fish learned with optimal efficiency. Of course, to deploy these functional rules, animals must possess the relevant perceptual and cognitive competences. Selection cannot favor a disproportionate tendency to copy the majority behavior in animals that cannot discern what the majority behavior is; nor can it favor payoff-based copying in species incapable of computing the returns to others. An animal cannot copy over long distances if it cannot see over long distances; nor can it imitate the fine motor patterns of other animals if they will not let it close enough to watch.[66] Such considerations allow us to imagine how natural selection favoring more accurate and more efficient copying could plausibly have spillover effects in shaping the cognitive, perceptual, and social characteristics of animals. Selection for more effective copying would have knock-on consequences for the evolution of brain and cognition. Evidence suggests that something along these lines happened in the primate lineage leading to humans, causing our ancestors to evolve a more general capability to copy others, with major ramifications for the evolution of mind. Our investigations of how that happened are described in the next two chapters.

CHAPTER 5

THE ROOTS OF CREATIVITY

In 1921, in a small village on the south coast of England close to South-ampton, a blue tit was first observed to peck open the foil top of a milk bottle delivered to the doorstep of one of the houses, and to drink the high-energy cream.[1] Whether it was truly the very first bird to steal the cream from bottles of milk is open to doubt. More likely, the bird spotted had copied a sneaky individual that stole in unobserved to plunder a free breakfast. Nonetheless, in a nation of bird watchers like Britain, no avian burglar could hope to get away with this pilfering for long without detection. Amateur ornithologists, followed by professional ethologists, noted the repeated appearance of the behavior as it spread to nearby locations. Soon dozens of other species of birds had picked up the habit. The British public was enthralled. Bird lovers ate their toast and boiled eggs glued to their windows, eager for a glimpse of the feathered bandits. Over the next thirty years, milk-bottle top opening was observed by an army of "twitchers" who, with characteristic obsession, carefully monitored the diffusion of this charming habit through dozens of towns and villages as it spread across the United Kingdom, and even into mainland Europe.[2]

This episode is perhaps the best-known example of a novel learned behavior spreading through an animal population. Subsequently, animal behaviorists carried out experimental tests of milk-bottle opening on captive birds,[3] and used mathematical and statistical models to analyze the diffusion of this habit.[4] The studies established that many individual birds were capable of solving the puzzle of how to peck open the foil caps, even without the opportunity to copy others. Bottle-top opening would seem to be quite an intuitive behavior for a bird. Researchers also found that the behavior spread easily because birds could not only pick up the habit through copying, but also through simply being exposed to the milk bottles that other birds had opened; this seemingly sufficed to put the idea in their heads. Apparently, this particular habit spread through a combination of multiple independent inventions at sites where milk bottles were introduced, followed by social transmission from bird to bird.[5]

Milk-bottle opening is an example of an animal *innovation*, defined as the devising of a novel solution to a problem, or a new way of exploiting the environment. The habit appears special only by virtue of its familiarity. In reality, many thousands of innovations have been devised by a broad variety of animals. Birds and mammals are known to incorporate new items or novel techniques into their foraging repertoires; whales, dolphins and birds introduce novel vocal elements into their songs; apes and monkeys concoct novel deceptive acts; primates and birds invent new tools; and countless other animals create novel courtship displays and social behavior.[6]

Animal innovations are highly diverse. They range from the ingenious (the orangutans that devised clever means of extracting palm hearts from trees with vicious defenses such as sharp spines and knife-edged petioles[7]), to the morbid (the herring gull that invented the habit of catching rabbits and killing them by dropping them onto rocks from height, or through drowning them in the sea[8]), to the enchanting (the group of Japanese macaques who started rolling snow balls and playing with them [figure 4][9]), to the plain disgusting (the rook that made a habit of eating pieces of frozen human vomit[10]).

My favorite example concerns a young chimpanzee called Mike. He was observed by primatologist Jane Goodall to shoot up the social rankings and become alpha male in record time by devising a thoroughly intimidating and noisy dominance display that involved banging two empty kerosene cans together.[11] Astonishingly, Mike achieved this without having a single fight. Also impressive are a group of ring-tailed lemurs, who were able to drink from an out-of-reach pool by dipping their furry tails in it while clinging from an overhanging branch and then squeezing water off their fur into their mouths.[12] A population of baboons independently invented the same habit.[13] What is more, those of us who were told off as kids for dunking our cookies in our coffee might be captivated by a population of Trinidadian birds called Carib grackles that have started dunking their food too.[14]

My laboratory has been investigating animal creativity and invention for two decades, and this chapter summarizes some of our findings. Our experiments convinced us that animals do exhibit behavior

FIGURE 4. Japanese macaques appear to enjoy playing with snow. The monkeys frequently roll snow-balls and sometimes youngsters even engage in snowball fights. By permission of Zoonar.

that can sensibly be termed "innovation," even if the consanguinity of nonhuman-animal and human innovation is a matter of debate.[15] Our investigations, and those of other animal innovation researchers, provide compelling evidence that humans do not have a monopoly on creativity. Many animals invent new behavior patterns, modify existing behavior to a novel context, or respond to social and ecological stresses in an appropriate and novel manner.[16] Of course, a vast difference exists between dipping food and inventing a microwave cooker, while banging cans together to send a message is a long way from developing e-mail. Why it should be that humans alone are capable of such truly spectacular innovation is the focus of this book. Undeniably, there is something distinctively creative about our species, and the issue of how this arose is addressed in later chapters. Nonetheless, I maintain that the study of animal innovation is pivotal to understanding human cognitive evolution. As we shall see, research in this field has generated some highly suggestive data that provide important clues with which to reconstruct aspects of the human story, particularly those related to the evolution of our enlarged brain. The innovation of other animals might not be impressive when juxtaposed against human achievements, but its study is central to comprehending the roots of human culture.

In recent years, researchers have demonstrated that animal innovation can be studied rigorously and systematically, that innovation can be distinguished from related processes such as exploration and learning, and that innovation plays important roles in the natural behavior of animals. Indeed, the ability to innovate can be critical to the survival of animals in changed circumstances.[17] For instance, innovative species of birds have been found to be significantly more likely than other birds to survive and establish themselves when introduced into new locations by humans.[18] In this age of human habitat destruction, a capacity for innovation is likely to be of critical importance to endangered species forced to adjust to impoverished environments.[19] Evidence is mounting that innovation plays important roles in animal ecology (e.g., where it facilitates the ability of the animal to expand its range) and evolution (e.g., where it can generate divergence between populations and is an important source of behavioral variation).[20]

In truth, many animals are enormously inventive, but the extent of animal innovation remained hidden until recently for a simple and obvious reason: one cannot recognize a behavior as novel until one has a good understanding of the "normal" behavior of the species. Only after capuchin monkeys had been studied in the wild for many years, could the first recorded use of a club to attack a snake with confidence be regarded as an innovation.[21] Likewise, only after decades of painstaking observation by chimpanzee primatologists could an extraordinary courtship display, which involved an adolescent male called Shadow flipping his upper lip over his nostrils to impress the females,[22] be recognized as genuinely novel. The adult females he sought to woo were dominant to him and responded aggressively to conventional courtship moves, but with his new display Shadow was able to convey his sexual interest devoid of aggressive overtones.

What was long established, at least in a subset of intensively studied species such as rats, cats, dogs, and pigeons, was that creativity is a natural aspect of the learning process. In the late nineteenth century, a prominent American psychologist, Edward Thorndike at Columbia University, conducted classic experiments on the problem-solving abilities of animals; these experiments led to the development of one of the most celebrated laws of animal learning, known as the "law of effect."[23] In one famous study, Thorndike confined cats to small boxes from which they could break out through mastery of an escape mechanism, such as pressing a button or pulling a string. The cats disliked confinement and responded by biting at the bars, thrusting their paws through openings, clawing at all in reach, flailing around wildly and incessantly, and generally trying all kinds of actions to escape. Eventually, the cat hit on something that worked, and got out. Thorndike repeated the confinement and noted how, over several trials, all the other unsuccessful actions gradually dropped out of the animal's performance, while the successful act was, as he termed it, "stamped in" by the experience. After many trials, when put in the box, the cat would immediately activate the escape mechanism in a calm and definite way.

Thorndike's experiment is renowned for establishing that animals learn by repeating actions that are followed by a positive consequence, while eliminating those followed by a negative outcome. However, the experiment also showed that this learning process frequently begins with the spontaneous generation of novel behavior,[24] which gradually becomes refined through experience to retain successful elements. A similar conclusion was reached by perhaps the greatest of all learning theorists, Burrhus Fredrick Skinner, the eminent Harvard psychologist who gave his name to the "Skinner box" in which much animal learning is now studied. Skinner emphasized how animals are naturally active, continuously emitting behavior, and that, depending on the circumstances and the animal's motivation, novel behavior would commonly be generated.

Not all animals are equally adept at innovation, however. As far back as 1912, the Bristol psychologist Conwy Lloyd Morgan speculated that behavior may be composed of a repetitive component that has occurred many times previously and a smaller component of novel behavior that is a creative departure from routine, and was prominent in so-called "higher organisms."[25] Nonetheless, research into the rules of animal learning strongly implies that the generation of novel behavior is a regular aspect of the way in which animals learn. Animal behavior studies in the modern era have confirmed these early learning theorists' conclusions. As we shall see, innovation is indeed widespread, and species do differ in their tendency to innovate.[26]

Animal behaviorists were nonetheless surprisingly slow to take up the challenge of studying innovation. After all, when a novel learned behavior spreads through an animal population as individuals learn from one another, typically a single individual will have initiated the process. Such diffusion requires two processes: the initial inception of the behavioral variant, which is *innovation*, and the spread of the novel habit between individuals, known as *social learning*. However, while many scientific books, conferences, and papers have been dedicated to animal social learning,[27] in comparison, the subject of animal innovation received scant attention right up until the twenty-first century. Although related topics, such as neophilia, exploration, and insight learning in animals have benefited from established traditions of research, and while innovation in humans has been subject to considerable investigation, animal innovation was largely neglected.

One exception was an important paper authored by primatologist luminaries Hans Kummer and Jane Goodall in 1985.[28] Kummer and Goodall surveyed the scientific literature on primate behavior, noting that there were extensive reports of primate innovation, although "of the many such behaviours observed, only a few will be passed on to other individuals, and seldom will they spread through the whole troop."[29] Several of these innovations derived from the ability of monkeys and apes to profit from accidental events, while others resulted from the ability to use existing behavior patterns for new purposes. Kummer and Goodall described how innovation was sometimes stimulated by an excess of resources—as occurred, for instance, when the animals were provisioned or held in captivity. In one case, a study compared the behavior of hamadryas baboons in Zurich zoo with a wild population in Ethiopia.[30] While all the motor and vocal signals seen in the wild were also observed in the zoo colony, 9 of 68 communicative signals seen in the zoo population were not found in the wild. This suggested that some zoo colony signals were innovations, which were typically elaborations on existing signals. More commonly, innovation was prompted by conditions creating a need, such as a period of drought or a social challenge. Later studies were to suggest that there is some truth to the old adage "necessity is mother of invention."[31]

What was particularly significant about Kummer and Goodall's article, however, was that it offered a suggestion for how animal innovation could be studied experimentally. At the time, many behavioral researchers felt that innovation was too rare to be easily investigated with experiments. How could researchers investigate a behavior that might only happen a few times a year? Kummer and Goodall offered a simple, pragmatic solution: "Systematic experimentation (such as the introduction of a variety of carefully designed ecological and technical 'problems') both in free-living and captive groups would provide a new way of studying the phenomena of innovative behaviors and their transmission through and between social groups."[32] This was to prove an important suggestion. In the last two decades, Kummer and Goodall's approach has been widely implemented in animal studies. Innovation can be induced in captive and natural populations of animals by presenting them with novel challenges, such as foraging puzzle boxes (ideally under controlled conditions), and by exploring the influencing factors, such as the innovator's age or the ecological context.

When my research group began studying animal innovation in the 1990s, Kummer and Goodall's article was far from the only treatment of animal innovation. Nonetheless, the scientific literature on this topic was scant. No satisfactory definition of animal innovation existed, and the few articles that had been published were mutually contradictory. As a step toward resolving these issues, I organized a symposium on the topic at the 2001 International Ethology Congress meeting, and invited the participants to contribute chapters to the book *Animal Innovation*,[33] which was published two years later. In the introductory chapter, my coeditor Simon Reader and I discussed some difficult issues concerning how animal innovation should be defined.[34] These are challenging, but here I skip the complications.[35] For our purposes, an animal innovation can simply be regarded as a new or modified learned behavior or manufactured resource not previously found in the population.

When prompted to name a human innovation, we might think of the invention of penicillin by Alexander Fleming, or the construction of the World Wide Web by Tim Berners-Lee. The animal equivalents are perhaps less dramatic. An example might be the invention of automobile-mediated nut cracking by Japanese carrion crows, who place walnuts in front of the wheels of cars at traffic lights and return to retrieve the nuts when the lights turn red.[36] Another cute example is the devising of novel nest decorations by starlings—birds famously fond of shiny objects—who raided a car-wash coin machine in Fredericksburg, Virginia, and made off with, quite literally, thousands of dollars in quarters.[37] What these innovations have in common is that they represent a newly introduced behavior or product.[38] As the milk-bottle opening example illustrates, animal innovations can subsequently spread by social learning, but the introduction of a novel behavior into a population through copying is not in itself innovation.[39] Nor, as Simon and I were keen to stress, should every quirky, random, or idiosyncratic behavior qualify as an animal innovation. What scientists call innovations must be both entirely novel and learned behaviors, and researchers only know that they are learned if they are produced repeatedly in a specific, functional manner.[40] Some researchers argued for a narrower use of the term—for instance, restricting innovation to cognitively demanding tasks. However, we felt that given the primitive state of knowledge on the topic, an inclusive definition would better benefit the field since an over-exacting definition risked hindering the collection of raw data, and amassing data is the most important goal of a young science. The subsequent growth of the field and widespread use of our definition arguably justify our stance.[41]

An initial question for animal innovation researchers to address was whether it was even appropriate to think of particular animals as "innovative." Perhaps animal innovations were simply induced by circumstances, and what we were calling innovators were simply those animals that happened to be subjected to changed or stressful environmental conditions, such as droughts or provisioned food. Conversely, if innovativeness was genuinely a property of individual animals, we needed to establish whether all individuals engaged in it, or if were there particular classes of individual or personality types who carried out most of the innovation.

Rachel Kendal,[42] a PhD student at the University of Cambridge, set out to address some of these questions with an investigation of innovation in zoo populations of callitrichid monkeys (marmosets, tamarins, and lion tamarins). The prevailing assumption in the primate behavior literature had been that young or juvenile primates were more innovative than adult individuals. However, the data underlying this assumption were not particularly strong. Researchers had perhaps been overly influenced by one or two prominent animal innovators, such as Imo, the famous juvenile macaque who devised sweet potato washing. The putative innovative tendency of the young was thought to be a side effect of juvenile animals' increased rates of exploration and play. This is plausible. Indeed, a compelling case can be made that play functions to generate creativity and stimulate innovation. Play may be an adaptation to get out of the rut and discover better solutions to life's challenges.[43] However, within callitrichids contradictory evidence for age differences in response to novel objects, foods, and foraging tasks had been found.

To shed some light on these issues, Rachel adopted Kummer and Goodall's recommendation, presenting novel foraging challenges, in the form of simple puzzle boxes that could be opened to retrieve highly desired foods; she offered these to family groups of callitrichid monkeys in zoos around the United Kingdom. Rachel set out to examine whether the age of the monkeys affected their responsiveness to novelty (neophilia), exploration, and innovation. The puzzle box tasks were given, over repeated trials, to more than 100 callitrichid monkeys in 26 zoo populations. The tasks required the monkeys to access food by, for instance, pushing open flap doors, reaching into holes, or lifting off the lids of the boxes. Rachel recorded the first individual to approach, contact, and solve each task, in each population, as well as a variety of other variables relevant to the spread of the solution.[44]

The study did indeed reveal systematic age differences in callitrichid innovation, but with older rather than younger monkeys significantly more likely to be the first to solve the tasks. While younger monkeys were among the first to explore and contact the tasks, adults were typically the ones who solved them first. Older individuals appeared better able to turn their manipulations of the apparatus into successful extraction of the food.[45]

When innovations spread through animal populations, the temptation is to assume that the animals must be learning from each other, and most times they are. However, researchers cannot just assume social transmission, because in theory, if the innovation was simple to master, each adopter could have learned it independently. Indeed, in the years that I have studied social transmission, I have come across several examples of novel behavior patterns diffusing through animal populations, which to all intents and purposes resembled socially transmitted behaviors; yet subsequent analysis revealed the animals each learned the task independently.[46] Distinguishing social from asocial learning is even more of a problem in natural animal populations, where the history of the animals concerned is rarely known. Major controversies revolve around this issue. For instance, the claim that animals such as chimpanzees or dolphins possess "culture" has been viewed as seriously undermined by the absence of clear evidence that habits such as nut cracking, foraging for termites, and sponging are socially learned.[47]

For this reason, my laboratory has dedicated much effort to devising new mathematical and statistical methods that allow the diffusion of innovations through social learning to be detected in natural populations of animals. The "option bias" method is one such tool.[48] The method works on a simple principle; that is, if there are more than one means of solving a task (that it, multiple "options"), then social transmission will generate "biases" in the pattern of option use, since individuals will copy each other's methods. Were researchers to come across a population of monkeys that were all solving a problem one way, when there was an equally easy alternative method available, they might infer from this bias in the option choices that the monkeys had copied the method from each other. In practice, researchers require statistical methods for working out how likely it is that a given level of skew in the options deployed could have arisen through chance or entirely through asocial learning. If the observed option bias is sufficiently large, then the alternative hypothesis, that the animals are each learning independently, can be rejected, and social transmission can be safely inferred. In the case of Rachel's monkeys, we were able to conclude that some of the puzzle box solutions devised by innovators had indeed subsequently spread through social learning.

All matters considered, Rachel's experiment suggested that the older monkeys' greater experience and physical competence had allowed them to innovate and solve novel problems more effectively than the youngsters, but that the young often acquired those same new foraging habits through social learning.[49] The greater life experience of adults seemingly enabled them to outperform younger individuals, although other developmental factors, such as improvements in manipulative skills, increased strength, and maturity with age may also have been important. There would appear to be a developmental watershed (at about four years, in the case of callitrichid monkeys), when prior manipulative experience generates sufficient competence in extractive foraging for individual callitrichids to begin to translate their manipulations into successful and efficient retrieval of the food without the help of others. In these monkeys, at least, innovation appears to require a certain amount of prior knowledge and skill.[50]

Age was not the only source of variation in innovative performance. The experiment also revealed consistent differences between species in their problem-solving capability. Rachel found faster responses, greater levels of success, more manipulation of the puzzle boxes, and greater attentiveness to both the task and the successes of others in lion tamarins, than in both tamarins and marmosets.[51] These findings fit with a previous study that had found species dependent on manipulative and explorative foraging tend to be less neophobic and more innovative than other species.[52] Extractive foraging is the act of locating and processing foods embedded in a substrate or casing, such as underground roots and insects, or hard shells of nuts, and fruits.[53] Lion tamarins are extensive extractive foragers, who use their claws to dig under the bark of trees to search for a variety of insects and other prey.[54] Marmosets are also extractive foragers, but are more specialized than the lion tamarins because they gain a large proportion of their nutrition from tree sap,[55] while all of the tamarin species that Rachel studied have been described as nonextractive foragers.[56] Hence, Rachel's experimental results of increasing innovation from tamarins to marmosets to lion tamarins fit the hypothesis that extractive foraging may have promoted the evolution of intelligence, in the form of an ability to respond flexibly to foraging challenges.[57]

Further insights into the factors that influence the invention and spread of animal innovations came from studies of birds. Neeltje Boogert, a Master's student at the University of St Andrews, investigated to what extent the pattern of spread of innovations in captive groups of starlings could be predicted by prior knowledge of a variety of individual and social-group variables,[58] such as association patterns (who spends time with whom), social-rank orders (who is dominant to whom), measures of neophobia (e.g., who is quick to explore a new area or object), and asocial learning performance.[59] Again adopting Kummer and Goodall's recommendation, small groups of starlings were presented with a series of novel extractive foraging tasks, and Neeltje recorded the time that it took for each bird to contact and solve each, as well as the orders of contacting and solving. We then explored which variables best explained the observed behavior.

Neeltje found, perhaps surprisingly, that asocial learning performance, recorded where each individual had been tested in isolation, predicted which birds would be the first solvers of the novel foraging tasks in the social group. In other words, one can foresee how innovative a starling will be on the basis of its previously measured learning performance. That innovative individuals should turn out to be good learners is of course very intuitive, but the results need not have panned out this way. We would not have been surprised if, say, dominant birds had monopolized the tasks, or if neophobia had proven a better predictor than learning ability, because timid individuals had failed to engage with the puzzle box. Often, intuitive predictions are not confirmed. For instance, Neeltje's initial findings suggested that association patterns did not predict the spread of solving, implying that birds were no more likely to learn from close associates than birds with whom they spent little time. This result had surprised us, particularly as other data suggested that some copying was taking place.[60] We came to the conclusion that perhaps the birds had not learned from their closest associates, because the relatively small size of the enclosures in the captive environment meant that each bird could readily see all group mates at virtually all times, leaving individuals able to learn from each other effectively.

Innovations may be more likely to spread along networks of association in larger groups living in more naturalistic environments, a prediction that has since been confirmed in populations of wild birds.[61] However, when we developed a more powerful statistical tool for detecting social transmission through animal networks, called *network-based diffusion analysis*,[62] and analyzed Neeltje's starling data again, we subsequently found evidence that the birds did learn from close associates after all. The same tools have subsequently been implemented to demonstrate network-based diffusion in birds, whales, sticklebacks, and primates.[63]

In my experience, many more people will accept the idea that primates and birds can innovate than are willing to entertain the suggestion that fishes have such a problem-solving capability. However, as we have seen, fishes are good learners and are convenient model systems for studying many aspects of social behavior. Moreover, forms of experiment can be conducted with fishes that would be practically impossible in more cerebral taxa, such as primates. This applies to the experimental investigation of innovation. That is because if the innovator is designated as the first individual in a population to solve a problem, a researcher would have to study large numbers of populations to start to see consistent patterns in innovation. Imagine that a researcher gives a novel foraging task to a population of baboons that a two-year-old, subdominant male solves first. What conclusion could be drawn? The researcher could not legitimately infer that innovators are typically male, young, and subdominant, since those characteristics might not be responsible for this particular innovator's success. The same task would have to be given to a large number of other baboon populations in order to start to see genuine patterns in the innovation that ensues. Were the researcher to find, say, that subdominant baboons consistently solved the task first, he or she would have grounds for concluding that there might be a genuine relationship between social rank and innovation, but that, perhaps, other factors prove not to be causally relevant. The trouble is no professional animal behaviorist has the resources to establish and study sufficiently numerous captive populations of baboons, so that kind of experiment simply could not be conducted with that particular species, or for that matter, any other primate. In contrast, establishing multiple populations of small aquarium fish is highly feasible, and that is where fishes come into their own.

Together with another of my PhD students at Cambridge University Simon Reader, now at McGill University, I established many populations of guppies,[64] a small tropical fish, in laboratory aquaria and presented them with novel maze tasks that could be negotiated to find hidden food. The mazes were composed of one or more partitions with small holes or compartments through which the fish had to swim to find the food source.[65] We designated the "innovator" as the first individual in each population to solve the maze and eat the food.[66] Sixty-nine such populations were established, each comprised of around 16 fish that were carefully chosen so as to vary in their sex, hunger level, and body size. The experiments were designed to ascertain whether any of these characteristics would consistently be found among the innovators.

We did indeed find consistent patterns. Innovators were significantly more likely to be females than males, more likely to be food deprived than not, and typically smaller rather than larger fish.[67] These patterns were not artifacts of differences among the fish in activity or swimming speed; innovators were neither the most active fish (since males are more active than females) nor those with the fastest swimming speed (since large fish can swim faster than small fish).[68] Rather, the observed patterns are best explained by differences among fish in their motivational state. The first individuals to solve the mazes are those driven to find novel foraging solutions by hunger, or by the metabolic costs of growth, or pregnancy. Small fish have higher metabolic costs associated with more rapid growth, and so need to acquire food at elevated rates relative to larger fish.[69] Guppies are live-bearing fish, and adult females spend much of their time pregnant, which exerts significant demands on their energy intake. Other experiments of ours have established that this sex difference in foraging performance disappears when sexually immature fish are tested.[70] Motivation, rather than cleverness or ability, is what explains patterns of innovation here. Fishes are typically reluctant to swim into unfamiliar holes and through dark compartments, since predators might be lurking behind them. The hungrier an individual, the more likely it will be to take risks and try out new solutions to find food.

To investigate further how motivational state affects innovation, we conducted an experiment, again using guppies, that explored the relationship between past foraging success and foraging innovation.[71] For two weeks, each day when the fish were fed, we would deliver food items one at a time and thus allow certain individuals to monopolize the food, thereby generating substantial variation between fish in foraging success.[72] We kept a precise record of the amount of food that each individual consumed over the two weeks and also weighed each fish before and after the experiment. Then we exposed each of the populations to three novel maze tasks, and recorded the time it took for each fish to complete each maze. Our prediction was that poor competitors (fish that had gained the least weight and obtained the fewest food items) would be more likely to innovate than the good competitors when presented with the novel foraging tasks. That was exactly what was found among male fish, where the time to complete the foraging tasks correlated positively with both weight gain and the number of food items consumed; however, no such correlation was observed in females. In spite of this, the females performed well. Female guppies appeared much more motivated to solve the foraging tasks than the males, irrespective of their past foraging success.

Evolutionary theory sheds light on these findings. Among many animals, females invest more in their offspring's development than males. Such differential parental investment often leaves males with comparatively little to lose from mating promiscuously and thereby having as many offspring as possible, but inclines females to be more cautious and choosy. This sex difference is known as "Bateman's Principle," after discover Angus Bateman.[73] In fact, the situation is a little more complicated, and a variety of factors are now known to influence the promiscuity and choosiness of the sexes,[74] particularly in humans.[75] Nonetheless, in many animals, a female's reproductive success is primarily limited by access to food resources, whereas a male's reproductive success is dependent on the number of mates that he can find.[76] This is particularly true in guppies, since females give birth to live young and pour a vast amount of energy and resources into their offspring's development. As a result, finding high-quality food has a much bigger impact on females' fitness than males'.[77] The more that female guppies eat, the more eggs are made, and the more offspring born.[78] Mating is rarely a priority for female guppies, as they can store sperm from previous matings for weeks. Most of the time they need food more than they need mates. That is why, irrespective of how well stocked their energy reserves are, female guppies always performed well in the problem solving task. They were constantly motivated to find food. Male guppies, on the other hand, spend vast amounts of their time pursuing females, displaying to them, trying to seduce them to mate, and, if they fail to impress, trying a sneaky mating attempt anyway. One study reported that in the laboratory,[79] on average, a male guppy displays to females seven times every five minutes, while observations in the wild have established that females receive a sneaky mating attempt every minute.[80] The male guppy that seeks out only sufficient food to keep it going while it chases females is the one that leaves the most descendants. Well-fed males have more important priorities than solving a foraging task, which is why the trade-off between foraging success and problem-solving performance arose in our experiment.

The aforementioned experiments, and many others like them, are the bread and butter of animal innovation research, and it was important that they were conducted to gain a basic understanding of the phenomena. However, in my judgment, current interest in animal innovation derives less from experimental investigations of animal problem solving and more from stimulating theoretical findings. Valuable though such experiments are, the investigation of animal innovation really took off with a major survey of over 2,000 examples of foraging innovations in birds[81]—the most complete survey of animal innovation ever conducted. This study was conducted by Louis Lefebvre, a biologist at McGill University in Canada, renowned for his creative research. Lefebvre noted that many scientific journals publishing articles on bird behavior carried "short notes" that reported when a species of bird was observed to do something unusual or novel. Seeing an exciting opportunity, Lefebvre led a team of researchers that compiled these reports, using keywords such as "novel" or "never seen before" to classify behavior patterns as innovations. They went on to investigate whether the size of the bird's brain predicted how innovative it would be.

There was a reason to anticipate such a relationship. Allan Wilson, a biochemist at UC Berkeley, about whom I will have more to say in the next chapter, had earlier proposed a "cultural drive hypothesis."[82] Wilson argued that the spread of behavioral innovations through cultural transmission led animals to exploit the environment in new ways, and thereby increased the rate of genetic evolution.[83] He suggested that the ability to devise novel solutions to life's challenges and to copy the good ideas of other animals would give individuals an advantage in the struggle to survive and reproduce. Assuming these abilities have some substrate in brain tissue, selection for innovativeness and social learning capability would favor larger and larger brains, which in turn would further enhance their innovation and social learning. Wilson speculated that this cultural drive had culminated in humans as the most innovative and culturally dependent species, with extremely large brains. If correct, cultural drive had played a central role in the evolution of the human brain.

Lefebvre and colleagues did indeed find the predicted relationship.[84] The rate at which different bird species engaged in innovation correlated positively with measures of brain size. Those species with the most reports of innovation tended to be the birds with the largest brains, while small-brained birds rarely innovated. Although such correlational studies are vulnerable to reporting biases,[85] Lefebvre and colleagues had devised statistical methods for evaluating and counteracting them. This work, and subsequent analyses, provided reasonable grounds to be confident that the innovation data represented a valuable, robust, and naturalistic measure of an aspect of behavioral flexibility.[86]

The pioneering work of Lefebvre's team inspired further analyses of the relations between innovation, ecology, and cognition in birds and primates. To test the idea that innovations might facilitate survival in novel circumstances, biologists Daniel Sol and colleagues took advantage of a series of natural experiments where humans had introduced bird species into new habitats; the first study was focused on New Zealand, but subsequently the same researchers went on to conduct a global analysis.[87] Innovative bird species were found to be more likely than noninnovative species to survive and establish themselves when introduced into new locations.[88] The study demonstrated that being innovative could aid survival, particularly in changed conditions. Another fascinating analysis revealed that migratory species of birds are less innovative than nonmigrants, and that the latter innovate most in the harsher winter months.[89] This implied that migratory birds were forced to travel because of an inability to adjust behaviorally to the tough winter months. Later studies showed that innovative species of birds were more likely to give rise to new species than less innovative birds.[90]

I remember reading Lefebvre's papers with considerable excitement. To me, his studies were a conceptual breakthrough. The findings not only implied that brain volume might prove a useful indicator of how innovative an animal was, but also provided the first compelling evidence that increased innovativeness might give animals an advantage in the struggle to survive and reproduce. Natural selection may have favored innovativeness as part of a survival strategy based on flexibility—that is, the flexibility to cope with unpredictable or changing environments and to alter behavior to outcompete others. Perhaps selection for innovativeness could be driving brain enlargement over evolutionary time.

Lefebvre's analysis also provided researchers with a new methodology. The experimental studies that we and others had carried out were starting to show patterns in innovative behavior, but inevitably such experiments can only investigate the behavior of a small number of species. How general these conclusions were was not clear. Our data suggested, for instance, that most innovation is carried out by adults rather than younger individuals, or that sex differences in innovation could be predicted using knowledge of parental investment patterns. However, Lefebvre had provided researchers with the means to test such hypotheses in a truly general way, across tens, even hundreds, of species.

Inspired by Lefebvre and his coauthors, Simon Reader and I took up the challenge of applying their methodology to primates. Unfortunately, primatologists did not have the same tradition as ornithologists of publishing innovations as independent short notes. Simon was forced to work systematically through literally thousands of scientific articles published in prominent primate behavior journals in order to collate examples of primate innovations. While doing so, he also collected data on rates of social learning. It was a Herculean effort on Simon's part to pull this data together and required years of effort, but he persevered. Eventually, he built up an extensive database of over 500 examples of innovations in primates, with a similar number of examples of social learning, all spanning 42 species.[91]

With eager anticipation, we conducted statistical analyses looking for patterns in the primate data.[92] The results were encouraging. A higher incidence of innovation in adult than in nonadult primates was reported across the entire database, just as our callitrichid monkey experiments had revealed. Consistent with our hypothesis that necessity was the mother of much animal innovation (derived from our fish experiments), across all primates Simon found more reported incidences of innovation in low-status individuals and fewer reports of innovation in high-status individuals than expected in either, given their numbers in the populations. Dominant animals typically have privileged access to resources, such as foods or mates, and hence, unlike subordinates, probably do not need to innovate to get what they want. Subordinate primates, on the other hand, were forced to get creative to survive; they did this by devising new foraging techniques, exploiting novel foods, and coming up with innovative means to secure matings. Often this required subordinates, more than dominants, to form strategic alliances.[93] We also found greater innovation in male primates and fewer reports of female innovation than expected, given their numbers in the population.[94] This sex difference was particularly strong in relation to sexual behavior and aggression. The latter finding can be explained in a similar manner to the sex difference in guppy innovation, except that in the case of the primates it is the males for whom the marginal benefits are greater, because the innovations allow them to access mates. In addition, Simon found that approximately half of the instances of innovation that had taken place among primates had followed some sort of ecological challenge, such as a period of food shortage, a dry season, or habitat degradation. Once again, this pattern fits with the necessity hypothesis. The primate database was lending confidence to the conclusions of our experiments.

However, the relationship between primate innovation rate and brain size was what I was most impatient to test. Compelling though it was, Lefebvre's study whetted the appetite for more. Only if the same relationship that Lefebvre had found in birds held in primates could Wilson's cultural drive really be central to human evolution. We also needed to establish whether rates of social learning covaried with rates of innovation and brain size, as Wilson predicted. Comparative statistical analyses are rarely straightforward to conduct. Sophisticated statistical methods need to be used to control for the fact that species are often closely related, and hence cannot be assumed to be independent data points in analyses. In our study there were other complications too. Some primate species might possess more reports of innovations simply because they had been studied more than other species. Clearly there are many more researchers studying, say, common chimpanzees than little-known species of nocturnal prosimians, such as aye-ayes or bushbabies.[95] We needed to correct our data for this "research effort." In addition, little consensus existed in the literature as to the most appropriate measure of brain size. Should we look at the absolute size of the brain, or focus on how large the brain is relative to body size? Should we focus on the whole brain, or just those regions thought to be important to innovation, such as the neocortex? Which potential biases needed to be controlled for? Simon and I ended up doing the analyses a number of different ways to ensure that our findings were robust.

Yet, irrespective of the details of the analysis, the results were extremely clear. We found that innovation rate correlated positively with a variety of measures of relative and absolute brain size, controlling for the phylogenetic relatedness of species, research effort, and other biases. Likewise, the reported incidence of social learning also co-varied strongly with brain size measures.[96] What is more, the incidence of innovation and social learning were tightly correlated. Big-brained species of primates invented more new behavior, and copied more, than small-brained species, precisely as Wilson had anticipated.[97]

These were significant findings. Despite extensive scientific interest in the evolution of intelligence spanning several decades, the intuitively appealing notion that brain volume and intelligence are linked had remained surprisingly untested. Certainly, important work on the relationships between neural volume and either cognitive capacity or behavioral complexity had been conducted previously, but these studies had tended to concentrate on rather specialized behavioral domains or brain regions, such as birdsong repertoire size and brain nuclei related to song,[98] or spatial abilities such as food storing in birds and the size of the hippocampus, which is thought to be the brain region in which spatial information is stored.[99] The only notable exception were the avian studies of Lefebvre and his colleagues, as described above. To our knowledge, despite ample evidence for links between neural measures and various aspects of behavior and life history in mammals,[100] no direct, unequivocal support for a link between brain size and general behavioral flexibility had hitherto been found. Yet here we had ecologically relevant measures of cognitive ability, the reported incidence of behavioral innovation, and social learning,[101] and found that brain size and cognitive capacity were indeed correlated.

What excited me more, however, was that Simon's analysis had provided emphatic support for one key aspect of the cultural drive hypothesis. The fact that brain size and innovation rate were linked in both birds and primates,[102] combined with our finding that rates of social learning and innovation covaried across primate species, really lent confidence to Wilson's explanation.[103] An account of human cognitive evolution was starting to emerge through the fog. Perhaps selection favoring innovation and social learning had driven primate brain evolution, and in the process selected for other cognitively demanding capabilities (such as tool use or extractive foraging techniques) in a runaway process that had climaxed in the human condition. Our experiments had revealed differences in innovativeness and social learning capability among individuals in fish, bird, and primate populations, so I was confident that there would be variation on which natural selection could act. We had also found differences among species, such as the superior performance of lion tamarins compared to tamarins and marmosets, which were consistent with the argument that species faced with challenging ecological circumstances would evolve enhanced problem-solving capabilities.

Cultural drive was an intriguing idea, but there were at least two problems that would need to be addressed before this explanation would really hold water. The first was highlighted by the increasing number of experimental reports of copying in invertebrates.[104] If fruit flies or damselfly larvae are capable of social learning with miniature brains,[105] why should primates need huge brains in order to copy? The same issue arose with respect to innovation, which we were finding in small fishes. An explanation was required for why selection might favor larger primate brains to support their innovation and social learning, when innovation and copying did not themselves inherently require extensive brain circuitry. The second problem was a long-standing bugbear of comparative phylogenetic analyses of brain evolution: many factors covary with brain size, so how could we be confident that any one relationship was causal? We were interested in the possibility that innovation and social learning were the critical factors that had favored primate brain evolution. However, another variable, such as selection to cope with the challenges of complex societies,[106] might have been responsible for the evolution of large brains in primates, and this enhanced computational power may only incidentally have been expressed in greater amounts of problem solving and social learning to generate the patterns we had found. To address these concerns and confirm the cultural drive hypothesis we would need to do a lot more work on primate brain evolution.

PART II

THE EVOLUTION OF THE MIND

CHAPTER 6
THE EVOLUTION OF INTELLIGENCE

The greatest disappointment of my academic career was that I never got to work with Allan Wilson. Wilson was a brilliant and visionary New Zealander, based at UC Berkeley, a leader in the field of molecular evolution. He was a pioneer of the method that used the similarity of different species' molecules to gauge the degree of relatedness between species, and thereby estimate the time since they shared a common ancestor. Wilson is most famous as the architect of the mitochondrial Eve hypothesis, which traced the mitochondrial DNA of all contemporary humans to a woman who lived in Africa around 200,000 years ago;[1] this laid the foundations of the now widely accepted "Out of Africa" model of human evolution.[2] However, Wilson was an extraordinarily creative and broad-minded scientist, and another idea of his called "cultural drive," had interested me in working with him. I had won a postdoctoral fellowship to join his lab in September 1991.[3] Tragically, just a month before I arrived, Wilson died of leukemia. He was only 56.

A few years earlier, Wilson had noticed an intriguing relationship between the rate at which animals evolve and their brain size.[4] Focusing on the lineage leading from early amphibians to human beings, Wilson plotted the relationship between an animal's brain size and the time since it had shared an ancestor with humans. He found that the animal brain had grown in relative size 100-fold in the last 400 million years. What is more, that expansion had accelerated over time; this suggested to Wilson that some sort of feedback mechanism must be involved. He resolved that the mammalian brain had driven its own evolution, and proposed a three-step hypothesis for how this might have occurred.[5] First, a new advantageous habit arises in an individual (i.e., an innovation occurs). Second, the new habit spreads through the population by way of social learning, with those individuals best equipped to copy others at an advantage as they are more likely to acquire the beneficial trait. Third, selection favors those individuals bearing "brain mutations" that endow an enhanced innovation or social learning capability. Wilson explained:

> Among the many kinds of mutations that could improve the brain's ability to innovate and catch on [i.e., copy], some will produce more neurons or dendrites and thus raise the relative size of the brain. The model enables one to understand in principle how the brain's relative size could go up exponentially with time.[6]

With each spread of a new habit through the population, Wilson argued, natural selection would favor improvements in a species' capacity to copy the discoveries made by others, resulting in bigger brains. New habits would also generate selection for changes to the animal's anatomy better suited to the behavior, leading to the fixation of novel mutations. Each increment in brain size, Wilson envisaged, would enhance the species' ability to generate and propagate new habits, making the spread of further innovations and the fixation of additional mutations even more likely. This runaway process, he believed, had driven brain evolution in a multitude of animals, particularly primates, but had climaxed in humanity—the brainiest, most creative, and most culturally reliant species of all.

Wilson anticipated that by constantly inventing and propagating innovations, large-brained species would frequently subject themselves to new selection regimes. These in turn would result in elevated rates of fixation of genetic mutations, expressed in the animal's body and brain. Thus, the cultural drive hypothesis led Wilson to expect a correlation between relative size of an animal's brain (weight of the brain measured relative to weight of the body[7]) and rate of anatomical evolution. This prediction was confirmed.[8] Rates of evolutionary change in body plan are found to correlate strongly with relative brain size in vertebrates. While rates of evolution of the molecules that make up animals' bodies depend largely on the mutation rate (which is independent of brain size), rates of anatomical evolution in animals also depend on the proportion of mutations that become fixed. In turn, this fixation rate, Wilson believed, hung on the rate at which new habits spread. He concluded:

Organismal evolution in the vertebrates may provide an example of an autocatalytic process mediated by the brain: the bigger the brain, the greater the power of the species to evolve biologically.[9]

In contemporary terms, their enhanced capacity for innovation and social learning confers on large-brained species a prodigious capacity for "niche construction,"[10] the ability of organisms to modify local environments and thereby affect the natural selection that ensues.

Cultural drive was a captivating idea, but was it correct? Many researchers were skeptical, and some evolutionary biologists contested the claim that evolutionary rates had accelerated in large-brained mammals.[11] In addition, there were several gaps in Wilson's argument when first proposed. For one, the critical relationship between brain size and social learning had been inferred rather than demonstrated, as based on a rather casual reading of the scientific literature. Wilson had merely noted that reports of social learning were more common in songbirds and primates than in smaller-brained vertebrates, such as amphibians and reptiles. For another, it was not known if animals generated the substantial amounts of innovation required for Wilson's argument to be credible, and if rates of innovation and social learning went hand in hand in all animals. A third concern was that the association between brain size and intelligence remained contentious. Intelligence is defined in many different ways, but in broad terms it refers the ability of an animal to solve problems, comprehend complex ideas, and learn quickly.[12] While it was widely believed that intelligence required a big brain, the mental abilities of only a small number of species had been comprehensively investigated. Conceivably, researchers had been biased to expect animals that are closely related to humans to be smart. These concerns could not be nonchalantly dismissed. While it was a minority position, prominent researchers had argued that, humans aside, all vertebrates are equally intelligent.[13]

Studies of animal innovation changed all that. First in birds,[14] then in primates,[15] the relationship between innovation and brain size was shown to be compelling, while experimental investigations demonstrated prevalent innovation in many animals.[16] By the time Simon Reader and I found that large-brained primates also engaged in more social learning than small-brained primates, that rates of social learning and innovation covaried across primate species, and that ecologically relevant measures of cognitive ability in primates were predicted by brain size, cultural drive was starting to look like a plausible hypothesis.

Nonetheless, the hypothesis would have to be fleshed out in greater detail and further nontrivial issues addressed before cultural drive could be regarded as comprehensively supported. First, the question remained of exactly how social learning could drive brain evolution when some animals managed to copy with tiny brains. Greater specification of the feedback mechanism by which cultural processes fostered the evolution of cognition was required if the argument was to be compelling. Second, many variables (e.g., diet, social complexity, latitude) had been shown to be associated with brain size in primates. In order to evaluate the hypothesis that cultural processes had played a particularly central role in the evolution of the human mind, we needed to establish whether social learning was a genuine cause of brain evolution. That required ruling out alternative explanations; plausibly, large brains had evolved for some other reason, and the greater computational power that big brains bestowed had been expressed in elevated rates of social learning and innovation. Third, talk of increases in "brain size" is rather simplistic. The brain is a complex organ with extensive substructure, and with particular features and circuitry known to be important to specific biological functions. We needed to establish how the brain had changed over evolutionary time, and whether the observed changes in size and structure were consistent with what the cultural drive hypothesis predicted. Had those brain regions associated with innovation and social learning in humans not increased in size over the last few million years, then cultural drive would appear a far less compelling explanation for the human condition. In this chapter, each of these issues are addressed in turn.

Consider first the challenge posed by invertebrate copying. Wilson probably never imagined that social learning would be demonstrated in animals like fruit flies and wood crickets. Human brains range between 1.25 kg and 1.5 kg, and contain an estimated 85 billion nerve cells. A honeybee's brain, on the other hand, weighs only 1 milligram and contains fewer than a million nerve cells.[17] If honeybees can copy with a brain the size of a pinhead, it becomes hard to envisage why natural selection for greater amounts of copying should favor bigger and bigger vertebrate brains.

This conundrum was another issue that, for me at least, was clarified by the social learning strategies tournament.[18] When the performances of the submitted entries were compared, the best performing strategies were found to be those most heavily reliant on social learning. By and large, the more that top strategies relied on social as opposed to asocial learning, the better they performed. However, the opposite relationship was found among poorer performers: the more copying, the worse their performance. These findings had demonstrated that it was not copying per se that was adaptive, but rather *efficient* copying. Put another way, copying only pays if it is done well. The winner of the social learning strategies tournament was the entry that copied with the greatest efficiency, carefully computing the optimal time to learn, weighting recently acquired information more heavily than older information, and making projections into the future about the likely utility of investing in further copying.

The tournament teaches us how to interpret correctly the relationship between the incidence of social learning and brain size found in primates. That relationship is unlikely to reflect selection for more and more social learning, since social learning is not inherently beneficial. What is more, animals do not require a big brain merely to engage in social learning, as studies of bees, ants, and flies compellingly demonstrate. Rather, if social learning is partly responsible for the evolution of larger brain size in primates, it is because natural selection has favored more and more efficient social learning. For Wilson's hypothesis to be correct, the key driver would have to be selection for more strategic, accurate, and cost-effective forms of copying, as well as for other cognitive, social, and life-history variables that enhance the efficiency of knowledge transfer between individuals, rather than selection for more copying. The greater amounts of copying exhibited by large-brained compared to small-brained primates were, almost certainly, just incidental outcomes of the former's greater proficiency at social learning. Copying ability, rather than copying frequency, was the causally relevant variable.

In the tournament, when agents copied the behavior of other agents, copy error occasionally arose. Such error typically led to poorer returns to the copier, as performed actions were usually high-payoff behavior, while error returned a randomly selected behavior. This finding makes intuitive sense. Individuals in the real world tend to perform those behavior patterns that have reaped dividends in the past, and accurate copying is more likely to lead others to acquire such behavior than inaccurate copying. In the tournament, the rate at which copying errors occurred was set by the organizers. All strategies that relied on social learning (i.e., that played the move OBSERVE) suffered lower payoffs if the error rate went up, and benefitted from higher returns when the rate declined. In reality, the rate of copy error depends not solely on the environmental context but also on the characteristics of the copier— for instance, on how good their perceptual systems are, how well they comprehend what the demonstrator is trying to do, and how accurately they can reproduce the demonstrator's behavior. Undoubtedly, were such individual differences in copying accuracy incorporated into the tournament, selection would have favored accurate (i.e., high fidelity) copying.

Drawing on these insights, let us reconsider Wilson's hypothesis, and see if we can spell out the putative feedback mechanism of cultural drive in a little more detail (see figure 5).

The tournament teaches us that natural selection will tend to favor those individuals who exhibit more efficient, more strategic, and higher-fidelity (i.e., more accurate) copying over others who either display less efficient or exact copying, or are reliant on asocial learning. As a consequence, functional capabilities and associated structures in the brain are expected to evolve to the extent that they enhance copying efficiency and fidelity, or alternatively, favor innovation. In turn, the evolution of such brain structures would plausibly result in growth in overall brain size, as well as an enhanced capacity for innovation and social learning, generating the feedback that Wilson envisaged. The question arises as to precisely which functional capabilities of the brain would enhance innovation and social learning.

Obvious candidates are enhanced perceptual systems. Selection for more accurate copying might be expected to favor greater visual acuity if, for instance, that enabled observing individuals to imitate fine motor actions such as food-processing techniques or tool manufacture with greater precision, or if enhanced perceptual abilities allowed individuals to copy over greater distances. Invertebrates such as bees are certainly

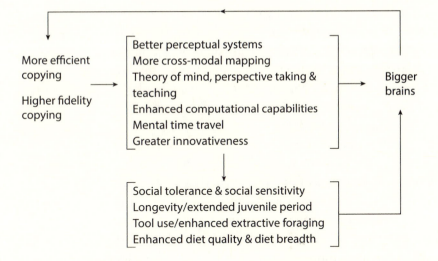

FIGURE 5. The cultural drive hypothesis. Selection for more efficient and more accurate social learning affects the primate brain and behavior. It favors a raft of cognitive capabilities, leads to increases in brain size, and feeds back to further enhance the efficiency and fidelity of social learning.

capable of social learning, but they typically need to be close by to copy one another because their visual range is limited. Copying from distance would bring a host of benefits, including greater choice of demonstrators, the vantage of safety, and being able to copy competitors and rivals who would not tolerate closer inspection. The ability to see, but also hear, with high resolution and over greater distances would likely not only enhance the precision with which copying occurred, but also open up new opportunities to learn from others. Hence, we might anticipate expansion of those regions of the primate brain associated with processing visual and auditory inputs, such as the visual and auditory cortex.

Copying, however, also requires integration across different perceptual systems, and mapping of sensory inputs to behavioral outputs. Imitation has long been of interest to psychologists and cognitive neuroscientists because of the enduring challenge of explaining what is known as "the correspondence problem," which is the challenge of understanding how the brain converts the perception of an observed act into a replica that is an enacted bodily movement.[19] To copy a fishing technique, for instance, the observer's brain must convert a stream of visual information about how others move their hands and arms into corresponding outputs specifying how the observer must also move its muscles and joints. The nature of this challenge is most apparent for what are known as "perceptually opaque" actions, such as copying facial expressions, where we cannot see the part of our own body that was moved. For this reason, selection for copying proficiency might plausibly favor the evolutionary expansion of circuitry linking the visual and auditory cortex to somatosensory and motor cortex regions of the brain.

A long-standing tradition in cognitive psychology has emphasized how, in order to imitate effectively, individuals must comprehend the goals and activities of the individual they copy.[20] Comprehension requires the ability to consider alternative interpretations or viewpoints of external objects and to construct alternative scenarios of how to interact with them; that includes considering the standpoint of other individuals. The ability to attribute mental states such as beliefs, desires, and knowledge to others, and to recognize that these may be different from one's own beliefs, desires, and knowledge, is known as "theory of mind." Effective copying might plausibly favor the evolution of a theory of mind and enhanced abilities for perspective-taking, with their associated neural underpinnings. Such abilities would allow the observer to comprehend the goals and procedures of the demonstrator more effectively, as well as allowing "teachers" to understand the state of mind and ability of their "pupils."

Animals cannot bias their copying toward high-performing individuals if they are unable to compute the payoff to others. Nor can they conform to the majority unless they can compute which behavior is in the majority.[21] Given the benefits of strategic copying, selection for the computational capabilities necessary to implement strategies should follow. Strategic copying should lead to the evolution of a focus on, or sensitivity to, the payoffs of others' behavior, the frequency of behavior within a sample, the consistency among demonstrators, and so forth, as well as the computational capabilities to process this information in efficient algorithms. For example, from an evolutionary perspective, the observation that six individuals have independently performed the same behavior is of greater relevance than that one individual has performed the same behavior six times, because the former gives far more of an indication about the prevalence of the behavior in the population.[22] Consequently, selection should favor attention to, and computation of, the frequency of individuals performing each behavior, rather than the frequency with which each behavior is performed.[23] More generally, the implementation of strategic copying demands efficient decision making, which is comprised of the integration of social and asocial streams of information into effective "rules of thumb," followed by computation of the most beneficial response. We have grounds to expect, then, that selection for efficient social learning will enhance the power and potency of all forms of learning, including asocial learning capabilities.

The social learning strategies tournament also leads to the expectation that effective social learners will possess a capability for "mental time travel." Like the winning strategy, *DISCOUNTMACHINE*, the optimal copier would be able to look back into the past and draw on historically effective solutions, taking account of whether each solution is relevant and biasing judgments toward more recent options. Equally, the optimal copier would be able to project forward into the future and conduct mental simulations of the likely consequences of adopting the behavior of others, including the payoffs yielded and risks incurred. Cultural drive, then, should favor the expansion of those ("executive") brain regions and circuits associated with computation, decision making, working and long-term memory, and mental simulation. In humans, these attributes are associated with the anterior part of the brain, known as the frontal and temporal lobes, and especially the prefrontal cortex.

Wilson's hypothesis also places emphasis on the capacity for innovation. An enhanced social learning capability would be virtually useless if no new behaviors were invented and there were no skills available to copy. The set of brain regions necessary for computation, memory, and simulation are also implicated in creativity and innovation—notably, the frontal lobe, and specifically, the prefrontal cortex.

Outside of the brain, selection for social learning might be expected to influence other aspects of social behavior and life history too. A heavy reliance on social learning should increase sociality, since the bigger the group and the more time spent in the company of others, the greater the opportunities for effective copying. Such selection for sociality should also favor a particular interest in, and sensitivity to, social cues, and also a social tolerance, particularly among kin.[24] The learning of a complex skill through imitation often benefits from the opportunity to spend long periods of time physically close to proficient individuals. Leading

FIGURE 6. Getting up close is critical to acquiring new foraging skills. At Suaq Balimbing in Sumatra, an infant orangutan (Lois) studies his mother's (Lisa) feeding behavior. By permission of Julia Kunz.

primatologist Carel van Schaik has argued compellingly that tolerance of other individuals is critical to the social transmission of tool use (figure 6).[25] Animals are thought to gain an evolutionary advantage from allowing relatives, particularly youngsters, to spend time in their company,[26] providing them with the opportunity to copy their skills and knowledge, as well as to scrounge scraps of food and play with abandoned tools.[27] These discarded elements are known to facilitate social learning in primates.

In restricted circumstances, discussed in the next chapter, individuals benefit from actively investing in the learning of their relatives, and therefore engage in teaching behavior. More commonly, a little social tolerance on the part of relatives will suffice to allow youngsters the opportunity to pick up the requisite skills. Studies of chimpanzees have established that learning complex tool-using behavior, such as nut cracking and fishing for termites, can take many years.[28] Young chimpanzees remain with their mothers until the age of around seven, probably because this allows them time to acquire complex foraging and social skills. A disproportionate amount of primate social learning involves infants and juveniles learning from their mothers,[29] which leads to the prediction that selection for social learning efficiency would favor longer periods of juvenile dependency. Moreover, the selection for innovation and social learning anticipated by Wilson could plausibly lead to longer lifespans for several reasons. Behavioral innovation occurs disproportionately in adults and appears to be heavily reliant on relevant prior experience.[30] Learning complex skills takes time,[31] and older individuals benefit from passing on their hard-earned knowledge to their descendants.[32] Greater longevity, in turn, provides further opportunities for animals to "cash in" on this knowledge.[33]

Finally, any enhanced capability for social learning must be expressed in some concrete way in the real world. As we saw in chapter 2, primates acquire diverse forms of knowledge through observation, including what to fear, what calls to make, and a variety of social conventions; however, foraging information is the most frequent form of socially transmitted knowledge. From the use of extractive foraging methods like digging out grubs from bark, to sophisticated tool-using techniques such as nut cracking or fishing for termites with sticks, monkeys and apes acquire all kinds of foraging skills and knowledge through copying. If social learning is what allows primates to pick up difficult-to-learn but productive food-procurement methods, species proficient in social learning should show elevated levels of extractive foraging and tool use, and should possess a richer, as well as a more diverse diet.

In summary, with the recognition that Wilson's hypothesis should operate through the natural selection of variation in social learning *proficiency* rather than social learning incidence, and drawing on the published literature on animal innovation and strategic copying, a credible feedback mechanism for cultural drive emerges. That mechanism allows specification of an extensive repertoire of functional capabilities in the primate brain,[34] as well as other structural, social, and life history characteristics—including longer lifespans, richer diets, and enhanced capacities for tool use—that are expected to evolve as a consequence of natural selection favoring innovativeness and efficient social learning. Larger brains are probably an incidental outcome of this process, since each of the functional capabilities will surely require the elaboration of one or more existing neutral structures or circuits. If correct, the natural selection of variation in innovation and social learning not only triggers the evolution of brain expansion, but feeds back to enhance copying efficiency further, fueling brain growth in an endlessly recurring and ever accelerating runaway process.

This reasoning leads to a key insight. If natural selection operating on social learning capabilities really has driven the evolution of intelligence in primates, then rates of social learning should not only covary with brain size but should also correlate with a host of measures of cognitive performance. Different measures of cognitive ability should cluster together in primates, with social learning aptitude at the heart of them. We set out to test this hypothesis.

At the same time, we wanted to address other controversies concerning the primate mind. Whether social cognition and social learning were adaptive specializations directly favored by natural selection, or instead incidental outcomes of selection for some other cognitive capability had been subject to debate.[35] Another controversy surrounded the question of whether or not primate cognition was organized in a modular fashion as an assortment of largely independent mental adaptations, with each designed to solve a specific problem; many evolutionary psychologists thought this was the case for humans.[36] The alternative possibility was that measures of primate cognition covary as a single dimension, one that is evocative of general intelligence. In addition, the comparability of the intelligence of humans and other primates remained highly contentious.[37] These debates would be clarified by knowledge of the extent to which diverse measures of social, ecological, and technical performance were correlated.[38]

Given that experimental studies have proven a productive vehicle for investigating the psychological abilities of several primates,[39] you might think that such controversies could be addressed through experimental investigation. Certainly, the little relevant experimental data that existed were consistent with Wilson's hypothesis.[40] However, there are obvious limitations to what can be achieved in this manner. Comparative experiments of learning and cognition are challenging. Laboratory tasks that are fair to all the studied species are notoriously difficult to design, leading to the concern that the findings reflect arbitrary decisions, such as the choice of task. These experiments are typically conducted in the laboratory, raising questions about the extent to which they reflect the behavior of animals in their natural environments. The real barrier for

us, however, was that experimental studies could not realistically provide data on large numbers of species, or on a broad range of cognitive
capabilities.[41] Most primate comparative experiments investigate just
two species, with the biggest including perhaps four or five, and these
typically assess the animals on a single performance measure. However,
given that there are more than 200 species of primates, realistically we
needed to look at the cognitive abilities of 50 or 60 species, using multiple measures of cognitive ability, to get a reasonable understanding
of the evolution of primate intelligence.

Instead, we took the approach that Lefebvre and his colleagues had
devised in their pioneering investigations of bird innovation.[42] We
compiled and analyzed data from the published scientific literature,
developing quantitative measures of the incidence of traits associated
with behavioral flexibility in natural primate populations.[43] This methodology circumvents the above problems with experimental studies,
and provides quantitative measures of performance for large numbers
of species across broad domains.

Simon and I recruited Yfke van Bergen, and with her help set out to
test Wilson's hypothesis. We needed to pull together a diverse set of measures of the cognitive abilities of primates that were applicable to a large
number of species. This would require a major effort. Simon and Yfke
went back to the journals and studiously compiled quantitative measures
of rates of tool use, extractive foraging, diet breadth, and diet quality
for over 100 primate species. We also drew on some existing measures.
For instance, in the great apes in particular, experimental studies had
suggested that deliberate deception sometimes occurred.[44] "Tactical deception" refers to behavior sufficiently deceptive that other individuals
misinterpret what is happening, to the advantage of the deceiver. Rates
of tactical deception had been compiled for many primate species,[45] and
had been shown to correlate with brain size.[46] Highly deceitful primates
were characterized as exhibiting "Machiavellian intelligence,"[47] named
after the scheming and unscrupulous political manipulations that Nic-
colò Machiavelli described in *The Prince*.

Various measures of primate group size were also available;[48] while
not a direct evaluation of cognitive performance, these had long been
thought to covary with social intelligence in primates. Primates often
form alliances, and siblings and members of matrilines will frequently

aid each other in disputes. Stealing food from a little guy could be disastrous if he is the son of the alpha female, so knowledge of social relationships is vital. Brain size may constrain the information-processing capability of a species to monitor a large number of such social relationships.[49] Primates living in large groups have to keep track of many more relationships than those living in small groups, which is thought to have created a need for bigger brains and greater computational power. Group size measures are regarded as a useful proxy of social complexity and, sure enough, they are typically good predictors of brain size in primates.[50] The issue for us was whether group size, rates of tactical deception, diet breadth, and all of our other measures would covary with social learning rates.

Eventually, we had eight separate measures that directly or indirectly assayed the cognitive abilities of a broad array of different primates, including the tendencies to (i) discover novel solutions to environmental or social problems ("innovate"), (ii) learn skills and acquire information from others ("social learning"), (iii) use tools, (iv) extract concealed or embedded food ("extractive foraging"), and (v) deceive others ("tactical deception"); these were five ecologically relevant measures of behavioral flexibility, while (vi) the number of different categories of food ("diet breadth"),[51] (vii) percentage of fruit in the diet, and (viii) measures of social group size, were thought to reflect the cognitive demands of exploiting and locating foods, and tracking social relationships.[52]

We investigated the relations between these measures using statistical methods. At one extreme, all could be found to correlate so tightly that a single factor would explain differences in the intellectual abilities of primates, while at the other extreme each of the measures might be entirely independent. The latter finding might imply that primate intelligence is organized into domain-specific modules that reflect species-specific ecological and social demands. The approach that we took, known as principal components analysis (PCA),[53] determined how many independent dimensions were required to describe the data. Strikingly, the answer was just one.

We first considered 5 cognitive measures (innovation, social learning, tool use, extractive foraging, and tactical deception[54]) in the 62 primate species for which data on all measures were available. The analysis revealed a single dominant component,[55] which explained over 65% of the variance in cognitive measures; all of the individual measures were tightly associated with the component.[56] In other words, 65% of the differences between primates in their performance on our five measures could be attributed to this single factor. All of our measures correlated strongly with each other,[57] with species that excelled in one of these cognitive domains typically excelling in all of them. A similar analysis that also incorporated diet breadth, percentage of fruit in diet, and group size produced broadly equivalent results, again generating a major component,[58] while further analyses confirmed that these cognitive abilities were not just correlated in the present, but had evolved together.[59]

Our analyses had uncovered a distinct dimension of cognitive ability in primates,[60] something that is highly evocative of what in humans is called "general intelligence." Human intelligence has been defined as "a very general mental ability that . . . involves the ability to reason, plan, solve problems, think abstractly, comprehend complex ideas, learn quickly and learn from experience."[61] The variables that went into our analysis (i.e., measures of problem solving, learning, creativity, tool use, deception) match the abilities emphasized in such definitions. The single factor that captured the bulk of the variation in primate cognitive abilities seemed appropriately characterized as "general intelligence," and hence we labeled it *primate g*.[62] Each primate species was positioned somewhere along this axis,[63] and hence a species' *primate g* score represents a measure of its general intelligence. Species such as chimpanzees that score high for tool use, innovation, social learning, diet breadth, and so forth, possess high *primate g* scores, while species like the South American bald-headed uakari,[64] which have never been seen to use tools or invent new behavior, receive low scores. More generally, high performers on our measure are species frequently reported to devise novel solutions, solve social and ecological problems, learn quickly and from experience, construct and manipulate tools, and learn from and deceive others; these are qualities attributed to intelligent humans. On average, the apes get the highest scores and the prosimians the lowest, which is what most primatologists would expect. Our statistical measure, then, constitutes a simple way to rate species according to their intelligence.[65]

Two sets of findings gave us confidence that this interpretation was legitimate. If *primate g* scores really do reflect the cognitive ability of the animal, then we would expect those scores to covary with brain size.

That is exactly what we found. Whether we focused on the absolute size of the primate's brain, brain size relative to body size, or other brain measures,[66] there are always strong associations of *primate g* with brain volume.[67] This pattern resembles that observed within humans, where general intelligence, known simply as *g*, shows a modest correlation with total brain and gray matter volumes.[68] Such findings suggest the "volumetric stance" among brain researchers is warranted, and that brain volumes are genuinely related to functionally relevant cognitive capacities.

Subsequently, we uncovered an even more compelling validation of our measure: *primate g* scores correlated strongly with species' performance in laboratory tests of cognition. A few years earlier, neurobiologist Rob Deaner at Duke University, and his colleagues, had published an impressive article that analyzed a large number of lab-based comparative studies of primate cognition.[69] Each study had presented different experimental tasks, and the species investigated also varied from one study to the next, making conventional comparisons between investigations hard to interpret. However, within a given study, species can be ranked according to their performance at the focal task. These ranks can then be compiled using sophisticated statistical methods to look for patterns across multiple experiments. In this manner, Deaner and his colleagues were able to produce an overall ranking of the performance of 24 primate genera in a large number of diverse tests of learning and cognition.[70] Again, the findings made sense. Deaner's measure placed orangutans and chimpanzees at the top, and marmosets and talapoins at the bottom. What was particularly encouraging for us, however, was that our *primate g* scores showed strong and significant correlations with Deaner and colleagues' ranked performance of primates in laboratory tests,[71] as well as with two other multispecies tests of comparative intelligence.[72] In other words, those species that were designated "smart" by our statistical measure of primate intelligence turned out to be precisely those species that had performed well in laboratory-based experimental tests of learning and cognition. These observations lent credence to our view that *primate g* was a genuine measure of primate comparative intelligence.

As described previously, numerous hypotheses had been proposed concerning the key drivers of primate brain evolution. As these had been presented as alternative explanations, researchers had worried that seemingly conflicting findings had not been reconciled.[73] In suggesting that selection may have favored general intelligence, rather than a single specialized ability, our analysis helped to alleviate this disquiet. Our *primate g* measure contained social (social learning, tactical deception), technical (tool use, innovation), and ecological (extractive foraging, diet breadth) components of intelligence, and hence supported a battery of hypotheses regarding the factors driving brain evolution,[74] but not the hypotheses that any one was the sole cause of brain evolution. The social intelligence hypothesis was the dominant view at the time, but our findings implied that social complexity might not be the sole driver. For instance, unlike *primate g*, group size did not predict the performance of primates in laboratory tests of cognition.[75] We began to believe that there was more to the evolution of primate cognition than selection for social intelligence alone.

We also examined the pattern of variation in *primate g* across the primate family tree, to determine whether enhanced cognitive abilities had evolved independently on multiple occasions, or whether "smart" primates were simply those closely related to humans.[76] The analysis revealed four independent convergent evolutionary events favoring high general intelligence across the primate lineages; these were in the capuchins,[77] baboons,[78] macaques,[79] and great apes.[80] Strikingly, these are precisely those primate groups that are renowned for their social learning and tradition. This pattern of results is exactly what would be expected if social learning really were a major driver of the evolution of brain and cognition in primates. Those primates with high *primate g* scores are those renowned for their complex cognition and rich cultural behavior.[81]

The major predictions derived from Wilson's cultural drive hypothesis had been confirmed. The analyses implied that diverse mental abilities, spanning multiple cognitive domains, had coevolved in primates to generate a between-species version of general intelligence. Seemingly, key cultural capabilities, notably social learning, innovativeness, and tool use, form part of a highly correlated composite of cognitive traits, with elements of cultural intelligence intimately wrapped up with many domains of cognitive performance. Such findings are inconsistent with the view, widespread within evolutionary psychology, that cognitive abilities evolve independently as separate modules,[82] and the results strongly imply general intelligence. Indeed, a recent study suggests that *primate g* is the principle dimension along which primate intelligence evolves.[83]

Our study had addressed the first of three concerns outlined at the beginning of this chapter by showing that social learning could plausibly drive brain evolution and the evolution of intelligence. However, at this juncture, two other concerns remained. First, social learning might well have driven brain evolution, but our findings were equally consistent with the suggestion that large primate brains had evolved for some other reason and were only incidentally expressed in elevated rates of social learning and innovation. Second, we needed to examine precisely how the primate brain had changed over evolutionary time, and consider whether the observed changes were consistent with the predictions of cultural drive.

Brain size in primates ranges from just 3 g in fat-tailed dwarf lemurs,[84] to around 1.5 kg in humans. Both absolute and relative brain sizes have increased in several primate lineages, including the apes,[85] largely due to increases in the neocortex, which comprises up to 80% of total brain volume.[86] The neocortex is the region of the human brain associated with problem solving, learning, planning, reasoning, and language, so it is the first brain structure that comes to mind when considering the neurobiological underpinnings of intelligence. The size of the neocortex scales allometrically with whole-brain size across primates,[87] which means that the neocortex gets bigger at the rate expected, given the size of the brain as a whole. However, that does not mean that the neocortex is equally important to all primates, because the correlation between neocortex and whole brain size is not one to one. To the contrary, the neocortex makes up a larger proportion of the brain in big-brained, compared to small-brained, primates.[88] Neocortices are also larger in absolute terms and are better connected to other brain regions in big-brained primates relative to their smaller-brained cousins.[89] Humans possess the biggest, proportionally largest, and best-connected neocortex of all primates. Thus, whether measured in absolute terms, relative to body size, or by comparing the relative size of the neocortex,[90] brain expansion is unquestionably one of the distinctive features of human evolution.[91] Yet, as we have seen, in spite of extensive research, controversy persists regarding the evolutionary explanation for this expansion, and its relationship to the evolution of cognition.

There were several reasons as to why the evolution of the brain and intelligence had remained unresolved for so long. One major issue was the poor quality of data on the primate brain. Researchers conducting comparative statistical analyses had been overly reliant on a relatively small number of primate species for which brain data had been available.[92] Typically, only a few brain specimens and often just a single brain were measured for each species, with the specimens frequently of unknown history. Often, for instance, whether the brain was from a male or female primate was unknown, which for sexually dimorphic species was a major source of error. Better data would be required if this concern was to be addressed. A second issue was the absence of a clear quantitative measure of cognitive performance applicable to multiple primate species. Researchers had typically just assumed that whole brain or neocortex size was a correlate of cognitive ability in primates, which of course left the true relationship unexamined. Fortunately, our analysis of primate intelligence had provided just such a measure, which meant this concern could now be addressed for the first time. A third concern was a historical overreliance on analyses in which a single variable, ranging from diet to group size, was found to predict brain size; this led researchers to make claims about the evolutionary significance of their favored variable.[93] Few analyses had considered multiple variables at the same time to see which was the most powerful predictor. These concerns meant that until very recently, lingering doubts remained over whether enlarged brain size, intelligence, and reliance on culture had actually coevolved in primates.

What was really required were appropriately controlled evolutionary analyses that included all the potentially relevant variables, and that set out to predict both variation in brain size and differences in cognitive performance across primates, with the use of good quality data. Motivated by these objectives, Ana Navarette and Sally Street,[94] postdoc and graduate student, respectively, at the University of St Andrews, began to collect more brain data and carry out the evolutionary analyses.[95] We set out to provide a definitive understanding of the evolution of enlarged brain size, cognitive abilities, and reliance on culture in primates. Brains obtained from the Primate Brain Bank at the Netherlands Institute for

Neuroscience were subject to magnetic resonance imaging (MRI scans) at the Utrecht Neuroimaging Centre.[96] Ana then used these scans to measure the size of various structures in the brains, and thereby was able to supplement existing published primate brain sources. By targeting our efforts both to increase the number of species represented in the database and increase the number of specimens per species, as well as using brains of known history, Ana's analyses substantially expanded the breadth and reliability of primate brain data.

Then, using our newly expanded primate brain database, and our recently validated measure of general cognitive ability (*primate g*), Sally and Ana conducted analyses that included all of the relevant socioecological, environmental, and life-history predictor variables to explore which factors truly explained variation in absolute and relative brain size and cognitive performance.[97] What factor would turn out to be the root cause of primate intelligence?[98] The analyses put all of the potential predictors into a statistical model, removing variables that didn't really help to explain the data and retaining variables that proved useful, until the simplest model that provided a good explanation for the data was left. When the results came in we were very excited.

Sally and Ana's multivariate analyses found that some combination of slow life history (e.g., long lifespan, extended period of juvenile dependence) and large group size was the dominant predictor of all brain size measures, as well as of cognitive performance.[99] In best-fitting models, diverse brain size measures (absolute brain size, relative brain size, relative neocortex size, relative cerebellum size) and assays of cognitive performance (*primate g*, social learning, innovativeness) always significantly increased with both group size and some measure of life-history length. Typically, life-history length was a stronger predictor of brain size measures than social group size. Alongside measures of social complexity, how long a species lived predicted how big its brain would be and how smart it would be, more so than diet, latitude, or any other variable. This fit with the expectation that cultural drive would generate selection to extend both the adult lifespan and the length of the juvenile period.

Our findings were evocative of an argument put forward by University of New Mexico anthropologist Hillard Kaplan and colleagues in 2000. These researchers had suggested that intelligence and longer lifespans had evolved in humans because our intellectual abilities allowed us to exploit highly nutritious but difficult-to-access foods, such as animals that have to be hunted, or plant parts that are difficult to extract (like palm fiber).[100] Brains are energetically demanding organs, but Kaplan argued that the energy and nutrients gleaned from embedded foods could "pay" for brain growth. On this view, increased longevity is favored because it allows more time later in life for individuals to cash in on the energetic bonanza yielded by the complex foraging skills acquired earlier.

We had not included humans in our analyses, so we could not assess the Kaplan hypothesis directly. However, given that foraging behaviors represented approximately half of recorded incidences of primate innovation and social learning, our findings strongly implied that the same argument was correct when extended to other large-brained, long-lived, and intelligent primates. These most clearly included the great apes, but also the capuchins, and perhaps macaques and baboons. As we have seen, these primate groups are renowned for their social learning and tradition, and the great apes and capuchins in particular are known for tool use and extractive foraging.[101] Extensive empirical evidence shows that the ability of many primates to acquire nutrient rich foods, such as nuts, termites, and honey, is critically dependent on social learning, as it clearly is for humans.[102] Extended periods of juvenile dependence would potentially create extra opportunities for such transgenerational knowledge transfer. Our findings fit neatly with Wilson's cultural drive hypothesis, in which selection for efficient social learning and innovativeness allowed for energy gains in diet, which in turn fueled brain growth and generated selection for extended longevity.[103]

High levels of knowledge, skill, coordination, and strength are required to exploit the high-quality dietary resources consumed by humans and other apes. Complex tool using and extractive foraging abilities require time to acquire, but in intelligent animals an extended learning phase during which productivity is low, can be compensated for by higher productivity during the adult period, provided there is an intergenerational flow of both food and knowledge from old to young. Mathematical analyses have established that because productivity typically increases with age, the investment of time in acquiring skill and knowledge leads to selection for lowered mortality rates and greater longevity.[104]

The currently dominant view is that the primate brain expanded to cope with the demands of a rich social life, including the aforementioned Machiavellian skills required to deceive and manipulate others, and the cognitive skills necessary to maintain alliances and track third-party relationships. The most important data supporting this hypothesis is a positive relationship between measures of group size and relative brain size.[105] In our analyses, group size remained as an important predictor of relative brain size, but also proved a significant secondary predictor of primate intelligence and social learning. However, group size was neither the sole, nor the most important, predictor of brain size or intelligence in our models. Combined with our earlier finding that social group size does not predict the performance of primates in laboratory tests of cognition,[106] this reinforced our view that there was more to primate brain evolution than selection for social intelligence.

The most parsimonious way to interpret these findings is to recognize multiple waves of selection for both bigger brains and greater intelligence in primates, with each operating at a different scale. Our analyses agreed with pre-existing research in recognizing the significance of natural selection for the social intelligence needed to cope with the complexities of social life, which was probably widespread across monkeys and apes.[107] However, we elaborated on this story to suggest that selection for social intelligence was followed by a later and more restricted, but nonetheless critical, bout of natural selection favoring cultural intelligence in a small number of large-brained, social primates—most notably the great apes.[108] In addition, our more recent analyses had found elevated rates of social learning in primate species with large social groups, which fit with the findings of theoretical analyses showing that large, well-connected groups are critical to the maintenance of cultural variation.[109]

Sally and Ana's analyses also revealed associations between primate intelligence and the relative size of two important structures in the brain, the neocortex and the cerebellum. The neocortex is the "thinking" part of the brain, and includes structures associated with both imitation (e.g., parietal lobe, temporal lobe) and innovation (e.g., lateral prefrontal cortex), which have indeed expanded during the course of human evolution.

The visual cortex also expanded, as has the circuitry linking the visual, and auditory, cortex to somatosensory and motor cortex regions of the brain,[110] as we predicted. The cerebellum is thought to play a key role in motor control, allowing precise movements of the limbs, for instance, as required for extractive foraging and tool use.[111]

Big animals tend to have big brains. Some whales, for instance, have brains more than six times the volume of the human brain, while three human brains would fit inside a typical elephant brain. Such comparisons, dubious though they are,[112] have historically led researchers to the view that the absolute size of the brain did not indicate an animal's intelligence. Larger animals may simply have bigger brains because they have more cells to process and control; bigger limbs to move may require larger nerves to move them. For this reason, studies of animal intelligence have tended to focus on relative brain size.

Recent thinking, however, suggests that absolute size of the brain, or of some key structures in it,[113] may be more relevant to an understanding of intelligence than hitherto thought.[114] Georg Striedter, an authority on brain evolution at UC Irvine, has shown that brains tend to change in internal organization as they increase in size, with the emergence of more subdivision and with larger regions becoming disproportionately connected and influential.[115] Consistent with this, projections from the neocortex to the brain stem and spinal cord are known to have become more extensive as hominins evolved,[116] with an unprecedented level of direct access to the motor neurons innervating the muscles of our jaws, face, tongue, vocal chords, and hands. This change in brain anatomy was predicted by psychologist Terrence Deacon at the UC Berkeley.[117] Deacon proposed a rule that specified if brain regions become disproportionately large, then, as they evolve, they tend to "invade" and become connected to regions that they did not innervate ancestrally. This increases the ability of the enlarged area to influence other brain regions and makes it more important to brain functioning. Deacon's rule is based on two fundamental principles of brain development. First, developing axons (the long slender part of a nerve cell along which impulses are conducted from the cell body to other cells) often compete with one another to connect with target sites. Second, this competition is generally won by those axons that participate in "firing" the target cells.[118] When a brain region evolves to become proportionally larger, its axons are given a competitive advantage over those from other regions; as the more axons a region can send to a target, the more likely excitation of the target cells will occur, allowing the connections to strengthen. These new connections might even "displace" older connections, particularly if the source of these old connections has decreased in proportional size.[119] The two largest structures in the human brain are the neocortex and the cerebellum, and the above reasoning leads us to expect they would become increasingly embedded in complex neural networks over which they would exert considerable influence as the brain gets larger.

What all this means is that selection for a larger neocortex or cerebellum automatically brings with it greater flexibility and dexterity, with these two brain regions exerting greater control over our arms, legs, face, and hands. Only a big neocortex would allow for the increased manual dexterity that is characteristic of humans.[120] To a large extent, the computational power that allowed our ancestors to manufacture and use tools brought with it the physical dexterity to move our hands very precisely. Likewise, the cognitive ability to learn a language went some way toward enhancing the flexible use of our mouth and tongue.[121]

The neocortex becomes an increasingly large proportion of the whole brain as brain size increases, which probably means that selection for enhanced neocortex functionality was brought about through selection on the brain as a whole, rather than just increasing neocortex size.[122] This fits with the finding that primate cognition evolves, to a surprisingly large degree, through selection for greater or lesser general intelligence.[123] Likewise, the cerebellum is known to play an important role in motor control, and to the extent that enlargement of the cerebellum is indicative of functional gains in the precise bodily movements necessary for tool use and extractive foraging, particularly the execution of complex behavioral sequences, increases in this region would be subject to positive selection. Again, neuroanatomical data supports this argument,[124] as does further comparative work conducted by Ana that shows "technical innovations" (e.g., primate innovations reliant on tool use) correlates more strongly and directly with brain size measures than nontechnical innovations.[125]

One further finding is worthy of note. A previous study had found that in mammals, the size of the adult brain is well predicted by measures of maternal investment, such as the duration of gestation or lactation.[126] The study implied that the relationship between brain size and longevity arose solely because in large-brained animals the life-span had been slowed to allow more time and resources to be ploughed into offspring development, including growing a big brain. However, Sally found that matters were different for primates. The associations of maximum longevity with brain size measures remained when gestation and lactation lengths were included as covariates in her analyses. Also of interest is the observation that the relationship between longevity and social learning, but not longevity and *primate g*, remains when measures of parental investment are included in analyses. In primates, unlike in mammals more generally, the relationship between brain size and lifespan is potentially indicative of a cognitive, rather than solely de-velopmental, mechanism. By virtue of possessing a big brain, and being able to use it to acquire all kinds of useful survival skills from others, some clever primates had apparently been able to extend their lifespans and live longer. In other words, in primates and primates alone, cultural intelligence facilitates survival.

Sally and Ana's findings excited us, because they seemed to make sense of a tremendous amount of comparative, statistical, and neuro-anatomical data. They implied that in some primate lineages charac-terized by large brains and complex social groups, a critical threshold in reliance on socially learned behavior had been reached. Once this threshold had been passed, mutually reinforcing selection for increased brain size, diverse cognitive abilities, and further reliance on social learning and innovation ensued, mediated by conferred increases in longevity and diet quality. Large brains are certainly not a prerequisite for social learning, but they may well support more efficient, higher-fidelity forms of social learning, thereby allowing copying over greater distances, high-precision imitation, cross-modal integration of per-ceptual and motor information, the computational power to imple-ment sophisticated social learning strategies, and more. Expansion of the neocortex and cerebellum likely enhanced learned motor skills,[127] and thus supported the ability to learn action sequences requiring fine motor coordination, such as nut-cracking in chimpanzees and nettle processing in gorillas.[128] Such sequence learning and coordinated action seems to be limited to only the largest-brained primates.[129] Consistent with this, those brain regions that have expanded during the course of human evolution include regions associated with social learning, imitation, innovation, and tool use.

We had focused on primates, but enlarged brain size, general cognitive abilities, and reliance on culture may have coevolved in other animals too, including some birds (e.g., corvids, parrots) and cetaceans (e.g., whales and dolphins).[130] Corvids such as rooks are renowned for their complex social cognition but also their tool use, causal reasoning, memory,[131] perspective-taking, innovation, and cultural transmission.[132] Like apes, they possess large brains for their body size; the same size, in relative terms, as chimpanzees.[133] The relative size of the forebrain in corvids is significantly larger than in all other birds,[134] with the exception of parrots, who are equally famed for their complex social cognition, innovation, tool use, and imitation.[135] Humpback whales, killer whales, and dolphins exhibit striking parallels to the patterns seen in primates, and also possess the complex social cognition, tool use, and innovation associated with unusually large brains and sophisticated social learning.[136] Indeed, it is hard to think of a "smart" animal that isn't a proficient social learner.

Current thinking in the study of animal cognition suggests that there has been convergent evolution of intelligence in apes, corvids, and cetaceans, and cultural drive is an outstanding candidate mechanism. However, the process as envisaged by Allan Wilson would have been autocatalytic,[137] with selection for more efficient social learning feeding back by favoring structures and capabilities that enhance cognitive ability along with technical skills, and with these capabilities in turn feeding back to impose further selection on brain and behavior. Here, I suspect that birds and cetaceans will each have experienced a major constraint not operating on primates, which may have hindered the development of tool use and extractive foraging, and limited the extent to which selection for technical intelligence could feed back on brain evolution. Cetaceans require their flippers to swim efficiently, which effectively confines tool use to what can be achieved with the rostrum, or mouth. Whales and dolphins cannot grip an object with one limb and manipulate it with another, preventing complex tool use and limiting their foraging techniques. Likewise, the requirements of flight put an upper ceiling on the evolution of large bodies and brains for most birds, because large birds struggle to get off the ground. In principle, bigger brains might have been possible in giant flightless birds, such as moas or elephant birds,[138] but their vestigial wings never evolved manipulative capabilities and their feet were adapted to running, which again imposed bounds on extractive foraging. Perhaps because of these constraints, those birds most dexterous in their tool use would appear to show limited behavioral flexibility.[139] In contrast, for primates, the runaway process would have been able to encompass selection for social, technical, and cultural intelligence, all enhancing each other. Such a mechanism could have operated across many groupings, particularly the apes, but it has clearly reached its zenith in humans, who are extreme in their encephalization, intelligence, tool use, and reliance on culture.[140] Perhaps Wilson had been on the right track. I was starting to believe that cultural drive might genuinely be a central plank of the explanation for the evolution of the human mind.

CHAPTER 7

HIGH FIDELITY

One of the most engaging aspects of academic life is the sheer relentlessness with which the finding of solutions to scientific challenges generates further questions. That was the case here. If cultural drive has operated on all the great apes and some monkeys too, then why haven't gorillas invented particle accelerators? Why haven't capuchins put a monkey on the moon, or devised a simian version of Facebook? For all the insight engendered, cultural drive had not explained why humans should be so much more accomplished than other primates. What was the secret of our success?[1]

One answer to this question attributes our species' attainments to chance. If cultural drive, or for that matter, any other mechanism fueling the evolution of intelligence has operated widely, not all species would be equally affected. For whatever reason, perhaps sheer luck, the brain size and intelligence of some species would be boosted more than others and, given the autocatalytic nature of the process, the cognitive prowess of one species could easily have "run away" from the others. Inevitably, the smartest species would be the one to look back and ask: "Why us?" This solution, however, appears unsatisfactory. Chance factors are hard to rule out, but identification of a trait or combination of traits that were uniquely possessed by our ancestors and gave them the edge is so much more compelling.

Another answer attributes our species' success to demographics. Once population size reached a critical threshold, such that small bands of hunter-gatherers were more likely to come into contact with each other and exchange goods and knowledge, then cultural information was less likely to be lost, and knowledge and skills could start to accumulate.[2] Such arguments are convincing, and population size and connectedness is undoubtedly an important part of the story. However, demographics cannot be all that counts, because there are numerous animals that live in large numbers, none of whom have invented vaccines or drafted declarations of individual rights. Large social networks will only support

150

complex culture in animals capable of generating complex culture in the first place.

The question remains as to what was different about the behavior, morphology, or circumstances of our ancestors that allowed our technology and culture to take off in such an extraordinary manner. Here, mathematical modeling has proven extremely insightful, and this chapter will describe three theoretical analyses that shed light on this issue. The chapter ends with an account of an experimental investigation in children, chimpanzees, and capuchin monkeys that reinforces the theoretical findings.

Early clues as to how the human success story is to be understood came from a mathematical analysis carried out by Magnus Enquist and his colleagues at Stockholm University.[3] Magnus was a theoretical biologist by training, but had become very interested in cultural evolution. He invited me to join a project that used mathematical models to explore how many "cultural parents" were necessary for stable cultural transmission.[4] A great deal about Magnus's findings surprised and intrigued me,[5] but one of the more prosaic findings had the most lasting influence. Magnus's team modeled how the fidelity of cultural transmission (that is, the accuracy with which learned information passes between individuals) affected the amount of time that a cultural trait would remain in a population. A positive relationship is to be expected, of course, because knowledge that is copied faithfully will remain in a population for longer than knowledge that is reproduced inaccurately. Ballet steps and musical scores have persisted for centuries precisely because protracted effort, including meticulous instruction and the drafting of written transcripts, has been put into ensuring accurate information transmission. What I hadn't anticipated, though it turned out to be highly significant, was that the mathematical function relating trait fidelity to longevity would be acceleratory, or exponential, in form (figure 7).

The shape of the curve is significant, because it means that for a given population size there is a threshold effect, whereby a small increase in the fidelity of social learning will transform cultural traits from being short lived to virtually immortal. That this could help to explain the difference in the volume of culture observed in humans compared to other animals was immediately apparent. The theoretical findings supported a

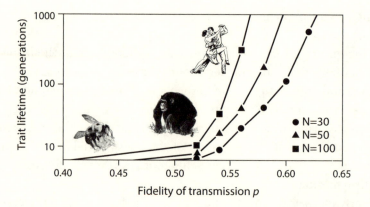

FIGURE 7. The longevity of a cultural trait (how long it remains in the population) increases exponentially with trait fidelity (how accurately knowledge and behavior are transmitted among individuals). This means that a small increase in fidelity can sometimes lead to a big increase in the amount and persistence of cultural knowledge. This relationship helps to explain why, with such high-fidelity transmission mechanisms, humans possess so much reliably inherited cultural knowledge. Based on figure 3c in Enquist et al. (2010).

verbal argument that had been made previously by psychologist Michael Tomasello at the Max Planck Institute for Evolutionary Anthropology in Leipzig. Tomasello had proposed that our species' unique capabilities for language, teaching, and efficient imitation had allowed us to transmit knowledge with a higher fidelity than observed in other animals, and that this transmission fidelity explained the existence of cumulative culture (what he termed "ratcheting") in humans, but not in other animals.[6]

I confess that initially I did not find Tomasello's argument particularly convincing, because it was not apparent to me why high fidelity should necessarily favor ratcheting.[7] After all, if everyone always copied perfectly then there would be no cultural change at all. Magnus's analysis resolved this for me by demonstrating that high-fidelity transmission would lead to cultural traits persisting for extremely long periods. In the more vertical region of the curve, a small increase in the fidelity of transmission made a big difference as to how long cultural traits remained in the population. Longer persistence times, in turn, created greater opportunities for improvements in technology to appear, since a trait must first be present before it can be refined, or improved. An additional consequence of having longer-lasting traits is that *more* culture will be preserved. High-fidelity transmission mechanisms potentially support vastly more cultural knowledge and behavior than low-fidelity mechanisms.

Magnus's analysis illustrated why Tomasello's reasoning was sound. The demonstration that the relationship between trait fidelity and longevity was exponential in character generated an important insight that allowed a great deal about the social learning and tradition of animals to be understood. For instance, while social learning is widespread in animals, including birds, fishes, and insects, the bulk of it is based on low-fidelity mechanisms. Countless animals learn through being drawn to visit objects or to interact with stimuli that they have witnessed other animals encounter; this involves simple forms of social learning known as "local enhancement" or "stimulus enhancement."[8] Sure enough, as Magnus's theory predicts, most animals capable of social learning either possess no traditions at all, or exhibit what are known as "lightening traditions"[9]—low-persistence patterns of behavior that appear and then disappear very rapidly.

A small number of animals are capable of more accurate or higher-fidelity forms of social learning. For instance, experiments demonstrate that chimpanzees can learn by replicating the motor patterns of others or through reproducing the movements of objects that others have brought about; these are known as "imitation" and "emulation."[10] It may be no coincidence that chimpanzee foraging and tool-using traditions can last for decades.[11] Charles Darwin described the nut-cracking habit of chimpanzees in *The Descent of Man*,[12] which establishes that this tradition has lasted for at least over a century. Moreover, with the exception of humans, chimpanzees also possess the most documented traditions of any animal, with 39 behaviors identified as habitual or customary in some populations and absent in others.[13] Finally, with our teaching, language, writing, and imitation, humans possess the highest-fidelity social transmission mechanisms and, sure enough, we possess vast amounts of very long-lasting culture. Magnus's analysis implied that the greater volume of culture associated with our species compared to others is a direct consequence of human-specific transmission mechanisms that significantly enhance the fidelity of our social learning.

We speculated that the same theory could also account for human culture's uniquely cumulative character, as Tomasello anticipated. More culture means greater opportunities for borrowing ideas from other domains and combining them to make new technologies that potentially fuel further innovation and refinement, thereby instigating cumulative culture. We had found that steeper curves arose in large compared to small populations, with cultural traits persisting longer in more populous groups.[14] Hence, a small increase in the fidelity of cultural transmission could have made a large qualitative difference to the character of human culture, and these differences would be magnified as human populations grew in numbers.

As we have seen, claims have been made for cumulative culture in animals such as chimpanzees and New Caledonian crows;[15] however, the evidence is limited, circumstantial, and contested.[16] In contrast, humans incontrovertibly possess complex technology that no single individual could invent alone. Could this difference really come down to transmission fidelity? The suggestion was far from new. In addition to Tomasello, fellow psychologists including Jeff Galef, Cecilia Heyes, and Andrew Whiten, and evolutionary biologist Richard Dawkins had all put forward related arguments.[17] However, although humans unquestionably possess both the cognitive abilities to engage in high-fidelity information transmission and a massively complex, cumulative culture, until recently there had been no formal theoretical investigation demonstrating how and why these might be linked. This hypothetical relationship, while plausible, needed to be proven.

Certainly, Magnus's theoretical analysis had allowed us to put forward a credible verbal argument that the increased existence times of accurately transmitted cultural traits could result in more opportunity for modifications or combinations to occur, and thus could lead to cultural ratcheting. The suggestion that animals would not get the opportunity to refine short-lived traits appeared plausible. Equally, the arguments that the long-lasting culture exhibited by humans would allow for refinement, and that the existence of many cultural traits would create opportunities for cross-fertilization and combination, made sense. However, it was one matter to speculate, and quite another to establish through mathematical modeling that the argument was rigorous.

Hannah Lewis, a postdoctoral researcher at the University of St Andrews with a background in mathematics, took up this challenge.[18] Hannah set out to explore the effects of high-fidelity cultural transmission on cumulative culture by manipulating trait longevity in a simulation model.[19] By systematically changing the rate at which traits are lost from the population, we could evaluate the hypothesis that cognitive changes that resulted in higher-fidelity transmission would lead to the kind of cumulative culture observed in humans.[20]

Clearly, processes other than trait loss must affect the emergence of cumulative culture. These include the rate at which new traits are invented from scratch (which we called "novel invention"),[21] the rate at which pre-existing traits are brought together into complex composites (henceforth "combination"),[22] and the rate at which established traits are refined or improved (henceforth "modification"). Novel trait invention, combination, and modification can all be regarded as different forms of innovation. These processes had been considered separately in models exploring the buildup of numbers of cultural traits,[23] or of trait complexity,[24] but there had been little general consideration of how these processes would affect cumulative culture. Moreover, how these forms of innovation interact and which would prove the most important had not been established.

Hannah's approach was beautifully simple. She constructed a mathematical model in which she manipulated the rates of novel invention, combination, modification, and loss of cultural traits to explore how these processes would affect the buildup of cumulative culture.[25] The model considered how traits were gained and lost across the entire population.[26] As a starting point, we assumed that there are a fixed number of traits that could appear within a group through novel inventions and that are independent of any other traits within the culture. We called these novel inventions "cultural seed traits."[27] Then, one of four possible events could occur. A new seed trait could be acquired by the group through novel invention (which occurred with probability ρ_1). Alternatively, two of the cultural traits present in the group could be combined to produce a new cultural trait (which occurred with probability ρ_2). A third possibility was that one of the traits present in the group could be modified or refined in some way to produce a new variant of the trait (which occurred with probability ρ_3). Finally, one of the traits present in the group could be lost (which occurred with probability ρ_4). Once an event had taken place, the culture of the group was updated (augmented with a new trait, or a lost trait removed) and the next event took place, with the process repeated, typically for 5000 such events.[28] The collection of traits in the population could thus change over time, either through the stepwise modification or loss of an existing trait, or through trait differentiation, where a single trait might give rise to more than one new element. Across multiple simulations, Hannah systematically varied the four rate parameters (ρ_1, ρ_2, ρ_3, and ρ_4) to explore how each process affected the buildup of cumulative culture.[29]

Hannah's results comprehensively established an important role for the fidelity of social transmission in the development of cumulative culture. The trait loss rate (ρ_4, the converse of fidelity) was by far the most important factor affecting the buildup of cumulative culture. Hannah's model strongly implied that transmission fidelity is the key factor affecting the appearance of ratcheting. If the fidelity of transmission is too low (i.e., ρ_4 is large, which means the loss rate is high), then it doesn't matter how much novel invention, combination, or modification takes place—cumulative culture simply cannot get started. Only when the trait loss rate is at, or less than, a threshold level of transmission fidelity, can cumulative knowledge begin to accrue.[30] Further small reductions in the loss rate, corresponding to increments in the accuracy of transmission, lead to big jumps in the cumulative nature of the culture (as represented by utility, complexity, and number of traits and lineages).[31] As the loss rate is decreased further, a threshold is reached beyond which it seemingly becomes impossible for cumulative culture not to develop, provided that some modest level of combination or modification occurs. Across thousands of simulations with different parameter values, trait fidelity (ρ_4) explained more of the variance in the buildup of cumulative culture than novel invention, modification, and combination combined.[32]

Hannah's findings strongly supported Tomasello's argument that the apparent absence of cumulative culture in other species resulted from their reliance on low-fidelity copying mechanisms, such as local enhancement, rather than the high-fidelity processes observed in humans.[33] For instance, teaching is rare among other animals,[34] as are instances of communication that enhance transmission fidelity (e.g., referential communication).[35] Imitation has been reported in chimpanzees and some birds,[36] but comparative studies show that imitation is quicker, more accurate, and more complete in humans when compared to other animals.[37] If most other animals do indeed rely on low-fidelity copying, then Hannah's work established that these learning mechanisms will not support cumulative culture.

Assuming high-fidelity transmission, the exact nature of the culture that develops depends on the mix of novel invention, trait modification, and trait combination, with different combinations leading to a stable, slowly changing culture at one extreme, or a rich, diverse, exponentially growing culture at the other. In terms of the creative processes, the rate of trait combination has the greatest effect, and the rate of novel invention has the least impact, on the buildup of cumulative culture. Our findings thus support the argument that trait combination is the major source of both human innovation and progress in cumulative culture.[38] Novel invention actually turns out to be relatively unimportant, which surprises many people. However, recent studies of human innovation report that innovation or discovery is often the result of chance, trait combination, and incremental refinement, rather than "genius."[39] Almost all human innovation involves the reworking or development of pre-existing technology.[40] Indeed, scholars of the history of technology criticize the "myth of the heroic inventor,"[41] lore that in our terms would attribute progress in cumulative culture to novel invention.[42]

Previously, the social learning strategies tournament had taught us that high levels of reliance on social learning routinely generated extreme longevity of cultural knowledge within the population. Now Magnus's and Hannah's analyses had taught us that high-fidelity transmission mechanisms automatically produced large amounts of long-lasting and cumulative culture. Put together, these theoretical results suggest that once our ancestors became heavily reliant on strategic and accurate forms of copying, they would almost inevitably develop a culture that very much resembled our own. We were starting to envision how something as monumentally complex as human culture could have begun to emerge.

One question that remained, however, was through which mechanisms high-fidelity information transmission had been achieved by our ancestors. One obvious answer is language, and an attempt to unravel the mystery of how language evolved is presented in the next chapter. However, a second outstanding candidate for a mechanism underlying high-fidelity transmission in humans is teaching. Indeed, teaching can be defined as *behavior that functions to enhance the fidelity of information transmission between tutor and pupil.*[43] My use of this term refers not just to the teaching of skills, but also to the passing on of knowledge. Both demonstrating to a pupil how to execute a food-processing technique, and telling the pupil about the necessary ingredients and where they can be found, I consider examples of teaching. Unsurprisingly, then, in the light of Magnus's and Hannah's findings, the attention of my research group turned to exploring the conditions under which teaching evolved in animals.

The evolution of teaching is a challenging, enigmatic, and beguiling topic in its own right. Alongside imitation, and leaving aside language, teaching is thought to be the primary mechanism through which humans pass acquired knowledge, skills, and technology among individuals and across generations. Teaching is widespread in human societies, and may well be a key human psychological adaptation.[44] As we will see in later chapters, teaching is the primary means through which individuals acquire an understanding of the norms, laws, and institutions of a society, and it is vital to many forms of human cooperation.[45] Yet the mystery remained as to why teaching should be either absent or exceedingly rare in virtually all other animals when it is manifestly so important to our own species.

Countless animals do acquire skills and knowledge from others, as we saw in chapter 2, but the experienced individuals who "transmit" the information generally do not actively facilitate learning in others.[46] Typically, animals just go about their own business and, while they may be copied by others, the demonstrators rarely go to the trouble of overtly helping others to learn. In fact, the commonly used term "demonstrator" is slightly misleading when applied to animal social learning because it may falsely imply that the transmitter's role is an active one.[47]

For many years it was thought that humans alone actively teach. Early studies of animal teaching were hampered by an anthropocentric viewpoint. The image of a teacher was of a traditional schoolteacher, and definitions of teaching stressed the intention of the tutor to educate.[48] This stance effectively restricted teaching to our own species, because such intentions are very difficult to infer in nonhumans. Progress was made in 1992, when two animal behavior researchers Tim Caro and Marc Hauser adopted a functional perspective.[49] That is, they characterized teaching as behavior that functions to teach, and has evolved specifically for that purpose. The major benefit of this approach was that observable criteria were used that could be applied to animals:

> An individual actor A [the tutor] can be said to teach if it modifies its behaviour only in the presence of a naïve observer, B [the pupil], at some cost or at least without obtaining an immediate benefit for itself. A's behaviour thereby encourages or punishes B's behaviour, or provides B with experience, or sets an example for B. As a result, B acquires knowledge, or learns a skill earlier in life or more rapidly or efficiently than it might otherwise do so, or would not learn at all.[50]

In the early 1990s, Caro and Hauser were able to point to just two putative examples of animal teaching,[51] in cheetahs and domestic cats,[52] although taken alone, neither unequivocally met their criteria. Mother cats with kittens and cheetahs with young cubs do not kill and eat their prey as other cats do. Instead, mothers bring the prey back to their offspring in dead, disabled, or intact form depending on the age of the young, apparently so that the latter can practice hunting. In the years that followed their initial work, refinements of Caro and Hauser's definition imposed additional criteria, such as feedback from pupil to tutor, or teaching restricted to the transfer of skills, concepts, and rules.[53] With such definitions, researchers have reported a small number of reasonably compelling cases of animal teaching, but patently not in the species that we had expected.

One interesting example is found in meerkats,[54] a small carnivorous species of mongoose that must survive in the harsh African desert through coordination and teamwork. Meerkat pups are almost entirely reliant on food provisioned by older group members, known as "helpers," which include both parents and other colony members.[55] Yet by three months of age, those same youngsters have become entirely self-sufficient at feeding and can handle a variety of difficult and dangerous prey, including lizards, spiders, and even scorpions with their deadly stings. Recent work shows that the helpers facilitate this transition to nutritional independence by gradually introducing pups to live prey.[56] Cambridge University researchers Alex Thornton and Katherine McAuliffe set out to establish whether this process meets the criteria of Caro and Hauser's definition of teaching.[57] This required them to determine whether the transfer of live prey by helpers occurred solely in the presence of the pups, at some cost to the adults, and led to the youngsters learning about food.

Adult meerkats normally consume prey immediately, but when young pups are present, they typically kill or disable mobile prey items before carrying them to a begging pup. Scorpions are often disabled by removing the sting, allowing pups to interact safely with the live prey. As pups grow older, they are increasingly given intact prey, stings and all. Thornton and McAuliffe established that whether or not helpers modify prey through sting removal before provisioning hung critically on the sound of the pups' begging calls, which changes with age. These researchers played recordings of the begging calls of old pups to groups with young pups, which led adults to bring back live prey, while playing the calls of young pups to groups with old pups caused an increase in the proportion of dead prey provisioned. In spite of the adult meerkat being fooled, the experimental manipulation demonstrates that the helpers normally adjusted their behavior to the age, and hence, competence of the pups. Indeed, the adults exhibited considerable sensitivity to the performance of pups, nudging prey items if the pups ignored them, retrieving escaped prey, and modifying the prey some more (for instance, disabling further) if pups were struggling. This provisioning strategy incurs costs, because considerable time is spent monitoring the pups as they handle live prey, and there is a nontrivial risk that the pups will lose the prey item. However, the strategy creates opportunities for the pups to acquire hunting skills.

Thornton and McAuliffe were also able to provide experimental evidence that the helpers' behavior promoted skill acquisition. Pups that were artificially given additional opportunities to handle live, stingless scorpions subsequently outperformed siblings that had been given dead scorpions, showing that the opportunity to practice on disabled but live scorpions facilitates skill acquisition.[58] Thornton and McAuliffe had demonstrated that the meerkat helpers' behavior meets all of Caro and Hauser's criteria, and was a genuine example of animal teaching. A clue as to why helpers teach is found with the observation that pups very rarely find mobile prey items themselves. Helpers can therefore actively facilitate the pups' acquisition of handling skills by giving them otherwise unavailable opportunities to practice handling prey. In the long term, adults benefit by reducing the costs of provisioning through hastening the time to independence, as well as through increased pup survival.[59] It is advantageous for adult meerkats to teach youngsters to feed themselves, because it saves the helpers time and effort in the long run.

Another compelling example of animal teaching is found in those ants that engage in "tandem running,"[60] a behavior that involves carefully guiding nest mates to new food sources or nest sites.[61] In most other ant species, individuals follow pheromone trails to food sources,[62] but tandem running has the function of teaching nest mates the locations of food. In tandem-running ants, knowledgeable individuals adjust their behavior to ensure that followers learn the route. Biologists Nigel Franks and Tom Richardson at Bristol University showed how, among tandem-running pairs, the leader ants only run ahead a few steps toward the food after being tapped by the antennae of a follower, which enables the pair to remain in close contact.[63] This slows the leader down considerably, resulting in a fourfold increase in the time that it takes leaders to reach the food. However, the followers find food substantially sooner when tandem running than when searching alone and, more importantly, learn the location of the food in the process. There is strong evidence that tandem running has evolved specifically to facilitate teaching. First, if knowledgeable ants merely need help to bring food back to the nest, it is considerably more efficient for them to carry naive ants directly to the food, as they sometimes do. Critically, however, when this occurs, route learning does not take place, largely because the ant is typically carried upside down and facing backward![64] Tandem running might be slower than carrying, but it has the advantage that the follower learns where to go.[65]

Using Caro and Hauser's functional criteria, or derivatives of it, animal behaviorists have now compiled evidence for teaching in a small and rather curious assortment of species; these include meerkats, ants, bees, and two species of birds called pied babblers and superb fairy wrens; with suggestive, but not yet conclusive, evidence in cats, cheetahs, and tamarin monkeys.[66] The functional similarities between teaching in, say, ants and humans should not obscure the fact that mechanistically, cases of teaching in other animals are entirely different from human teaching, and are reliant on entirely separate psychological and neural processes.[67] Indeed, current thinking suggests that instances of teaching function to enhance the fidelity of information transmission through adaptive refinements of those forms of social learning that are already present in the animal.[68] This thinking leads us to expect quite distinct teaching mechanisms in different species. Also important to recognize is the fact that animal teaching has not been subject to a great deal of scientific attention, leaving the taxonomic distribution of this character uncertain. Cases of animal teaching remain controversial, with some researchers wishing to impose more stringent criteria before a behavior is classed as teaching.[69] Nonetheless, as examples of animal teaching trickled in, we were struck by the observation that no compelling evidence for teaching had been found in nonhuman apes, dolphins, elephants, or other large-brained mammals celebrated for their intelligence. Why this should be intrigued me.

Animal teaching studies left us pondering a series of challenging questions. First, we wanted to comprehend why teaching was not more widespread in animals when it seemed so beneficial. Second, we needed to make sense of the curious taxonomic distribution associated with teaching. If ants and bees were capable of teaching, why should intelligent animals, such as chimpanzees, not teach? Third, in contrast to humans, teaching in other animals appeared restricted to isolated traits, which led us to wonder how and why a very general capability for teaching evolved among our ancestors. One might imagine that these issues would have been subject to widespread interest across multiple academic disciplines, including evolutionary and behavioral biology, psychology, and anthropology. Yet, although related topics, such as the evolution of social learning,[70] learned communication,[71] and learned cooperation,[72] had been the focus of extensive research, until recently there was no formal theory of the evolution of teaching.

One of the hottest research topics in evolutionary biology currently is the evolution of cooperation. Any behavior that provides a benefit to another individual (the "recipient"), and that was selected because of this positive effect, is defined as cooperation.[73] Many cases of cooperation in nature involve the donor passing on physical resources such as food or help to the recipient, but the same definition would apply if the donor instead provides useful information. Hence, most instances of teaching in humans, and all known cases of teaching in other animals, meet definitions of cooperation, with the tutor's behavior being favored because it promotes the acquisition of fitness-enhancing information in the pupil. In principle, individuals could be taught skills or knowledge that damage their chances of surviving and reproducing (to take class *A* drugs, for instance), but such cases are rare exceptions. Teaching is generally a form of cooperation. Nonetheless, patterns of teaching in nature are not currently well explained by theories of the evolution of cooperation.[74] Rather, certain unique features specific to teaching mean that an understanding of its evolution requires a specialized treatment. These include the fact that taught information can be acquired through means other than teaching (e.g., through trial and error, or through social learning), and that the dynamics of information transfer differ considerably from the dynamics of the spread and accumulation of physical resources. This meant that if we wanted to understand the evolution of teaching, we would need to develop our own theory.

Once again, we looked to mathematical modeling for help. Two mathematically minded members of my research group, a PhD student named Laurel Fogarty and a postdoctoral researcher called Pontus Strimling, worked together closely to devise a model that would explore under what circumstances teaching might be expected to evolve.[75] I found their analyses extremely insightful, helping to explain the bizarre distribution of teaching behavior in nature, the absence of teaching in many large-brained mammals, and the widespread use of teaching in humans. Laurel and Pontus came up with a model in which the evolutionary fitness of individuals hung critically on whether or not they possessed a valuable piece of knowledge or skill, such as meerkats knowing how to tackle a scorpion safely, or our ancestors knowing how to manufacture a functional digging tool. In the hypothetical mathematical world that Laurel and Pontus constructed, individuals that possessed this knowledge would be more likely to survive and reproduce than individuals without the skill. We assumed that young individuals (henceforth "pupils") could acquire this information in three possible ways. The skill could be learned by (1) trial and error (that is, asocial learning) with probability A; (2) copying another individual (social learning) with probability S: or (3) being taught by a relative (henceforth "tutor"), at some cost to the tutor, which occurred with probability T. Only individuals with an evolved ability to teach would be able to act as tutors. Laurel and Pontus then explored the circumstances under which individuals carrying an initially rare genetic mutation that conferred an ability to teach would have a fitness advantage over nonteachers, such that teachers would on average leave more offspring; in this way the mutation would increase in frequency until the population was dominated by individuals capable of teaching. This approach allowed us to explore the circumstances under which a capacity for teaching would evolve through natural selection.

Some of Laurel and Pontus's findings were very intuitive. As one might anticipate, the more closely related the tutor and the pupil were, the more likely teaching would be favored. The reasoning here, known as "kin selection," is that behaviors can spread—even if they reduce the chances that the individual performing them survives and reproduces—in circumstances where those behaviors provide benefits to the survival and reproduction of close relatives (inclusive fitness).[76] Laurel and Pontus found that teaching would evolve where its costs were outweighed by the inclusive fitness benefits that resulted from the tutor's relatives being more likely than other individuals to acquire the valuable information. Obviously, that means that the probability of teaching being favored decreases as the costs of teaching rise.

Other findings, however, were less obvious. Particularly instructive was the relationship between the fitness difference of teachers and nonteachers (Wd), and the probability that individuals could acquire the skill through means other than teaching (i.e., through asocial or inadvertent social learning, $A + S$). In the latter case, high values of $A + S$ can be thought of as corresponding to easy-to-learn skills (i.e., behaviors that could readily be picked up through trial and error, or through copying), while low values would represent difficult skills. When these variables are plotted against each other graphically, the resulting functions generate a characteristic n-shaped curve. Animal teaching is more likely to evolve for tasks of intermediate difficulty than for either easy or difficult skills.

Consider first the situation with skills that are easy to learn. Here teaching is not favored by selection because there is no advantage to adults investing in a costly means to ensure that their relatives (for instance, their immediate offspring) acquire a skill that they are highly likely to pick up anyway. This finding neatly explains why teaching is not more common among smart animals. Chimpanzees, for example, are very good at trial-and-error learning and particularly adept at social learning, at least compared to other animals. The high probability with which chimpanzee youngsters will pick up foraging skills such as nut cracking or fishing for ants through a combination of trial and error and copying, greatly reduces the probability that it would be worth adults investing in the costs of teaching them how to do so. Going to a lot of trouble to teach youngsters skills that they are going to learn without your help is not economical. Hence, the youngsters being good at learning actually raises the bar, and makes it more difficult for teaching to evolve in that species. That is why no simple relationship exists between the intellectual ability of the animal and the presence of teaching.

Now consider the case of traits that are difficult to learn. Here too, teaching is not favored by selection, but this time for a different reason. Few individuals in the population have managed to acquire these traits, precisely because they are difficult to learn, which means that most teachers will not possess the requisite knowledge to instruct their pupils. For other animals, at least, there is no advantage in their investing in a costly means of information transmission if they don't have useful skills or knowledge to pass on. Thus, among nonhuman animals, tasks of intermediate difficulty are most likely to be taught. If teaching is to be favored, the skill taught should not be so simple that it is easy to be picked up without teaching, but at the same time, not so difficult that few individuals possess the skill and can pass it on. Highly valuable skills with a big impact on survival are more likely to meet the restrictive criteria under which teaching evolves than skills that don't greatly increase fitness. However, Laurel and Pontus found that there typically needs to be a very substantial increment in fitness associated with the skill if teaching is to pay; that is, the knowledge taught would really have to make a difference to the pupil's chances of surviving and reproducing. There are relatively few traits that satisfy these constraints, which explains why teaching is probably rare in nature.

Armed with these findings, we can now understand why teaching should be found in such an odd assortment of animals. The incidence of teaching only appeared puzzling because we had the wrong intuitions. We had expected teaching to be exhibited by clever animals that are good at social learning. In fact, with some caveats that we will come to, smart animals rarely need to teach, because most of their skills can be picked up through copying or trial and error. What cases of animal teaching have in common are a high degree of relatedness between tutor and pupil, factors that reduce the costs of teaching, and an otherwise difficult-to-learn skill that confers a substantial fitness benefit. If, as it would appear, animals rarely engage in teaching, it is because these criteria are difficult to meet.

Yet the puzzle remained of how humans could teach so widely. Humans don't just teach the odd, isolated high-fitness skill to close relatives, but impart all sorts of material and often to entirely unrelated individuals—from tango dancing to calculus—much of which seems to have no impact at all on the pupil's survival. Laurel and Pontus had identified the conditions under which teaching evolved, yet our own species appeared to violate those rules.

Once again, we went back to the drawing board to reflect further on the differences between animal and human teaching. We became impressed by the fact that compared to other animal teaching, humans were manifestly capable of imparting highly complex knowledge, be it automobile mechanics, the use of computers, or advanced mathematics. We knew this was because humans alone have undergone cumulative cultural evolution, which has in essence reduced our reliance on information acquired alone, through trial and error, and allowed the amassing of knowledge and skills that are learned through social guidance and direct instruction. Valuable information—concerning, for instance, industrial practices or technological manufacture—had become available for individuals to teach through the accumulated efforts of many past individuals. We began to wonder whether cumulative culture might be the critical factor that made the difference. Unlike other animals, for whom tasks that are difficult to learn cannot be taught because few individuals possess the relevant knowledge, humans are able to teach very complex skills; we can do this because that knowledge is widespread in the population, having gradually accrued over time. This led us to the hypothesis that the generalized capacity for teaching exhibited by humans might have coevolved with the cognitive traits that underlay cumulative culture.

Laurel and Pontus responded by extending their model to allow for cumulative knowledge gain. This was implemented by allowing individuals that had acquired the first piece of information to gather further knowledge that improves or refines it. This further knowledge simultaneously made the acquired skill more beneficial but also more difficult to learn.[77] The results were resounding. Cumulative knowledge gain did indeed make teaching more profitable, and as a consequence more likely to evolve.[78] The relative fitness of teachers compared to non-teachers was virtually always higher in a cumulative setting compared to a noncumulative setting.[79] What is more, cumulative culture favored teaching precisely for the reasons that we had anticipated; difficult-to-learn information had been made available in the population at a higher frequency through prior accumulation, and this was then available to tutors to pass on to their pupils.[80]

The teaching conundrum had been solved. Teaching evolves where the costs are outweighed by the inclusive fitness benefits that result from the tutor's relatives being more likely to acquire the valuable information. Teaching is not favored when the pupil can easily acquire the information on their own or through copying others. Nor is it favored when imparting traits that are difficult to learn, as teachers generally do not possess the information to pass on to their relatives. These restrictions typically lead to few circumstances under which teaching would be efficacious. Models that allow for cumulative cultural knowledge gain, however, suggest that teaching evolved in humans despite, rather than because of, our strong imitative capabilities, and primarily because cumulative culture renders otherwise difficult-to-acquire valuable information available to teach. The analyses suggest that human teaching and cumulative culture evolved together, through mutual reinforcement.

A small subset of nonhuman animals does appear to satisfy the stringent conditions for the evolution of teaching, and our findings help explain what this curious assortment has in common. The possibility of high relatedness among female workers in the social insects may help to explain why teaching is observed in some tandem-running ants and some social bees,[81] but comparatively rarely in vertebrates.[82] There is something else that many teaching species have in common, however; this is cooperative breeding, a social system in which helpers provide care for offspring produced by other group members that breed.[83] Strikingly, the most compelling cases of animal teaching occur in cooperatively breeding species, such as ants, bees, meerkats, and pied babblers. Humans too have been characterized as cooperative breeders, because children often receive care and resources from more distant relatives, or nonrelatives.[84] The significance of cooperative breeding here is that helpers engage in costly,[85] and prolonged,[86] provisioning of the young that they help rear. Plausibly, the alleviation of heavy provisioning costs, as well as the sharing of teaching costs across multiple tutors, significantly lowers the per capita cost to an individual teacher; both help to render teaching economical.[87] Teaching may be favored only when the tutor's costs are unusually low,[88] and this is likely to be more frequent in cooperative-breeding species because the costs of teaching can be offset against those of provisioning.[89] This finding is relevant to understanding human teaching too. Humans are characterized by a very long period of juvenile dependency, which corresponds to a remarkably long period of time during which adult humans must provision their offspring. This prolonged juvenile dependency may well have contributed to the extensiveness of teaching observed in our species, by providing an economic incentive for us to teach our children how to fend for themselves.

The extent of animal teaching may, as yet, have been underestimated, and new examples may be uncovered in the future.[90] Nonetheless, the generality and pervasiveness of human teaching offers a striking contrast to teaching in other animals. In part, this follows from our capacity for cumulative culture, but there is a second factor, also unique to our species, that may be important as well. The fidelity of human teaching is likely to be unusually high relative to teaching in other animals.[91] There are a number of reasons for this—the most obvious is our capacity for language. However, there is also teaching through imitation, through manual shaping of the pupil's body by the tutor, and through passing on subtle clues. The latter is known as pedagogical cueing, where factors like eye contact, joint reference, and verbal cues (i.e., "watch this") help pupils learn to direct their attention. Moreover, humans have considerable powers of mental state attribution, which allows tutors to adjust their teaching to the pupil's state of knowledge.[92] All these abilities, which are examples of the kind of cognitive traits that might plausibly have evolved through Wilson's cultural drive mechanism, combine to enhance the fidelity of human information transmission.

Not all of the issues surrounding teaching have yet been clarified. In modern human societies teaching is a profession in which individuals are paid to provide guidance and instruction to nonrelatives, and Laurel and Pontus's model does not explain how such institutionalized teaching arose. Nor have we yet discussed the origins of language, which I believe is intimately tied to the evolution of teaching. These are topics we will return to later in the book. Another complication are reports that teaching is not common in some hunter-gatherer societies.[93] However, such claims refer to an absence of direct instruction, and neglect the prevalence of more subtle forms of teaching, such as attention-grabbing verbal cues, or the directing of the pupil's gaze to relevant objects.[94] Recent thinking within anthropology supports the argument that teaching is widespread in humans,[95] in spite of earlier claims to the contrary, but that human teaching takes on a variety of forms from direct verbal instruction to more subtle scaffolding and cueing. There is no escaping the fact that every one of us has been taught valuable knowledge, skills, and lessons throughout our childhood, and teaching is core to knowledge transfer and personal development in virtually every profession, from training accountants to coaching IT skills. All human beings live in a world structured by laws, regulations, and complex institutions, the vast majority of which would be impossible to understand without teaching.

I end this chapter with a study that provides clear experimental support for the link between high-fidelity transmission mechanisms and cumulative culture. The experiment was carried out by Lewis Dean, a graduate student at St Andrews.[96] The study, which was published in the journal *Science*,[97] investigated the cognitive processes necessary for cumulative cultural learning, as well as the proposed links between teaching, language, and cumulative culture.

As noted previously, the evidence that other animals exhibit cumulative culture is circumstantial and contested.[98] A scientific debate has ensued, in which researchers have discussed why the culture of humans, but not the socially learned traditions of other animals, frequently ratchets up in complexity and diversity. The debate has spawned a large number of distinct hypotheses concerning the cognitive capabilities, or social conditions, thought to be necessary for cumulative culture to arise. These explanations include a hypothesized critical dependency of cumulative culture on aspects of social cognition deemed to be exclusive to, or substantially enhanced in, humans, including teaching, language, imitation, and prosociality (i.e. helping others).[99] However, other explanations stress features of social structure that mitigate against the spread of superior solutions in animals other than humans.[100] These include scrounging, which can hinder social learning and demotivate resource production;[101] the tendency of dominant individuals to monopolize resources, thereby preventing subordinates from learning;[102] and a lack of attention to low-status inventors.[103] A further hypothesis is that satisficing, or conservative behavior, is what hinders ratcheting in nonhumans.[104]

As no extensive and rigorous experimental investigation of the capacity for cumulative cultural learning had been conducted,[105] we decided to carry out such an investigation.[106] We designed a puzzle box specifically to test the ability of individuals to engage in cumulative cultural learning. The puzzle box was a simulated foraging task that could be solved at three stages of difficulty, with success at stage 2 building on stage 1, and success at stage 3 building on stage 2. Progress at each stage accessed an ever more desirable reward. To solve the first level, individuals merely had to slide a door to one side to reveal a chute, through which low-grade rewards (a carrot for the nonhuman primates, or small stickers for the human children) were delivered by the experimenter. Stage 2 required a button to be pushed, allowing the door to slide further to reveal a second chute, containing a higher-quality reward (an apple, or a medium sized sticker). Stage 3 required the turning of a dial, which allowed the door to slide further still, revealing a third chute containing high-quality rewards (grapes, or a large sticker). All stages could be completed through two parallel options, allowing us to investigate cooperation, tolerance, and social learning within the task, while presentation in social groups allowed solutions to each level to spread among individuals.

We then presented appropriately scaled versions of the puzzle box under a variety of conditions to small groups of children, chimpanzees, and capuchin monkeys.[107] More specifically, the participants were eight groups of three- to four-year-old children from three nursery schools in Scotland, eight mixed-age groups of chimpanzees at a research facility in Texas,[108] and two groups of capuchin monkeys at another facility in Strasbourg, France.[109] Our study had two objectives. First, we wanted to establish whether children, chimpanzees, and capuchins were capable of cumulative cultural learning (i.e., whether higher-level solutions would arise and spread through the groups).[110] Second, if they were not, we wanted to understand why not; that is, we wanted to sort between the alternative hypotheses for why humans seemingly alone possess a cumulative cultural capability.

The results of Lewis's experiments were very clear. After 30 hours of presenting the task to each of 4 chimpanzee groups, only 1 of 33 individuals reached stage 3.[111] A further experiment revealed that performance was not greatly enhanced by the introduction of chimpanzee demonstrators who had been pretrained to solve the task at the highest level. The trained chimpanzees repeatedly demonstrated the solution in a highly proficient manner, yet this solution did not spread to others in their group. A similar pattern was observed in the capuchins, where after 53 hours, no individual reached stage 3 and only 2 individuals reached stage 2. These findings stand in stark contrast to what we witnessed in the human children. Despite a far shorter exposure to the apparatus (only 2.5 hours), five of the eight groups had at least two individuals who reached stage 3, with multiple solvers at stages 2 or 3 in all but 2 groups. The children, but not the chimpanzees or capuchins, had provided us with clear evidence for cumulative cultural learning.

What were the children doing differently from the other two species? They were teaching each other how to solve the task. A total of 23 unambiguous instances of teaching by direct instruction (i.e., referencing part of the puzzle box) were observed exclusively in the children; all of these involved task-relevant communication (e.g., "push that button there") and approximately one-third involved gesture (i.e., pointing to the relevant part of the puzzle box). Of course, we knew that teaching is rare in animals, so perhaps these findings were not too surprising. For this reason, we went on to consider teaching precursors, or subtle processes similar to teaching, such as "scaffolding,"[112] which might not satisfy definitions of teaching but might nonetheless inadvertently aid learning in the other primates. One such possibility was that chimpanzees or capuchins might facilitate learning in others by allowing infants to steal food the adults had extracted. A second was that they might enlist the offspring's interest in the task through food calls. However, when we examined rates of provisioning, we saw no tolerated theft in mother-infant pairs of capuchins, while among the chimpanzees, the mothers were actually stealing extracted food from their offspring! Moreover, our examination of food calling established that in both nonhuman species there was no difference in the rate of recruitment of others to the puzzle box before, versus immediately after, a call. In fact, low rates of food calling were observed. In stark contrast, those children who received verbal instruction strongly outperformed those who did not.

We also compared the rate at which individuals from each species performed a matching manipulation of the puzzle box to that observed being performed by another individual departing the box. For instance, had the last contact of the preceding individual at the box been to push down the button on the left, we examined whether the observer that followed also pushed that button. Matching could constitute copying the actions of others (i.e., imitation) or making the same manipulanda move in the same way (i.e., emulation). We found that children alone performed more matching than nonmatching manipulations and that they produced a significantly greater proportion of matching actions than both chimpanzees and capuchins. Other analyses revealed evidence for chimpanzee social learning at stage 1 but not at higher stages.

In addition, we witnessed striking evidence that the children were helping each other solve the task. There were no fewer than 215 altruistic events when a child spontaneously gave a retrieved sticker to another (half of all children exhibited such altruism), but not a single instance of the voluntary donation of food in either the chimpanzees or capuchins occurred. Such altruism likely signifies that the children, but not the other two species, understood that their motivations and goals were shared. There were other signs that the children approached the task cooperatively. The proportion of occasions when two children performed the task together was far greater than among chimpanzees or capuchins, which indicated greater tolerance of others and collaboration among children.

In contrast, there was little evidence that the capability for cumulative culture was affected by either the social structure of the species or aspects of nonsocial cognition.[113] We found no support for the five hypotheses that cumulative culture is absent in chimpanzees or capuchins because in these animals (1) the social transmission of superior solutions is hindered by scrounging, (2) dominant individuals monopolize key resources, (3) there is a lack of attention to low-status innovators, (4) these animals satisfice, or (5) there was an inability to discriminate higher-quality from lower-quality rewards. Nor can the results be easily dismissed as an artifact of captivity testing, because wild chimpanzees and capuchins have been subjected to long-term studies that reveal no unambiguous evidence for cumulative culture.[114] Likewise, our animals could be described as "dysfunctional" because they had performed effectively in previous studies demonstrating social learning and tradition of noncumulative tasks.[115] Rather, the children's use of teaching, imitation, and language explained the differences in performance between species.

The link between high-fidelity transmission mechanisms and cumulative cultural learning was cemented by our finding strong positive and significant correlations between an individual child's performance at the puzzle box and how much teaching and verbal instruction they received, how much imitation they engaged in, and how much prosociality they benefited from. All children who successfully solved stage 3 received at least one form of this social support, and 86% received at least two. Conversely, children who did not benefit from social support performed poorly.

These data not only provide clear and strong evidence for a cumulative cultural capability in the children but also strongly link their elevated performance to their social cognition. A package of social cognitive capabilities that encompassed teaching, which was largely carried out through verbal instruction, as well as imitation and prosociality, was critical for elevated performance. The study implicates this package of processes as important for cultural transmission to ratchet. The children responded to the apparatus as a social exercise, manipulating the box together, matching the actions of others, facilitating learning in others through verbal instruction and gesture, and engaging in repeated prosocial acts of spontaneous gifts of the rewards that they themselves had retrieved. In contrast, the chimpanzees and capuchins appeared to interact with the apparatus solely as a means to procure resources for themselves in an entirely self-serving manner, largely independent of the performance of others, and exhibited restricted learning that appeared primarily asocial in character. Our study provided strong support for the view that cumulative culture requires psychological processes that are present in humans but are absent or impoverished in chimpanzees and capuchins.[116]

Human cultural traditions accumulate refinements over time, thereby producing both technology and other cultural achievements of astonishing complexity and diversity unprecedented in the rest of nature. Although numerous hypotheses had been proposed for this phenomenon, the explanation had for many years remained elusive. Lewis's experiment provided a clear answer to this conundrum, again lending strong support for the position advanced by Michael Tomasello,[117] and echoed here: the mindboggling complexity of human culture hangs critically on our capacities for teaching, language, and imitation. What our studies had added to Tomasello's arguments was an account of how and why these abilities had evolved.

The theoretical analyses and experiments described in this chapter serve in combination to clarify how important accurate information transmission is to the origins of culture. Humans alone possess cumulative culture because humans alone possess sufficiently high-fidelity information transmission mechanisms, including an unusually accurate capacity for imitation, teaching, and language. It sounds paradoxical that teaching should both explain the advent of human cultural complexity and be the product of it, but that is exactly what we should expect if a feedback mechanism, like Wilson's cultural drive, is operating. In fact, evolutionary feedback mechanisms help explain a lot more about the human condition than I have touched on so far, including that quintessentially human attribute, our language.

CHAPTER 8

WHY WE ALONE HAVE LANGUAGE

No longer mourn for me when I am dead
Than you shall hear the surly sullen bell
Give warning to the world that I am fled
From this vile world, with vilest worms to dwell.
Nay, if you read this line remember not
The hand that writ it; for I love you so,
That I in your sweet thoughts would be forgot,
If thinking on me then should make your woe.
O, if, I say, you look upon this verse,
When I perhaps compounded am with clay,
Do not so much as my poor name rehearse,
But let your love even with my life decay;
Lest the wise world should look into your moan,
And mock you with me after I am gone.

—WILLIAM SHAKESPEARE, SONNET 71

Powerful expressions of emotion and self-sacrifice, such as this Shakespearian sonnet, move many of us. In a world in which people frequently clamor to make their mark or leave a legacy, that someone could love another so desperately that they would choose to be forgotten rather than have their memory cause pain and suffering to their loved one is very poignant. There is an anguished bellow of conflicted emotion perfectly captured in Shakespeare's words to which we relate. Yet the sonnet encompasses far greater specificity of information transfer than any anguished bellow ever emitted by an animal. No other primate expresses heartfelt sentiments with such clarity. No other species conveys after-death contemplations, or anticipated societal judgments. "What a piece of work is a man," indeed.

Shakespeare would certainly not have caste humanity as "infinite in faculty" had our species not evolved language.[1] Without language there would be no sonnets, plays, theatre, literature, history and, of course, no Bard of Avon. If the eminent Soviet psychologist Lev Vygotsky is correct, and he surely is, without language there would also be precious little complexity to human thought. Vygotsky argued compellingly that the development of human reasoning is mediated by signs and symbols, and hence is critically contingent on language.[2] He established a connection between speech and the development of mental concepts and cognitive awareness that has remained prevalent within psychology to the present. A similar view was championed by the great American linguist Noam Chomsky, who emphasized that language is as much a system for structuring our thinking about the world as it is a vehicle for communication.[3] When we think, we typically do so linguistically.

No account of the evolution of human cognition and culture would be complete without an explanation for the origins of language. Yet, in spite of its unquestioned importance, the reason why human language evolved remains an enigma. Certainly, there are no shortage of candidate explanations. Selective scenarios for the emergence of natural language are bounteous.[4] Language evolved to facilitate cooperative hunting.[5] Language evolved as a costly ornament that allows females to assess male quality.[6] Language evolved as a substitute for the grooming exhibited by other primates when groups got too large.[7] Language evolved to promote pair bonding,[8] to aid mother-child communication,[9] to gossip about others,[10] to expedite toolmaking,[11] as a tool for thought,[12] or to fulfill countless other functions or purposes.

The problem is that there are too many explanations for the evolution of language. This is a domain rife with speculation and teeming with plausible stories for the genesis of our most cherished faculty, although most such stories are somewhat weakly supported. The wealth of diverse historical narratives is enough to leave some researchers deeply skeptical as to the value of such theorizing.[13] Difficulties arise primarily because the origin of language was a singular event, a unique adaptive response in an isolated lineage. Of course, evolutionary biologists commonly study the emergence of unique traits, such as the giraffe's neck or the elephant's trunk. In those instances, however, there are other long-necked or long-nosed creatures that can be used to shed light on the factors that favored the focal trait. In contrast, there is nothing remotely like language in the animal kingdom. People speak of "the language of the bees" and the "chatter of dolphins," but, in spite of intensive research, science has yet to reveal strong similarities between the communication of humans and other animals. That does not mean a comparative perspective is not useful. To the contrary, through the study of animal communication a great deal has been learned about the origins of language, including what exactly is unique about human communication and how its neural underpinnings and vocal apparatus have changed.[14] However, the singularity of language does leave its origins far more challenging to unearth.

When the natural communication systems of primates are examined, for instance, no straightforward increase in complexity from monkeys to apes to humans is observed. Many researchers characterize great ape communication systems as more limited in range than those of monkeys.[15] For example, monkeys, but not other apes, have functionally referential alarm calls,[16] although whether monkey calls are truly referential like human language remains contested.[17] This particular ape-monkey difference makes biological sense. Great apes are larger and stronger than monkeys, and hence are less vulnerable to predation. Apes almost certainly didn't evolve referential alarm calls because they had comparatively little to be alarmed about. Indeed, there is little that is learned at all in the vocal communication of nonhuman apes.[18] Apes do possess gestures to initiate play, for instance, or when infants signal they wish to be carried—many of these gestures have learned elements.[19] However, apes seemingly do not use their gestures referentially, nor do their gestures exhibit any symbolic or conventionalized features.[20] Hence, the communication of our closest living relatives typically consists of single, unrelated signals that cannot be used outside of highly restricted, functional contexts; they are rarely joined together to make more complex messages, and can only relay information about the here and now.[21]

Admittedly, some apes were taught to use meaningful hand signs, and much has been made of this. However, there is virtually unanimous agreement among language experts that the strong claims—for instance, that Washoe the chimpanzee or Koko the gorilla have learned full grammatical language—are not supported by the data.[22] What such studies do demonstrate is that apes can be taught to learn the meaning of a large number of signs, and to communicate using these; but any assertion that a nonhuman ape has acquired the rules of grammatical structure remains highly contentious.[23] One can train a rat or a pigeon to form an association between a cue and an action, and likewise there is little in the ape sign language literature that cannot be explained by simple rules of associative learning, and perhaps a little imitation.

Revealingly, all talking ape utterances are devastatingly egocentric in character. Give an ape the means to talk through signing and it will say things like, "Gimme food" or other expressions of the animal's own desires. For instance, the longest recorded utterance of Nim Chimpsky, the chimpanzee taught sign language by Herbert Terrace of Columbia University, was, "Give orange me give eat orange me eat orange give me eat orange give me you."[24] Chimpanzees, bonobos, and gorillas seem to make rather poor conversationalists. In contrast, just a few months after uttering their first words, two-year-old children are producing a variety of complex sentences, comprising verbs, nouns, prepositions (e.g., on, in, by), and determiners (e.g., the, their, my) in the correct grammatical relations and on a diverse range of topics. Even very young children are able to communicate about the past and the future, as well as about distant objects and locations.

The evolution of language seemingly required a major shift in referential communication away from specific, isolated, unlearned signals that concern concrete events in the present, to a general, flexible, learned and socially transmitted, infinitely combinable, and functionally unconstrained form of communication—the latter is entirely absent in the animal world. Linguist Derek Bickerton summed up the paradox of language evolution beautifully when he wrote, "Language must have evolved out of some prior system, and yet there does not seem to be any such system out of which it could have evolved."[25]

To compound matters, human language exhibits a multitude of distinct, contemporary uses that we all witness on a daily basis. There is a limit to what an animal can do with a long neck or a prehensile trunk, and this provides clues as to its original function. However, language can be used to woo prospective mates, demonstrate dominance, coordinate teams, reassure a child, teach a pupil, deceive a rival, lay down laws, sing a song, and in numerous other ways. Once complex language evolved, it would have been rapidly coopted for all kinds of uses utterly unconnected to its original function. Distinguishing between a genuine selective scenario and the subsequent exploitation of what is surely one of human beings' most flexible characteristics has proved extremely difficult.

That does not mean that we can never know why language originally evolved; there are, in fact, ways to impose discipline on the explanatory accounts and sort between them. Criteria can be implemented that allow researchers to judge the relative merits of alternative historical narratives on the original function of language. However, to my knowledge such criteria have never been fully compiled, and hence have not yet been applied in unison. To explore how and why language evolved we need to ask what caused our ancestors to take the first steps away from the communication systems that other animals, particularly other primates, possess.

The approach that I follow was pioneered by Szabolcs Szamado and Eors Szathmary,[26] and Derek Bickerton[27]—between them, they sketched six criteria for determining the validity of competing language evolution theories. To these I add one further criterion of my own.[28] Seven benchmarks are thus generated with which to evaluate alternative explanations for the original adaptive advantage conferred by early forms of language. Although individually most of the criteria are not particularly constraining, this compilation of criteria is important because when considered together, they comprise a tough standard against which almost all of the proposed accounts fail. In fact, to my knowledge, only one functional account of the origin of language meets all of these criteria, an explanation that follows naturally and directly from the line of argument presented in this book. Let us consider these seven constraints in turn.

First, *the theory must account for the honesty of early language*. Research into animal communication has established that for signals to be honest and reliable, they will often need to be costly to produce; otherwise they can easily be faked.[29] Many forms of communication in nature are thought to involve costly signaling between individuals. The costs associated with producing signals act to ensure that signaling remains accurate, and receivers are thought to respond to signals as long as the information being transmitted is reliable on average.[30] Cost-free signaling that is accurate and honest can evolve, but primarily where no conflicts of interest between the participants arise.[31] Human beings are thought to signal status through conspicuous shows of wealth or risky activities such as hunting,[32] but the production costs of speech or gesturing themselves are comparatively trivial. Human language constitutes a uniquely economical and flexible signaling device, allowing humans to engage in "cheap talk" in an unprecedented range of circumstances. However, if words are notoriously easy and cost free to produce, why should anyone believe what others say, and what is the incentive to learn thousands of words if one cannot be confident that any convey an accurate message? This constraint implies that researchers should favor theories that propose a context for the evolution of early language in which there was either no conflict of interest between the signaler and receiver, or one in which the reliability of the signals could easily be assessed.[33]

Second, *the theory should account for the cooperativeness of early language*. In many acts of linguistic communication, the transmitter imparts information that benefits the receiver. As a consequence, the receiver is then able to exploit this information alongside the transmitter, perhaps even in direct competition with them, with little to ensure that the receiver will reciprocate on a future occasion. This raises the issue of what is in it for the transmitter. When considered alongside the honesty criterion, the cooperativeness criterion becomes more significant. Had language evolved to deceive or manipulate, then the benefits of transmitting knowledge to others might be easier to envisage. On the other hand, if early language was honest, and the transmitted information benefited the receiver, then language was effectively a form of cooperation. While the production costs of the signal may be modest, the time costs and the competition generated may mean that the costs are often not trivial for the transmitter. The successful theory must explain why, at the time of the origins of language, an individual would go out of its way to help another by passing on accurate information.

Third, *the theory should explain how early languages could have been adaptive from the outset*. Bickerton frames this constraint in the form of a "ten word test." He suggests that there must have been a time when the first protolanguage that was devised had ten words or signs,[34] or fewer, and hence a challenge for any account of language evolution is to explain what could usefully be said with so few words. The exact number of words is immaterial. The point is that language had to be adaptive from the outset, otherwise it is hard to envisage how it would have been favored by selection. Ideally, a successful theory would demonstrate experimentally that a selection pressure is generated from the beginning, and that it favored more and more complex forms of communication.

Fourth, *the concepts proposed by the theory should be grounded in reality*. Any successful account must be able to explain how the words and symbols that were devised acquired meanings tied to the real world (known as "the symbol grounding problem").[35] There has to have been some parallel pathway by which early words could acquire their meaning—for instance, through pointing, imitation, or some form of representation. No symbol can function unless there is something that links it to the entity in the real world that it represents. However, if an individual can say "bird" and point to a bird, sketch a bird in the sand, or imitate a bird at the same time, they can potentially transfer meaning to the symbol. The third and fourth criteria collectively represent a serious challenge for any theory of language origins. Szamado and Szathmary point out that "most theories do not consider groundedness and do not say anything about the possible first words," while others propose implausibly abstract first terms.[36]

Fifth, *the theory should explain the generality of language*. Language is characterized by the range and power of generalization that it confers. The topic of conversation is not linked to the observable present; humans can transmit information about the past and the future, as well as events or objects distant in space. Assuming that this capability to generalize was favored by selection, a successful theory would specify an exact context for these generalizations. According to Szamado and Szathmary, the hypotheses that language originally evolved to replace grooming; to facilitate group-, mate-, or parent-child bonding; or to allow for song fail to meet this criterion.

Sixth, *the theory should account for the uniqueness of human language*. A compelling theory of language evolution must not only explain why language was acquired by humans, but also explain why the context that favored language in humans either did not arise or did not favor the evolution of language in any other species. Edinburgh linguist James Hurford expressed this particularly clearly in stipulating: "for any set of circumstances proposed as individually necessary and collectively sufficient to explain the evolution of language, one has to show that this combination of circumstances applies (or applied) to humans and to no other species."[37] For Szamado and Szathmary, this criterion alone rules out most proposed theories of language origins.[38] Most contexts put forth to explain language evolution (e.g., mate choice, pair bonding, parent-offspring communication) can be found in other animals too, yet none has evolved a communication system remotely like language.

Seventh, *the theory should explain why communication needed to be learned*. Leaving aside the role of evolved structure in language acquisition, human language is learned, and learned socially. Moreover, relative to the communication systems of other species, human languages change rapidly, primarily not through changes in gene frequencies, but at an entirely different level—a cultural or linguistic level.[39] Charles Darwin, in *The Descent of Man* (1871), was the first person to point out that languages evolve through a process of differential "survival and preservation of certain favoured words" that resembles natural selection. Given that nonhuman primate communication is largely unlearned,[40] and changes at rates little different from other biologically evolved characters, the following question arises: What was language needed for that required it to be both socially learned and rapidly changing?

Mathematical theory sheds light on this issue.[41] Theoretical analyses suggest that cultural transmission is favored in changeable and variable environments. Exposed to slowly changing conditions that vary little, populations can evolve appropriate behavior through natural selection, and learning is of little value. At the other extreme, in rapidly changing or highly variable contexts asocial learning pays, provided the environment retains some semblance of predictability (although some theory suggests that social learning can be of benefit in changing conditions too[42]). Social learning is generally favored at intermediate to fast rates of change, because individuals can acquire relevant information without bearing the costs of asocial learning, but with greater flexibility than is the case with unlearned behavior. Within this window of environmental variability, vertical transmission of information (social learning from parents) is thought to be an adaptation to slower rates of change than horizontal transmission (social learning among unrelated individuals of the same cohort), with oblique transmission (offspring learning from nonparent adults) somewhere in between.

One implication of this body of theory is that if the calls of other apes are largely unlearned, the content over which they communicate must be relatively stable. Conversely, humans have socially learned communication, which implies that the matters that we talk about must change at a significantly faster rate, a rate that no evolved forms of communication could track. This raises the question of what our ancestors needed to talk about that was changing at such a rate that biological evolution could not keep up. Moreover, why did this dynamic feature of the environment not have an impact on the communication of other species?

To my knowledge, there is no established hypothesis for the evolution of language that meets all seven of these criteria. That is encouraging because it implies that collectively these criteria constitute a tough hurdle that any credible, selective scenario for the evolution of language must overcome. Yet I believe a credible candidate explanation does indeed exist, one that does pass these tests.[43] The explanation emerges naturally from the data and theoretical findings presented in the preceding chapters. Let me rehearse the argument.

We have learned from experimental studies that copying is widespread in nature, but that animals are highly strategic in their use of information acquired from others. The social learning strategies tournament explains this by demonstrating that there is a selective advantage to copying, provided it is implemented with accuracy and efficiency. This leads us to expect natural selection to have favored more efficient and higher-fidelity forms of social learning, which may have had an impact on brain evolution. Comparative data across primates support this suggestion, with strong associations among social learning, innovativeness, and brain size in primates, and with social learning covarying with several traits thought to be indicative of intelligence, and with performance in laboratory tests of cognition. The findings imply that a "cultural drive" process may have operated in restricted primate lineages. This, in turn, raises the question of why humans alone should exhibit a culture that ratchets up in complexity. Theoretical studies answer this question by showing that high-fidelity information transmission is necessary for cumulative culture, but then pose the supplementary question of how our ancestors achieved high-fidelity transmission. The obvious answer is through teaching, by which I mean not just explicit tutelage, but also a host of more subtle processes, including the use of eye contact, joint reference (coordinating attention between an object and another person), and utterances designed to help pupils learn what they should pay attention to or to what a symbol refers. Defined in this way, teaching is rare in nature, but universal in human societies. Mathematical analyses reveal stringent conditions that must be met for teaching to evolve, but show that cumulative culture relaxes these conditions. This implies that teaching and cumulative culture coevolved in our ancestors, creating for the first time in the history of life on earth a species that taught their relatives across a broad range of contexts.

A cumulative culture setting is not the only factor identified that increases the likelihood teaching will evolve. Increments in the probability of teaching also occur when the costs of teaching are low or can be offset against the costs of provisioning, when teaching is highly accurate and effective in transmission, and when there is a strong degree of relatedness between tutor and pupil. Given that an animal is teaching, adaptations that reduce the costs of instruction without diminishing effectiveness, or that enhance effectiveness without increasing costs, ought to be favored by selection. Should a character appear that simultaneously increases the effectiveness and reduces the costs of teaching, then we would predict it would be subject to strong positive selection, but crucially, *only in a population of teachers*. The more teaching contexts in which the character could be applied, the greater its selective advantage.

Language is such a character. First, language is an incredibly cheap way to teach. Telling someone where to find a food patch is far easier than taking them there. Instructing a child that the red berries are poisonous is much more straightforward than somehow getting this across through other means. A simple "yes" or "no," or "this way, not that way," will allow a tutor to provide helpful guidance for a pupil acquiring a new skill at very low cost. Second, language is an exceedingly accurate way to teach. A precision is brought to information transfer by language that is virtually impossible to achieve through any other means. This exactitude, combined with the efficiency with which tutors can cue their pupils to relevant events as they occur and provide instructive guidance when skills are being learned, means that teaching through language greatly enhances knowledge transfer. Simple utterances that carry messages like "pay attention," "dig here," "like this," "faster," "this way," provide invaluable clues, and help the learner to focus on what actions need to be imitated or precisely where new skills need to be applied. Once an individual is committed to teaching, language is by far and away the most efficient means to do so. That is why almost all teaching in human societies occurs through use of language.[44] In addition, language extends the range of phenomena that can be taught, and allows conveying abstract concepts, the understanding of which may significantly enhance the pupil's performance. Language enables instruction about the past and the future, and about events or objects some distance away. Language opens up new domains to teaching.

Over and above these considerations, the setting fits too. The theoretical work described in the preceding chapter specifies that teaching is far more likely to be advantageous among close relatives than unrelated individuals, and two million years ago our ancestors were living in small, kin-structured groups.[45] Humans may have evolved as cooperative breeders.[46] Eminent anthropologist Sarah Hrdy at UC Davis has presented evidence that women would have struggled to raise children on their own in ancestral hunter-gatherer societies and were heavily reliant on "allomothering" helpers.[47] Relatives other than the mother, such as the father, grandparents, and older siblings, would have spent time with the infants, leaving mothers free time to gather food. The unusually long period of juvenile dependence in our species helps make the teaching of life skills to children economical, because the costs of instruction can be offset against years of feeding and caring. The faster children can be taught to fend for themselves, the lower the burden of childcare.

The latest thinking on the evolution of early *Homo* suggests that increases in brain size were coupled with increased toolmaking and stone transport, dietary expansion, and greater developmental plasticity (the flexible adjustment of development to environmental conditions).[48] This means that there would be plenty to teach, because our hominin ancestors subsisted on a broad omnivorous diet and were reliant on a large number of extractive foraging and tool-using skills.[49] This period in human history was the dawn of cumulative culture, when our ancestors first began manufacturing stone tools, using the flakes to butcher carcasses for food and in a variety of other ways. In other words, it was the beginning of the phase in which (according to our analysis of the evolution of teaching) cumulative culture would help make teaching widely adaptive. Here, then, is a setting in which teaching among close relatives could be beneficial across a broad range of contexts.

Perhaps, then, language originally evolved to enhance the efficiency and scope of this teaching. Natural selection may have favored early language among our ancestors because it made their teaching so much more economical and effective. As a hypothesis, it sounds plausible enough, but as we have learned, there are too many *prima facie* plausible accounts for the selective scenario that favored language. The key question is whether the hypothesis meets the seven criteria outlined above. Let's consider them in turn.

To begin with, if language evolved to teach relatives,[50] we would indeed expect it to be honest. No new conflicts of interest between the signaler and receiver arise in a teaching scenario. The whole function of teaching is to ensure accurate knowledge transfer so that relatives can acquire fitness-enhancing skills and information, such that the pupils' increased rates of survival and reproduction increases the inclusive fitness of the tutor. Deception or inaccuracy on the part of the tutor would typically not provide such benefits. Hence, no problem arises in meeting this first challenge. Conflicts between parents and offspring, or close relatives, might well occur in other contexts, but provided early communication was initially restricted to that which was taught, the interests of both parties are broadly aligned. Of course, if parents wish to see resources evenly distributed among their descendants, while individual offspring want more than their fair share, disputes between parents and offspring can occur.[51] The same reasoning would apply to teaching if overly demanding children sought to monopolize their parents' attention. However, some level of resolution to this conflict must have evolved for teaching to be practiced, and the emergence of language as a cheaper, more efficient means to instruct would not create any further conflict. Plausibly, parents and offspring might differ in how much instruction they would like to see imparted, but what teaching does occur is expected to be honest, because inaccurate instruction is simply wasting a parent's time and effort. If early language evolved to promote teaching, then language would be expected to be honest.

Likewise, the cooperativeness of early language is readily understood. If language evolved for teaching, it emerged in a context that was already collaborative. No difficulty arises in explaining why it benefits a tutor to impart valuable information if by doing so he or she teaches life skills to a relative in a manner that ultimately increases the tutor's inclusive fitness.

How language could have gotten started in a teaching context, and how symbols acquired their meanings (the third and fourth criteria) are also easy to envisage. Simple attention-grabbing commands would do little to get most messages across, but they have been proven to help to facilitate social learning. One of the challenges of imitation is that the demonstrator's behavior is a constant stream of actions, which means that what should be copied is not always apparent to the naive observer; nor is it often obvious where the relevant actions start and stop. Here, a simple verbal cue (or even a nonverbal one) can be invaluable. This has been demonstrated through developmental psychology experiments, where extensive data now shows that adults cue the learning of babies and young children with simple vocalizations. Such cues are known to generate referential expectations in infants, triggering a tendency to follow the gaze of adults as they orient; an example is a shift in the adult's gaze toward the particular object with which the adult interacts, thereby facilitating joint attention.[52] Infants will check the facial expressions of adults responding to unfamiliar objects and use this to guide their approach or avoidance behavior.[53] The use of all such cues, and the resulting gaze and joint attention, are thought to contribute to the infant learning about both the properties of objects and how they can be manipulated, as well as the meaning of words.[54] At the same time, pointing, other gestures, and movement can ground teaching utterances to provide meaning to unfamiliar terms. A tutor can exclaim "here" at the same time as pointing to where the stone must be hit. He or she can mime digging with a stick while uttering the word "dig." Utterances equivalent to "no, this way" can be emitted while manually shaping the pupil's body movements. Experimental findings demonstrate that this is not only plausible, but regularly happens when children learn new skills.[55] Hence, in the context of teaching, the grounding of language in reality, its value when comprised of just a handful of words, and its gradual expansion become possible to envisage.

Then there is the requirement that the theory should provide an explanation for language's power of generalization. Here too there is a natural link to the teaching context. Teaching through language, once started, could be applied to all kinds of proficiencies that are difficult to learn, including a variety of extractive foraging procedures, food-processing methods, and hunting skills. Paleontologists' investigations provide insights into the diets of our ancestors; these include examination of the tiny ("microwear") scratches on fossilized hominin teeth, as well as analysis of chemical traces left in fossil bones and teeth ("stable isotope analysis"). Such studies suggest that the diet of *Homo habilis* and later hominin species was highly versatile, with our ancestors eating a broad range of foods including fruits, tougher material like woody plants, and various animal tissues.[56] The earliest bones marked by the cuts and hammering of stone tools, which date back at least to 2.6 million years ago, are associated with the butchery of large animals, and provide direct evidence of meat and marrow eating by species like *Australopithecus garhi*, *Australopithecus afarensis*, and *Homo habilis*. Around 1.9 million years ago, *Homo erectus* appears in the fossil record. This was a species that created large cutting tools like hand axes and cleavers, as well as campfires with hearths used for cooking food and fending off large predators. The broadening of the hominin diet is associated with increasing reliance on difficult-to-access but nutrient-rich foodstuffs that required extraction from a substrate and some form of processing. Often, this food processing required not just tool use but prior technological manufacture. As a consequence, there would be more and more difficult skills for individuals to acquire, and hence additional opportunities for beneficial teaching. The teaching of foraging, hunting, and scavenging methods; tool manufacture; food preparation and food-processing skills; fire maintenance; and collective defense (some of which required the coordinated actions of multiple individuals) would particularly have benefitted from verbal instruction. There is a short step between teaching coordinated foraging actions in the present, and planning future coordinated foraging. In this way, a modest protolanguage would be expected to become increasingly elaborate, and to generalize in a number of dimensions.

The sixth criterion, that the theory should account for the uniqueness of human language, is also readily met. No other animals, aside from humans and perhaps their immediate hominin ancestors, evolved language because humans alone were engaged in extensive teaching. Only recent hominins participated in extensive teaching because they alone possessed cumulative culture, a feature that our theoretical analyses tell us promotes the evolution of teaching. Without cumulative culture, other animals were faced with stringent conditions for the evolution of teaching, conditions that barely, if ever, were met. In the absence of widespread teaching, other animals could not have experienced selection for traits like language that reduce teaching costs or promote teaching efficiencies. Only in hominins did language, teaching, and cumulative culture coevolve in a runaway, autocatalytic process initiated by selection for strategic and high-fidelity social learning.

All that remains is to explain why early language needed to be learned. Here it is relevant to note that chimpanzees and orangutans both have extensive tool-using repertoires, as well as behavioral traditions that exhibit considerable interpopulation variation.[57] The richness of extant ape culture suggests that the apparent constancy of Oldowan and Acheulian stone tool traditions is probably misleading.[58] Members of our genus are likely to have constructed richer and more geographically diverse cultural repertoires than contemporary apes, including both stone and nonstone tool-using traditions; and local, population-specific, learned, and socially transmitted foraging repertoires. This viewpoint is strongly supported by recent archaeological evidence.[59]

Also relevant is the observation that humans transmit large amounts of learned knowledge across generations, far more than any other animal.[60] As preceding chapters describe, the traditions of other animals are largely based on simple forms of social learning and rarely persist for long.[61] A comparative perspective thus implies that our genus has evolved increasing reliance on information acquired from the previous generation. Theoretical analyses of the evolution of culture, described above, teach us that a shift toward increased transgenerational cultural transmission is a sign of greater constancy in our ancestors' environment. Yet paradoxically, there is no evidence to suggest that environments have become more constant over the last few million years—rather, the opposite. Moreover, had environments really become more constant, other animals would also be expected to show more transgenerational cultural transmission than they do.

A more compelling hypothesis is that our ancestors constructed the environmental conditions that favored hominid reliance on culture,[62] building niches in which it was advantageous for them to transmit more information to their offspring.[63] This is not as farfetched a suggestion as might first appear. In fact, all organisms construct important aspects of their local environments through their activities in a process known as "niche construction."[64] For instance, countless animals dig burrows, build nests and mounds, spin webs, and manufacture egg sacks, pupa cases, and cocoons.[65] Humans, of course, are renowned as "champion niche constructors," and our niche construction is more potent than that of any other species largely because of our capacity for culture.[66] That human activities, particularly agriculture, deforestation, urbanization, and the manufacture of transport systems, have driven dramatic changes in our environment and in the process modified the natural selection acting on our species and countless others is now well recognized.[67] This environmental modification also influences the constructor's reliance on culture. The more an organism controls and regulates its environment and that of its offspring, the greater is the advantage of transmitting cultural information across generations.[68] For instance, by tracking the movements of migrating or dispersing prey, our ancestors increased the chances that a specific food source would be available in their environments; that the same tools used for hunting could be used; and that the skin, bones, antlers, and other materials from these animals would be on hand for the manufacture of additional tools. Such activities create the kind of stable social environment in which technologies such as food preparation or skin-, bone-, and antler-processing methods would be advantageous from one generation to the next, thereby increasing the likelihood that these methods will be repeatedly transmitted across generations. Once started, cultural niche construction may also become autocatalytic, with greater culturally generated environmental regulation leading to increasing homogeneity of the social environment as experienced by old and young, which would favor further social learning from parents and other adults.[69]

The emergence of language can be understood in a related manner. Vocal learning in the food grunts of chimpanzees has recently been reported,[70] and some ape gestures are learned and flexibly used;[71] however, nonhuman apes exhibit largely unlearned vocalizations.[72] A shift from unlearned to learned vocalization suggests an increase in the rate of change of features of the environment that select for primate com-

munication. However, an explanation based on independently changing external conditions, such as fluctuating climates, is not particularly credible for several reasons. First, any such account violates the
uniqueness criterion for language: an external source of selection ought
equally to have favored extensive learned communication in other primates, including other apes. Second, the requirement for increasing
rates of change in the external selective environment for language con-
tradicts the requirement for increasing stability in the external selective
environment for culture. Third, the scale of climatic change is probably too slow. If, however, language initially evolved as an adaptation
to cope with self-constructed elements of the environment, such problems are alleviated. Which features of the self-constructed ape environment change and diversify sufficiently rapidly to require learning to
track them? From a comparative perspective, the most obvious features
to fit the bill are cultural practices, particularly tool use, extractive foraging, and material culture. Cultural practices are typically transmitted
among close relatives, used to exploit difficult-to-access but nutrient-
rich foodstuffs, and are challenging to learn; these features make them
precisely the kind of traits that would benefit from teaching.

Human culture exhibits dramatic and relentless growth in complexity and diversity with time,[73] but such increments appear largely absent
from nonhuman ape culture.[74] Hence, at some stage in the last two
million years, our ancestors began to generate cultural variants (e.g.,
tools, foraging techniques, social signals, courtship rituals, medicative
treatments, gestures) at such a rate that they could no longer communicate about their world without being required to constantly update
and elaborate communicative signals and meanings. If each new tool,
foraging technique, display, or treatment has to be learned, and if, as
the comparative evidence suggests, cultural variants such as tool use
are typically learned by young apes from their mothers and older siblings,[75] then conceivably language may have coevolved with cultural
complexity as a means of facilitating and enhancing socially transmitted
life-skill acquisition in young hominins.[76]

Language originally evolved to teach, and specifically to teach close
relatives. That, at least, is my thesis. Difficult though it is to be totally
confident about any account of the selective scenario that originally favored language, this explanation clearly has many virtues. The account

explains the honesty, cooperativeness, uniqueness, and "symbol grounding" of language, as well as how it got started, its power of generalization,
and why it is learned. The explanation meets all seven of the criteria
required of a successful account of language origins—something that,
to my knowledge, no other hypothesis has done.[77] Human language is
unique (among extant species) because our species uniquely constructed
a sufficiently diverse, generative, and changeable cultural world that had
to be talked about.[78]

One curious feature of the debate over language evolution is that
there remains considerable doubt as to whether language is an adaptation.[79] Given that language is so manifestly complex and functional,
and exhibits many of those "design" properties associated with adaptations, it is tempting to assume the neural mechanisms that allow
language must have been favored by natural selection specifically for
communication. However, there is a long-standing tradition of describing language as a "spandrel," the term within evolutionary biology that
describes a character that evolved as a side effect of selection for some
other capability. Chomsky, for instance, adopted this position,[80] maintaining that language is a side effect of a large and complex brain and
the enhancements in human thinking these afford. In contrast, evolutionary psychologist Steven Pinker is perhaps the most prominent
critic of this stance.[81] From the above account, it will be apparent that
I am suggesting language is indeed an adaptation; specifically, it is an
adaptation that functions to increase the accuracy, reduce the costs,
and increase the scope of teaching.

Naturally, the selective scenario for language only begins here, and
is likely to have been coopted and amplified in a wide variety of other
ways to fulfill diverse alternative functions. Tecumseh Fitch, an evolution of language expert at the University of Vienna, has argued that
language originated to facilitate communication among close kin,[82] and
I endorse this position. In the kin-structured groups exhibited by our
ancestors, the initial selection for language was probably an adjunct to
the teaching of young by parents or siblings. However, from that point,
early language could subsequently spread to teaching more distant relatives—an expansion particularly relevant to activities such as collective
foraging, scavenging, and hunting, all of which required coordinated
activity among multiple individuals. The evolution of teaching model

described in the preceding chapter suggested that both a significant fitness increment associated with the skill taught, and a high degree of relatedness between tutor and pupil, are required for teaching to be favored. Here, the direct benefits of ensuring that relatives possess relevant skills and knowledge in the form of enhanced foraging returns, would compensate for the reduction in the degree of relatedness among more distant relatives. Complex, coordinated actions are often difficult to carry out without a means to teach, or tell, individuals what their specific roles should be. In this regard, language would prove an extremely powerful coordination tool.[83]

Subsequently, with language, teaching could be extended to support other established cooperative processes, such as mutualistic exchanges, indirect reciprocity, and group selection. Both reciprocal altruism and mutualistic trade (at least, the trade of distinct, desired commodities) are surprisingly rare in other animals outside of the context of kinship.[84] This may be because trade requires some capacity to agree on a rate of exchange, something that would be very difficult without at least protolanguage.[85] Similarly, the efficient functioning of indirect reciprocity probably requires gossip.[86] Linguistically taught social norms allow humans to institutionalize the punishment of noncooperative individuals, for instance, through policing or socially sanctioned retaliation, which further enhances cooperative endeavors.[87] I suspect that evolutionary biologist Mark Pagel is correct to suggest that "language evolved as a trait for promoting cooperation,"[88] but I maintain that the origins of language began with a highly specific form of cooperation—namely teaching. Other cooperative contexts could certainly have exploited a pre-existing linguistic capability, generating selection for enhanced linguistic skills. Such selective feedback would likely have made a big difference both to the scale of human cooperation that ensued, and to the potency of human language,[89] and plausibly helps to explain how early language extended into domains in which honesty could not be assumed, and vigilance against malevolence or incompetence was required. However, other cooperative contexts struggle to meet the *honesty* and *adaptive from outset* criteria described above, and hence cannot be the answer to how language got started.

I am far from the first person to suggest that language began with the creation and use of meaningful signs. Many other scholars have championed this argument, with perhaps the most prominent being Terrence Deacon, an anthropologist at UC Berkeley.[90] From a comparative perspective, the suggestion makes sense, because the use of symbols but not syntax is a feature of animal communication.[91] As described above, symbolism is present in the natural communication systems of many animals; a variety of apes have been taught to recognize and use symbols—whether gestures or lexical objects—with meanings that are broadly equivalent to words, and have even learned to string these together in simple combinations. However, there is little compelling evidence that any nonhuman possesses an understanding of syntax.[92]

As the number of socially transmitted food types, extractive foraging skills, processing methods, gestures, patterns of coordination, and labeled threats increased, learning their associated symbols would itself start to become a challenging task, and a major source of selection acting on our ancestors. Like others,[93] I suspect that our ancestors constructed a world sufficiently rich in symbolism to generate evolutionary feedback in the form of self-modified selection pressures that favored structures in the mind functioning to manipulate and use those symbols with efficiency.[94] Indeed, this feedback, variously cast as manifestations of the Baldwin effect,[95] or of niche construction,[96] is simply a more specific representation of the general "cultural drive" argument laid out in the last two chapters of this book. That is, selection for more efficient and higher fidelity forms of social learning has favored the evolution of specific structures and functional capabilities in the brain, and in the process has driven the evolution of brain and intelligence. The syntax that we witness in contemporary human language is only possible because of a long history, spanning perhaps two million years, of symbolic manipulation in protolanguage, which created selection pressures that, in turn, brought about significant changes in the hominin brain.[97]

As the sheer volume of symbols that our ancestors were required to learn the meaning of and string together in unambiguous messages increased, the demand for rules and conventions specifying usage patterns was created (and these are important aspects of syntax). If words are simply strung together without syntax, then ambiguities over their collective meaning rapidly arise and create a heavy processing burden on the receiver of the message. For instance, the utterance "bear man eat" is consistent with a bear eating a man, a man eating a bear, or the bear and the man both eating other food. Syntax alleviates this burden by breaking up the message hierarchically and recursively into meaningful and readily comprehensible chunks, phrases, and clauses that the brain can easily and quickly process. Syntax introduces rules that eradicate ambiguities. With this syntax came not just full-blown language, but an almost infinite flexibility in usage. Words have highly restricted meanings until they are strung together; but in combination, and underpinned by a mutually understood set of combinatorial rules, they are capable of communicating highly complex messages.

Language probably began as a means of reducing the costs of teaching complex foraging skills, but at some point it must have been coopted to teach linguistic symbols too. Once early language itself became something that was frequently taught (although often implicitly, without overt tutelage), it would in turn have generated selection for effective means of teaching language to children; this includes "infant-directed speech," also known as baby talk, or "motherese."[98] Children are known to hear some linguistic structures selectively and to ignore others, a phenomenon that may have generated selection for language structure that is "child friendly."[99] Infant-directed speech is typically slower and higher in pitch than regular speech, and uses shorter and simpler words. Studies have shown that infants prefer to listen to this type of speech, that it is more effective in getting and keeping their attention, and that it helps them learn words faster, compared to standard speech.[100] The suggestion is commonly made that language learning by children is spontaneous, or "instinctive,"[101] as if adults play little role in their learning. Such arguments underestimate the important ways in which adults facilitate language learning in children. Experimental studies show that the children who learn language fastest are those who receive the most acknowledgement and encouragement of what they say; who are given time and attention to speak; who are corrected, questioned, and spoken to in a child-friendly manner; and who are exposed to syntactically complex speech at the right time.[102] Infant-directed speech is found in most, but possibly not all, societies,[103] which suggests that it may be a widespread, socially learned tradition. That, however, does not preclude the possibility that important elements of infant-directed speech, such as children's sensitivity to its linguistic features, or adults' tendency to engage in behavior that elicits rewarding responses (e.g., smiles), have been favored through a biological evolutionary process.

As protolanguages began to increase in complexity from their rudimentary foundations, they would have generated strong selection for cognitive adaptations that facilitated language learning and information transmission. For instance, compared to other primates, humans appear particularly adept at inferring the meaning of what others are saying.[104] While this ability is partly attributable to the aforementioned pedagogical activities of the transmitter, an enhanced capability of the receiver to extract meaning through observation of others is also highly likely. What is more, the selective feedback from the use and manipulation of linguistic symbols to the human mind likely extends far beyond the acquisition of a capacity to extract meanings and comprehend syntax. Chomsky described language as the main engine of thought, and there is now little doubt that humans possess a mind uniquely fashioned to acquire and process information linguistically. Indeed, language not only facilitates our own thought, but provides us with a means to understand what others are thinking ("metacognition"). Such metacognition further upgrades the human capacity to teach. For instance, through the use of language a tutor can ask of a pupil, "Do you understand that?" The tutor thereby gains a critical insight into the pupil's state of mind. Indeed, I suspect that it was in the context of teaching, probably through language, that theory of mind (the ability to attribute beliefs, desires and knowledge to others) initially evolved. In all these ways, language is a powerful example of niche construction.[105] It is a dramatic change that our ancestors brought about in their (conceptual) world, and it is a transformation that helped make us human.

As the conventions of linguistic structure varied from one (proto) language to another, and changed over time, the rules of syntax would themselves have to be learned. Whether they are learned through a dedicated language acquisition device as envisaged by Chomsky,[106] or through some more general process mechanisms such as Bayesian learning, is a moot point within the field.[107] Either way, I believe that the symbol-rich cultural world constructed by our ancestors was a major source of selection for enhancements in language learning. That selective feedback would have operated at two levels. These include a gene-culture coevolutionary dynamic, where human cultural activities generate natural selection favoring enhanced language learning and transmission capabilities. But a cultural evolution dynamic also functions, whereby human cultural activities feed back to affect the learned properties of the language.

The notion that features of languages could be fashioned through cultural transmission is perhaps less intuitive than the idea that our ancestors' minds were molded by natural selection. Yet, if linguistic structures are to persist over time they must repeatedly survive the process of being learned, expressed, and adopted by others. Utterances that are difficult to learn or speak would be at a disadvantage relative to those that are more intuitive; as a result, languages will adapt over time. The Edinburgh linguist Simon Kirby and his colleagues labeled this process *cultural selection for learnability*.[108] Many researchers have been impressed by the ease with which children acquire languages and have assumed that this precocity implies that the human brain must possess a specialized language-learning capability.[109] However, children may appear preadapted to decipher the rules of syntax in part because languages have evolved rules that are easy to learn.[110] The cultural evolution of language has been studied through mathematical modeling, and researchers have established that key properties of language could evolve in this manner.[111] For instance, compositionality—the idea that the meaning of a complex signal is a function of the meaning of its parts—can arise through a process of cultural evolution, and typically more rapidly and at lower cost than through biological evolution.[112] Likewise, both transmission-chain experiments and mathematical models show how languages that are propagated culturally evolve in such a way as to maximize their own transmissibility, becoming easier to learn and more structured over time.[113] This research is important, because it shifts some of the explanatory burden away from natural selection for language-specific cognitive adaptations, and makes the challenge of explaining the origins of language more manageable. Once our ancestors evolved a socially transmitted system of symbolic communication, many other features of language came for free.[114]

By and large, my argument in this chapter is a functional one. We have been addressing the question: *What was the original function of language?* However, there is another evolutionary question that we can ask too, a question about the evolution of mechanism: *How was it computationally possible for hominins to learn language?* The challenge of language learning becomes substantially greater with the advent of syntax, because it required not only learning the meaning of symbols, but organizing them into appropriate strings in which order was critical, and comprehending higher-order patterns of meaning in that sequence. How can humans manage this?

Here, the coevolutionary argument presented in the previous paragraph is, I think, only part of the story. I believe humans were predisposed to be highly competent manipulators of strings of elements, because many of their ancestors' tool-manufacturing and tool-using skills, extractive foraging methods, and food-processing techniques had required them to carry out precise sequences of actions.[115] These types of skills only succeed if the action units are carried out in the right order. Many foods can be rendered more nutritious and less toxic if appropriately processed.[116] For instance, many wild and domesticated plants in the human diet, particularly the pulses and cereals such as rice and wheat, cannot be properly digested when eaten raw. Legumes, likewise, are highly nutritious foods but contain mildly toxic substances that are difficult for humans to digest unless they are first detoxified by soaking and cooking. The carbohydrates in root and seed foods, such as starch, require chopping, slicing, mashing, and heating to change into forms that humans can more easily chew and digest. Often these food gathering and processing actions require long sequences of actions, commonly comprising hierarchical, and sometimes recursive, elements, where subsets of the sequence must be repeated. This is not just true for contemporary humans, but even applies to some foraging methods used by apes, such as nettle consumption by gorillas,[117] and so is unlikely to be a very recently acquired computational skill.

A theoretical study carried out by Andrew Whalen and Daniel Cownden,[118] graduate student and postdoc, respectively, in my laboratory at St Andrews, has investigated the learning of such sequences. Andrew and Daniel found that long sequences of actions are unlikely to be learned through trial and error, but social learning hugely increases the chances that individuals will acquire the appropriate sequence. Sequences in which the reward comes right at the end of a long string of behavioral elements, and where the final rewarding action must be preceded by a long series of nonrewarding or aversive actions, are particularly challenging to learn without help from others. The same holds where there are a large number of options at earlier stages. For acquisition of such tasks, social learning is vital. Such sequences may be learned through imitation; indeed, one form known as "production imitation" refers precisely to the learning of sequences of actions through observation.[119] Many tasks such as basket weaving or pot making, may be more readily learned through "sequence emulation," where rather than attending to the motor pattern of the demonstrator, the observer attends to the sequence of object movements effected by the demonstrator's actions.[120] Irrespective of the mechanism, an evolutionary history of acquiring long sequences of actions through social learning, as well as the computational capabilities to process hierarchical and recursive subroutines, may well have left humans cognitively predisposed to learn and process long strings of symbols. Indeed, many linguists believe that the key distinguishing feature of language is hierarchical syntactic structure.[121] I suspect that our ability to process strings of elements into hierarchically organized sets arises through hundreds of thousands of years of selection for the computational skills necessary for complex food processing. Language could probably only have evolved in an exceptionally proficient social learner, an observation that again speaks to the uniqueness criterion outlined above.

I end this chapter by describing an experiment with findings that both support the argument that toolmaking, teaching, and language coevolved, and illustrate the kind of selective scenario in which early language was likely to have been favored.[122] What is particularly relevant here it is that the experiment provides a demonstration that a specific learned skill, one known to have been widely practiced by our ancestors, would certainly have generated a selection pressure for enhanced teaching and language. That skill is stone toolmaking.

For over two million years, hominins were highly skilled at shaping stones such as flint, chert, or obsidian into cutting and hammering tools, a process known as "knapping." Competent knappers were capable of producing large numbers of sharp flakes and a variety of other tools from a single cobblestone core by striking it with a hammerstone. Analyses of recovered flake tools have shown this was far from the crude bashing of stones together to generate a sharp edge, but comprised systematic flake detachment, during which the manufacturer would maintain precise flaking angles and would even repair the core stone if damaged.[123] This complexity, along with present day tool-making experiments,[124] implies that even the oldest lithic (i.e., stone tool) technology, known as Oldowan, was learned and required considerable practice.[125] Furthermore, the technology's continual existence and wide geographic spread, along with hints of regional traditions,[126] strongly imply that tool-making methods were socially transmitted; however, the underlying psychological mechanisms remain poorly understood.[127]

If I am correct, and feedback mechanisms have played important roles in shaping the evolution of the human mind, language, and intelligence, stone tool manufacture makes for an interesting test case. The appearance of this technology 2.5 million years ago at the dawn of *Homo*, and its repeated use for millions of years, means that stone tool manufacture and use is a particularly strong candidate for an activity that might have imposed selection on human cognitive evolution. The ecological niche that early hominins occupied is widely thought to have been extremely challenging,[128] and tool-making skills are known to be difficult to acquire.[129] Archaeologists have presented evidence that fitness benefits were likely associated with the ability to construct and use effective cutting tools, and to transmit those skills rapidly.[130] Hence, a coevolutionary relationship between toolmaking and cognition, including teaching and early forms of language, would seem plausible.

Oldowan stone tool production could have been one of those difficult-to-acquire skills, described above, that generated selection for more complex and accurate forms of knowledge transfer, including language. In turn, increments in the fidelity of information transmission might have aided the acquisition and spread of more complex tool-making technologies, such as the Acheulean, which might have generated selection for further increases in the complexity of social transmission, and so on. Consistent with this hypothesis, paleontological and archaeological remains show that changes to hominin morphology, including increased overall brain size, follow the advent of Oldowan toolmaking.[131]

Experiments with contemporary humans are one way to explore this hypothesis further, as these have provided important insights into the cognitive and motor processes supporting lithic technology.[132] However, until recently there had been virtually no research on the social learning of making stone tools.[133] Such experimentation would be instructive, because the transmission mechanisms could plausibly constrain the technology that can be propagated, yet researchers disagree as to the mechanisms that toolmaking requires.[134] Positions range from the hypothesis that chimpanzee-like emulation or imitation was sufficient to transmit knapping technology,[135] to the view that toolmaking required a major development in hominin cognition,[136] such as language.[137]

We decided to carry out a large-scale experimental study testing the capability of five social learning mechanisms to transmit Oldowan knapping techniques across multiple transmission steps.[138] By establishing the rates of transmission resulting from different mechanisms, we set out to determine which forms of communication might plausibly have been selected for as a result of reliance on stone tool use. The project was carried out by a team of over 10 members of my research group, led by graduate student Tom Morgan, and under the guidance of archaeologist Natalie Uomini at the University of Liverpool.

In our experiment, adult human participants first learned to knap stone flakes using a granite hammerstone and flint core, and then were tested on their ability; next, they helped others learn this skill. The process was repeated to explore whether and how tool-making knowledge was transmitted along chains of participants. Experimental subjects were allocated to one of five conditions that varied according to the type of information that could be passed from "tutor" to "pupil."[139]

1. *Reverse Engineering*: Subjects were provided with a core and hammerstone for practice, but only saw the flakes previously manufactured by a "tutor", without meeting the tutor or seeing the flakes being made.

2. *Imitation/Emulation*: In addition to having their own core and hammerstone, pupils also observed a tutor making flakes, but could not interact with them.

3. *Basic Teaching*: In addition to demonstrating tool production, tutors could also manually shape the pupil's grasp of their hammerstone or core, slow their own actions, and reorient themselves to allow the pupil a clear view.

4. *Gestural Teaching*: Tutors and pupils could also interact using any gestures, but no vocalizations.

5. *Verbal Teaching*: Tutors and pupils were also permitted to speak.

For each condition, we carried out 4 short transmission chains with 5 participants learning in sequence and 2 longer chains with 10 participants. Experimenters trained in stone knapping acted as tutor to the first participant. Nearly 200 adult human participants took part in the experiment, producing over 6,000 pieces of flint, each of which we weighed, measured, and assessed for quality using a novel method that we developed and verified.[140] It was a mammoth research effort (perhaps the biggest project ever carried out in the field of experimental archaeology). However, the effort paid off in generating remarkably helpful findings.

Across numerous measures, Tom and Natalie consistently found that teaching and language, but not imitation or emulation, enhanced the acquisition of stone knapping skills relative to the reverse engineering condition. For instance, the quality of the flakes generated showed clear improvement with gestural or verbal teaching, with language nearly doubling performance relative to reverse engineering,[141] but imitation/emulation generated no improvement in performance. The number of viable flakes produced showed a similar pattern, with substantial improvements in performance only occurring with gestural or verbal teaching.[142] Nor was there evidence for an increase in the rate of manufacture of viable flakes with imitation/emulation; only verbal teaching generated such an increase, and only verbal teaching led to a clear increase in the volume of core reduced.[143] Finally, although there was no evidence that imitation/emulation increased the probability that a viable flake would be produced per hit, gestural teaching doubled and verbal teaching quadrupled this probability. Across the six measures taken, there is strong evidence that verbal teaching increases performance relative to gestural teaching.

Teaching, but particularly teaching through the use of language, therefore greatly facilitated the rapid transmission of flaking, whereas there is little evidence that imitation or emulation did so. As expected, in all conditions performance decreased along chains as knapping information was gradually lost in transmission. However, with teaching, transmission was sufficiently improved that performance declined steadily along chains,[144] whereas without teaching, the drop in performance was so severe that performance immediately fell to baseline levels. Set in a more naturalistic context, with longer periods of interaction between tutor and pupil, and interspersed with periods of individual practice, teaching would certainly generate stable transmission.[145]

Tom and Natalie's study demonstrated that transmission of Oldowan technology is enhanced by teaching and, in particular, by language.[146] This fits with the suggestion that Oldowan stone tool manufacture would have generated natural selection favoring increasingly complex teaching and language. Critically, we found a continuous improvement in the rate of transmission of toolmaking with increasingly complex forms of communication.[147] Such data imply that stone toolmaking would have generated selection for improved communication, starting with observational learning and continuing to the emergence of full-blown language. This process was probably already underway during the Oldowan and likely continued well after, with the transmission of more complex technologies also benefitting from more sophisticated means of communication. Indeed, the evolutionary feedback could have lasted for millions of years, with more complex communication allowing the stable and rapid transmission of increasingly multifaceted technologies; this in turn would generate selection for even more complicated communication and cognition, to then continue in perpetual cycles.

A second significant finding was that the rate of transmission of Oldowan toolmaking is, at best, very minimally enhanced by imitation or emulation.[148] We cannot, of course, rule out the possibility that a benefit of observational learning might have appeared had we conducted our study over a longer learning period. However, any such benefit would clearly have been less than what would be derived through teaching across a similar timespan. While it is hard to say with absolute confidence that lithic technology could not be transmitted through imitation, our findings support other work in implying that observation alone is a very inefficient means of acquiring stone tool-making skills.[149] Clearly, some information can be acquired through imitation—the requirement to strike the core with the hammerstone and some notion of the force required, for instance. However, it seems that the requisite rapid striking action hinders the transmission of subtler information crucial to knapping, such as details of precisely where to hit the core, at what angle, and how best to first orientate the core. Here, teaching by slowing down the striking action, pointing to appropriate targets to hit, demonstrating core rotation, or manually shaping the pupil's grasp—particularly if aided by verbal instruction—provides immediate benefits to the pupil. Written transcripts from our verbal teaching condition show that abstract knapping concepts,[150] such as the platform angle,[151] were indeed effectively transmitted between individuals in the verbal teaching condition. Plausibly, the use of arbitrary labels such as "platform angle" in verbal instruction facilitates transmission. Such labels break the task into constituent parts, can be used to identify the important elements, and provide a clear framework with which pupils can go on to teach others. In other words, language not only allows acquisition of the skill itself, but also enhances the learner's ability to transmit the skill to others.

What was particularly exciting was that Tom and Natalie's findings shed light on one of the most enduring puzzles of human evolution—the apparent stasis of Oldowan technology.[152] For a very long period of some 700,000 years, these stone tools did not change very much until Acheulean technology appeared, around 1.7 million years ago.[153] The study implied that Oldowan technological change could well have been restricted by the fidelity of social transmission—a suggestion supported by the slow spread of Oldowan technology across Africa.[154] We obviously cannot conclusively identify what form Oldowan transmission might have taken, but our data point to imitation or emulation as strong candidates. While imitation and emulation generate higher-fidelity transmission than simpler forms of social learning,[155] the relatively poor transmission of stone-tool manufacture that we observed with these mechanisms might well, in naturalistic contexts, have been too slow and imprecise for innovations to be conveyed reliably. If so, the technology may plausibly have been unable to become much more complex until more effective communication evolved.[156]

Our findings implied that the spread of the more regularly shaped Acheulean tools probably would have required a capacity for teaching at least equivalent to the "basic teaching" condition in our experiment. Rudimentary forms of language are possible, but the suggestion that modern language evolved during the Oldowan seems unlikely, given how slowly technology evolved thereafter. More plausibly, the transmission of Acheulean technology was reliant on a form of gestural or verbal protolanguage.[157] Acheulean hominins were probably not capable of generating complex grammars, but they may well have learned and strung together a small number of symbols. Our findings imply that simple forms of positive or negative reinforcement, or directing the attention of a learner to specific points on the core, as was common in the gestural teaching condition, are considerably more successful in transmitting stone knapping than observation alone. (This finding neatly illustrates how a teaching explanation for early language could meet Bickerton's "ten word rule.")

The study supports the hypothesis that a gene-culture coevolutionary dynamic between tool use and social transmission was ongoing in human evolution, starting at least 2.5 million years ago and continuing to the present. The simplicity and stasis of Oldowan technology are indicative of a restricted form of information transmission, such as observational learning, that only allowed the communication of the broadest concepts of stone-knapping technology. This mechanism was sufficient to support limited knowledge transfer among individuals with prolonged contact, but insufficient to propagate innovations more rapidly than they were lost, and would have contributed to the stasis in the Oldowan technocomplex. However, hominin reliance on stone technology would have generated selection for increasingly complex communication that allowed more effective spread of tool-making methods. Under this continued selection, teaching, symbolic communication, and eventually language would have been favored, allowing the transmission of abstract flaking concepts. Our results suggest that hominins possessed a capacity for teaching, and probably protolanguage, as early as 1.7 million years ago.

I reiterate that selection for language was not brought about by tool-making per se, but solely by those aspects of tool manufacture and use that required teaching. More precisely, I submit that language originally evolved to teach close kin, and that toolmaking was just one of several life skills that our ancestors taught their relatives. However, while manufacturing stone tools would not have been the only technology taught by our ancestors, it almost certainly was an important one, because the livelihood of hominins was critically dependent on this technology for millions of years. Minimally, stone toolmaking is a good example of the kind of technology that would have required teaching, and would have generated selection for forms of teaching with reduced costs and enhanced accuracy. The study also demonstrates beautifully how early language could have begun to generalize from nouns, actions, and prepositions to abstract concepts.

Derek Bickerton has made a case that language evolved to help our forebears scavenge for large, dead prey items as a group, fighting off competition from predators.[158] I think this too is plausible since, as Bickerton describes it, this coordination would have required teaching to be carried out effectively. This would encompass teaching individual scavenging roles, instructing about the whereabouts of food, telling others what prey item has been found, coaching collective defense, coordinating large teams of individuals, and so forth. The hypothesis is attractive, not least, because it provides a further illustration of how language might have been generalized to refer to events distant in space. However, it was not the scavenging but the teaching that was critical to the evolution of language. Likewise, Michael Tomasello has argued that human language originally evolved to coordinate collaborative foraging activities,[159] and again I have sympathy with this position, because it is hard to imagine collaborative, coordinated foraging among hominins without teaching. Once again, however, I emphasize that it was not the foraging but the teaching that was critical, and such teaching would likely have been expressed in domains that extended beyond foraging. Almost certainly, a broad array of different traits, skills, and knowledge were taught by our ancestors, each exerting different demands on their linguistic capabilities.

Language experts will, of course, point out that, even if my hypothesis is correct, a great deal about language evolution would remain a mystery. I have only hinted about how the generative computations underlying language evolved, how systems of semantic representations developed, how phonological representations emerged, how the interfaces between these coevolved, and how all of this internal machinery was externalized in linguistic communication, whether expressed acoustically or visually. Nor have I said anything about the evolution of vocal learning. That is all true, but I nonetheless believe that the account is of value in removing some of the mystery of the origin of language. The analysis places the origins of language in a broader context through which we can understand the evolution of diverse aspects of human cognition. Alfred Russel Wallace, codiscoverer of natural selection, famously failed to accept that selection could account for human evolution, partly because he could not imagine how a trait like language, and the other unique features of human cognition, could evolve.[160] I would like to believe that, had he known of the material in this chapter, he might have reached a different conclusion.

CHAPTER 9

GENE-CULTURE COEVOLUTION

In the classic sci-fi fantasy *Planet of the Apes*, space traveler Ulysse Mérou becomes trapped on a terrifying planet in which gorillas, orangutans, and chimpanzees have usurped control, having acquired human language, culture, and technology by imitating their former masters. Cast out of their homes, human beings rapidly degenerated into brutal and unsophisticated beasts. Much of the sinister realism of Pierre Boulle's 1963 novel stems from the author's impressive knowledge of scientific research into animal behavior. Of course, in reality, here on planet Earth other apes could never actually acquire human culture solely through imitation, because that capability requires an underlying evolved propensity. In our case, that propensity was fashioned by millions of years of gene-culture coevolution.

In the last chapter, I described how the manufacture and use of stone tools may have played a vital role in human evolution by generating coevolutionary feedback between cultural practices and genetic inheritance, and thereby contributed to the emergence of language. Our tool knapping study supported the hypothesis that a gene-culture coevolutionary dynamic between tool use and social transmission was ongoing in human evolution, starting at least 2.5 million years ago and continuing to the present.[1] Indeed, this entire book is one long advocacy for the significance of evolutionary feedback that encompasses a cultural drive mechanism initiated by natural selection that favored accurate and efficient copying. That selective feedback propelled the evolution of cognition in some primate lineages, and ultimately was responsible for the awesome computational power of the human brain.

Any such account, however, rests on the plausibility of gene-culture coevolution. If the arguments are to prove compelling, at this juncture some weighing up of the evidence for such evolutionary interactions would be beneficial. Perhaps there are means to detect gene-culture coevolution, or historical traces left in the human genome or brain that are testament to its legacy. This chapter examines the evidence that our cultural activities have influenced our biological evolution, again drawing on a cocktail of theoretical and empirical findings. It begins by relating findings from theoretical studies, which show through mathematical modeling that gene–culture coevolution is, at least in principle, highly plausible. Then the anthropological evidence for gene–culture coevolution is surveyed. Here, compelling and well-researched case studies provide incontrovertible evidence that gene-culture coevolution is a biological fact. Finally, some genetic data are presented— specifically, studies that have identified human genes subject to recent natural selection, including genes expressed in the brain. Many such genes (strictly, "alleles," or gene variants) have increased extremely rapidly in frequency over a few thousand years, and this unusually swift spread, known as a "selective sweep," is taken as a sign of their having being favored by natural selection.[2] The relevance of such studies stems from the fact that the geneticists who carried them out have concluded that the sweeps are almost certainly a response to human cultural activities. Collectively, these three bodies of evidence make a compelling case that culture is not just a product, but also a codirector, of human evolution.

The argument that genes and culture can coevolve was first championed over 30 years ago by pioneers of the field of "gene–culture coevolution," a branch of mathematical evolutionary genetics.[3] These researchers treated genes and culture as two interacting forms of inheritance, with offspring acquiring both a genetic and a cultural legacy from their ancestors. The two streams of knowledge flow down the generations, but far from independently. Genetic propensities, expressed throughout development, influence the cultural traits that are learned, while cultural knowledge, expressed in behavior and artifacts, spreads through populations and modifies how natural selection affects human populations in repeated, richly interwoven interactions.

Gene–culture coevolutionary models build on conventional evolutionary models. The latter track how the frequencies of genetic variants change in response to evolutionary processes such as natural selection or random genetic drift, but the former differ by incorporating cultural transmission into the analyses. This allows exploration of how learned behavior or knowledge coevolves with alleles that affect the expression or acquisition of the behavior, or whose fitness is affected by the cultural environment. The approach has been used to explore the adaptive advantages of reliance on learning and culture,[4] to investigate the inheritance of behavioral and personality traits,[5] and to examine specific topics in human evolution, such as the evolution of language, or cooperation.[6] When my plans to work with Allan Wilson at Berkeley in 1991 were tragically curtailed, I started to collaborate with Marc Feldman, a geneticist at Stanford University and leading authority on gene-culture coevolution. I was delighted to get the opportunity to learn gene-culture coevolutionary methods from arguably the world's greatest expert.

One of the first projects that I worked on with Feldman, his student Jochen Kumm, and psychologist Jack van Horn (now at the University of Southern California) was a model of the evolution of human handedness. I'll describe the study in detail, partly because it provides a useful illustration of a gene-culture coevolutionary analysis, but also because it connects with the topic of the preceding chapter.

The conundrum that prompted our study was the question of why everyone isn't right handed. Experimental studies had established that globally, approximately 90% of people are right handed.[7] This estimate is loosely consistent across the world, although it does vary to some degree among societies.[8] Yet, in no society in the world are left-handed people in the majority, and this has led researchers to conclude that right handedness must have been favored by natural selection throughout recent human evolution. However, if it is advantageous to be right handed, why isn't everybody? What processes might be preserving left-handers in human populations? The most common answer to this question is that genetic variation underlies variation in handedness. This explanation suggests that left- and right-handed people have different genotypes, and that this genetic variation had been preserved in human populations through natural selection.[9] However, if genes cause people to be right- or left handed, with different combinations generating different patterns of hand usage, then the more closely related a pair of individuals (i.e., the more alike their genes), the more similar should be their handedness, and this is not the case. To quote two leading authorities: "knowledge of a person's handedness tells us virtually nothing of the handedness of that person's twin or sibling."[10] Genetic models would predict that identical twins ("monozygotic" twins), who have the same genes, would be more alike for handedness than fraternal ("dizygotic") twins, who share only 50% of their genes; yet the two types of twins have essentially the same concordance rates. If the handedness of a thousand pairs of identical twins were measured, 772 pairs on average would possess the same pattern of handedness, while the corresponding figure for fraternal twins would be 771.[11] Handedness does not exhibit strong heritability, which seriously undermines a purely genetic explanation.[12]

Moreover, exclusively genetic models of handedness fail to account for some well-established cultural influences on handedness. Left-handed people have experienced a long history of discrimination in many parts of the world,[13] including parts if the Middle East and Far East, and some eastern European countries; they are found at lower frequency in such societies that have associated left handedness with clumsiness, evil, dirtiness, or mental illness.[14] As might be expected, societal attitudes have an impact on patterns of handedness. Studies of schoolchildren in China, where left handedness has historically been frowned upon, report only 3.5% used their left hand for writing (the figure for Taiwan is only 0.7%), compared with a 6.5% estimate for children originating from the same localities but living in the United States,[15] where the pressure to use the right hand is relaxed. Because the worldwide dominance of right-handers suggests a role for genes, but cross-cultural variation reveals a cultural influence, handedness appears to be well suited to a gene–culture coevolutionary analysis. My colleagues and I set out to explore the evolution of handedness to try and make sense of the observed patterns.

We constructed a model that assumed that the probability of becoming right or left handed was influenced by the combination of alleles found at a single genetic locus.[16] The two possible gene variants at this locus were a "dextralizing" (or "right-shift") allele that predisposes its carrier toward right handedness, and a neutral allele that put handedness down to chance.[17] This is not to suggest that we believed only a single gene influenced handedness, but rather we focused on a single hypothetical gene as a means of exploring how any genetic variation would respond to selection. (In fact, our model implies a series of selective sweeps of dextralizing genes throughout human evolution, each ratcheting up the proportion of right-handers.) Cultural factors were also assumed to affect handedness, primarily operating through a parental influence. This assumption seems reasonable given that handedness is usually fully developed by the age of two to three.[18] Hence, in our model handedness depended on the individual's genotype (i.e., whether the individual possesses zero, one, or two dextralizing alleles) and the handedness of their parents. We also considered various forms of selection operating on the population, including selection favoring right handedness directly, or indirectly via selection on another lateralized structure (e.g., like left cerebral dominance for language).

The findings of the analysis were extremely straightforward; irrespective of the starting frequency of right handedness, the magnitude of the selective advantage to right-handers, or the degree of dominance of the two alleles, all initially genetically variable populations converged on a single evolutionary trajectory and continued to evolve until the dextralizing allele was completely fixed, with the chance allele eliminated. No genetic variation in handedness had been preserved, which ostensibly undermined the explanation that genetic differences explained variation in handedness.[19] Nor was it likely that human populations were still evolving toward this equilibrium, because that would require right handedness to be increasing over time when the data suggested the opposite.[20] How then could the existence of left-handers be explained? A further possibility was suggested by our analysis: human populations may have reached the final equilibrium state predicted by the model, but left-handers could nonetheless remain in the population if the influence of the dextralizing allele was sufficiently weak.

We decided to explore this possibility by collating data on patterns of handedness in families and using this data to estimate the values of the model parameters, including the impact of the dextralizing allele. We pulled together 17 studies that gave the frequencies of right- and left-handed offspring born to parents categorized into 3 groups; both parents were right handed, one was right handed and other left handed, or both were left handed. As the number of left-handed parents increased, these 3 groups gave rise to increasing proportions of left-handed children. We then carried out an analysis in which we used the data set on patterns of handedness in families to estimate the best-fit values of the parameters in our model.[21] With these values, the model gives a good fit to 16 out of the 17,[22] and across all studies combined. Similar analyses applied to the same data by the leading genetic models give a substantially poorer fit. Our model had given a good fit to more studies, and a poor fit to fewer studies, than any other model.

The analysis suggested that all humans are born with a predisposition to be right handed. Other factors being equal, 78% of people would be right handed (or a single child would have a 78% chance of growing up to be right handed). All other factors are not equal, however, as parents exert an important influence on patterns of handedness. Two right-handed parents increase the probability that their child will be right handed, by an amount that averages around 14%, to give an overall probability of a right-handed child being born to right-handed parents of 92%. Likewise, two left-handed parents decrease the probability by the same proportion, leaving a 64% probability of a right-handed child being born to two left-handed parents. Parents of mixed handedness roughly cancel out each other's influence.

We went on to conduct various tests of our model. One such test was to derive an overall expected frequency of right-handers, which reassuringly came out at 88%, close to the observed value. A second, stronger test was to collate studies giving the frequency of right-right, right-left, and left-left pairs of identical and fraternal twins, and compare the expected proportions in each category from our model with the observed data. Using the same parameter values derived from the familial data set, our model generated expectations that were a good fit to 27 out of the 28 twin data sets, and across all studies combined. Once again, our model outperformed all other published models.

The study strongly suggested that patterns of inheritance and variation in human handedness are the outcome of a gene-culture interaction. A history of selection of handedness has created a universal genetic predisposition toward right handedness; our genes load the die to favor the right, but not so strongly that right handedness is obligate. However, patterns of hand use also show a parental influence—specifically, an advertant or inadvertant parental tendency to shape their child's handedness to resemble their own. This parental influence probably represents a combination of direct instruction (i.e., parents telling their children to use their right hand), children imitating their parents, and parents unintentionally shaping their child's hand use (for instance, repeatedly placing spoons or crayons in a particular hand), although inherited epigenetic effects are a further possibility.[23]

An interesting connection arises between this study and the experimental analysis of stone tool manufacture described in the preceding chapter. When our ancestors made flake tools, they tended to rotate the core stone as they broke off flakes, and the direction of rotation turns out to be a reliable indicator of the handedness of the knapper.[24] Archaeologists sometimes find multiple flakes from the same core at a single site, and can thereby reconstruct the knapping process. This allows them to use patterns of flint knapping, supported by skeletal data, to infer patterns of handedness in ancient hominin populations. The data provide evidence for increasingly strong biases in handedness with time, with 57% of the earliest stone tool makers (living 2.5–0.8 million years ago), 61% of Middle Pleistocene hominins (0.8–0.1 million years ago), and 80–90% of Neanderthals (0.3–0.04 million years ago), being right handed.[25] Whether other primates, such as chimpanzees, exhibit handedness is a contentious topic that has generated conflicting findings, but any population-level bias in handedness among apes must be quite weak.[26] Individual animals might have strong hand preferences, but across the entire population there is, at best, only a modest tendency to prefer the same hand. Hence, archaeological and comparative data support the suggestion that right handedness has increased over time, almost certainly through the selection of (dextralizing) handedness distorter genes, which have been repeatedly favored over hundreds of thousands and perhaps even millions of years.

Knapping is a skill that takes time to master and requires strength and precision; these are properties that tend to generate hand specialization. One would imagine that the social transmission of knapping skills would encourage pupils to use the same pattern of handedness as their tutor—for instance, holding the core and hammerstones in the equivalent hands, thereby generating a bias in hand usage. Perhaps right-handers originally reached ascendency through chance, or perhaps there was some pre-existing bias toward the right. Either way, with each selective sweep favoring a dextralizing allele, the proportion of right-handers would be ratcheted up. This increase in right handedness would occur not just because of the immediate effect of the alleles, but also because with the increasing frequency of right-handed parents, the proportion of children exposed to a parental bias favoring right handedness would increase. Both indirectly (by introducing new behaviors that benefited from hand specialization) and later directly (by constructing an environment suited to the right-handed majority), hominin cultural processes increasingly reinforced selection favoring right handedness, right up to the present.

Studies like this cannot prove that gene-culture coevolution occurred, but they do demonstrate that the feedback mechanisms discussed in the preceding chapter are logically sound and highly plausible. If the cultural practices of our ancestors could generate selection for dextralizing alleles, then they could also plausibly generate selection for genetic variation expressed in larger brains, or advanced cognition. Such a conclusion would hold even if my particular theory of handedness were eventually disproven. Many such gene-culture coevolutionary analyses have now been conducted, collectively generating a number of important insights into human evolution. The studies conclusively demonstrate that genes and culture could coevolve, and that in such interactions culture would become a potent codirector of evolutionary events. Cultural processes are every bit as influential as genetic processes in gene–culture coevolution, and many circumstances arise where cultural transmission overwhelms, or reverses, natural selection, or where the observed patterns of selection depend intimately on the details of cultural transmission.[27]

A consistent finding of these models is that cultural processes can dramatically affect the rate of change of gene frequencies in response to natural selection, sometimes speeding up genetic evolution, and other times slowing it down. Recent estimates of the evolutionary response of human genes exposed to culturally modified conditions reveal extraordinarily strong natural selection. One of the best-studied cases is the coevolution of dairy farming (and associated consumption of dairy products) with alleles that allow humans to digest lactose, the sugar in milk. In most humans, the ability to metabolize lactose disappears in childhood, but in some populations lactase activity persists into adulthood. This "lactose tolerance" is known to result from a mutation at a single genetic locus. Both comparative analyses and ancient DNA extracted from 7,000-year-old human remains reveal that dairy farming appeared before the spread of lactose-tolerance alleles, generating conditions that made the production of lactase by adults advantageous.[28] Only subsequently did the genetic variant that promoted adult lactose tolerance spread. This evolutionary response is now known to have happened independently in at least six separate dairying populations, with different mutations being favored in each case.[29] Lactose-tolerance alleles have spread from low to high frequencies in less than 9,000 years since the inception of dairy farming and milk consumption, generating an estimated selection coefficient of 0.09–0.19 in one Scandinavian population[30]—one of the strongest responses to natural selection ever detected. Such observations, combined with the recency with which many human genes have been subject to selection,[31] imply that natural selection deriving from human cultural activities may be unusually strong. The reasons for this are twofold. First, unlike the vagaries of the weather, climate, or several other nonanthropogenic processes that generate inconsistent selection from one year to the next, human activities are purposeful and goal directed. Through the dissemination of cultural knowledge, large numbers of individuals in a population engage with their environment in a consistent, directed manner—generation after generation manufacturing the same tools, eating the same foods, and planting the same crops. The result is consistency in the patterns of natural selection that such activities generate. Although their constancy varies from trait to trait, evidence suggests that culturally modified environments are capable of creating unusually strong natural selection, because these conditions are highly consistent over time.[32] Second, many gene variants are favored as a result of coevolutionary episodes, such as predator-prey or host-parasite interactions, triggered by evolutionary changes in another species. When changes in one genetic trait are the source of selection for changes in a second, the rate of response in the latter depends in part on the rate of change in the former, which, as a rule, is not fast. In comparison, if a cultural practice modifies selection acting on human genetic variation, then the greater the proportion of individuals in the population that exhibit the cultural trait, the stronger the selection on the gene. As a consequence, the rapid spread of a cultural practice often leads quickly to the maximally strong selection of the advantageous genetic variant, which rapidly increases in frequency. Cultural practices typically spread more quickly than genetic mutations, simply because cultural learning typically operates at faster rates than biological evolu-tion.[33] What does the speed with which a cultural trait spreads depend upon? Answer: the fidelity of cultural transmission. The very factor that is critical to the emergence of complex cumulative culture in humans is also a major determinant of evolutionary responses to that culture.

For illustration, consider a mathematical analysis of the coevolution of dairy farming and adult lactose tolerance conducted by Stanford ge-neticists Marc Feldman and Lucca Cavalli-Sforza.[34] In their model, adult lactose tolerance was controlled by a single gene, at which two types of alleles arose, with one promoting adult lactose tolerance and the other intolerance. The model showed that whether or not the lactose toler-ance allele achieved a high frequency in the population depended cru-cially on the probability that the children of milk drinkers themselves became milk drinkers; that is, it depended on the fidelity of cultural transmission. If this probability was high, then lactose-tolerant individ-uals would possess a significant fitness advantage, which resulted in the spread of the lactose tolerance allele within a few hundred generations. However, if a significant proportion of the offspring of milk drinkers did not imbibe dairy products, then unrealistically strong natural selection favoring lactose tolerance would be required for the allele to spread.

Gene–culture coevolutionary models commonly report more rapid responses to natural selection than conventional genetic models, aris-ing from the combined effects of the consistency and rapid spread of cultural practices.[35] This is one reason why many geneticists are now arguing that culture has "ramped up" human evolution.[36] High-fidelity information transmission both supports cumulative culture and pro-motes gene-culture coevolution, but the selective feedback does not end there. The construction of a dairying niche allowed humans to disperse into areas that otherwise would have been uninhabitable, and provided the raw materials for the emergence of diverse technologies needed to process dairy products. This generated an opportunistic expansion of the dairy farming niche into new regions throughout Europe and Africa. That dispersal, like earlier hominin dispersals from Africa, was critically contingent on culture and was the catalyst for further gene-culture coevolution.

Nearly two million years ago, and perhaps earlier, ancient hominin species such as *Homo erectus* moved away from the tropics and began to populate different regions of the world. The subsistence of these early humans in many regions of Europe or Asia was contingent on their ability to manufacture stone tools suited to the butchery of large animals, control fire, make clothes and shelter, coordinate group hunting, and use other technologies. Through these dispersals, our ancestors modified the natural selection acting on themselves in numerous ways. For example, the new environments had different climatic regimes, which selected for genes expressed in skin pigmentation, heat tolerance, and salt retention, all of which show signs of being subject to recent selection.[37] Humans need salt to transport nutrients, transmit nerve impulses, and contract muscles within their bodies. In the African regions where humans first appeared, temperatures were high and available salt must have been limited and quickly lost through sweat. People better able to cope with heat stress, or with superior salt retention, would have had a significant survival advantage. However, this advantage would decrease as humans spread to cooler climates, where salt retention could lead to disease, and hence the selection of alleles that reduced heat tolerance and salt retention.

Another example is provided by the movements of early Polynesians. During their settlement of the Pacific, beginning around 1800 BCE and continuing for two millennia, these people experienced long ocean voyages that subjected them to severe cold stress and starvation. Apparently genes expressed in enhanced energetic efficiency and fat storage among the pioneers, which helped individuals to survive famine, were favored by natural selection. Unfortunately, in contemporary environments in which food is plentiful, these same genes predispose their carriers to obesity and related health problems, such as diabetes.[38] This explains why a type 2 diabetes–associated allele thought to lead to a "thrifty" (i.e., fat-storing) metabolism is prevalent at a high frequency in present-day Polynesians.[39]

Perhaps the most compelling illustration of the sheer multiplicity of feedbacks that can arise from cultural activities is provided by the agricultural practices of Kwa-speaking populations in West Africa.[40] For hundreds, perhaps thousands, of years, these people have cleared forests through slash and burn techniques to create fields in which to plant their crops, which are typically yams.[41] Tree removal, however, had a striking and unintended consequence, because with no roots to soak up all that rainwater, the agricultural practice greatly increased the amount of standing water. These puddles were perfect breeding grounds for malaria-carrying mosquitoes, such as *Anopheles gambiae*, which need sunlit pools to breed in and thrived in the new conditions.[42] The mosquitoes are vectors for the protozoan parasite *Plasmodium falciparum*, which causes malaria. Mosquito bites transfer the parasite into the human bloodstream, from where it travels to the liver and invades red blood cells; these rupture within 72 hours and release more new parasites into the blood.[43] Today, there are several hundred million clinical cases of malaria worldwide and roughly 800,000 deaths each year,[44] the majority of them in sub-Saharan Africa.

As an aside, lest you think that people in modern postindustrial societies could not be so foolish as to construct an environment rife with disease in this manner, let me point out that modern car tire manufacturing is also promoting disease vectors today. Mosquitoes infest pools of rainwater that collect in tires, which are typically stored outside; in this way, tire export contributes to the spread of malaria and dengue around the world.[45] In fact, the transition to urban living and associated increases in population density, as well as the spread of pathogens through long-distance trade, and pathogen exposure through animal husbandry and irrigation, have long been thought to have promoted the spread of infectious diseases.[46] Historically, these links have been difficult to prove, because most infections do not cause skeletal damage that can be detected in ancient specimens, while the beginnings of urbanization typically predate writing.[47] However, a recent genetic study took an innovative approach by reasoning that, should an association between urban history and disease exist, then populations living in regions with a long period of urban settlement should have evolved disease resistance to a greater extent than other populations.[48] Variants of the *SLC11A1* gene are known to be associated with susceptibility to tuberculosis (TB) in humans,[49] and are also linked to other infectious diseases such as leprosy, leishmaniosis, and Kawasaki disease.[50] As predicted, the study found a highly significant correlation between the frequency of those alleles that confer resistance and the duration of urban settlement. Populations with a long history of living in towns are now better adapted to resisting infections that thrive in the urban environment.[51]

A similar story arises with the Kwa. By inadvertently promoting malaria, they generated conditions in which alleles that confer resistance to the disease would increase in frequency through natural selection. One such allele is the hemoglobin S allele (*HbS*), generally known as the *sickle-cell* allele because it causes red blood cells to become stiff and take on a sickle shape. Those humans who carry two copies of the sickle-cell allele suffer from sickle-cell anemia, itself a life-threatening disease. Sickled red blood cells are stiff, and as a consequence tend to clog up small blood vessels. This is particularly problematic for those individuals who have two *HbS* alleles, and thus suffer from severe sickle-cell disease.[52] Most children with sickle-cell disease die before the age of five.[53] However, individuals with just one copy of the allele ("heterozygotes") experience comparatively mild sickling; this actually provides some protection against malaria, because sickled cells are recognized by the spleen as they flow through and are removed, flushing the parasite out with them. The result, as many of us learned at school, is a classic case of "heterozygote advantage, where individuals with one copy of *HbS* survive better than those with either two copies or none. Long periods of crop cultivation have intensified natural selection on the *HbS* allele, causing it to increase in frequency. The fact that neighboring Kwa-speakers, with different food-procurement practices, do not show the same increase in *HbS* supports the conclusion that a cultural practice (clearing fields to grow yams) has triggered genetic evolution.[54]

Between 1550 and 1820, some 12 million people were transported from West Africa, first to the West Indies and then on to the Americas, as a result of the slave trade.[55] These people carried their *HbS* alleles to the New World, where they again provided them with protection from malaria. With the eradication of malaria in portions of the Western Hemisphere by the mid-twentieth century (itself dependent on further cultural practices, such as the use of pesticides to kill mosquitoes and the discovery of medical treatments for malaria), one might predict that the *HbS* allele would start to decrease in frequency. Consistent with this expectation, people of African origin in North America today exhibit a relatively low frequency of *HbS* compared to their African forebears.[56] However, African-American ancestry includes people from other regions of Africa, which complicates the inferences that can legitimately be drawn. A neater comparison occurs in the West Indies where, some three centuries ago, Dutch settlers imported slaves from West Africa to the island of Curacao and neighboring mainland South America. On Curacao, unlike the swampy and infested mainland, malaria has been eradicated. Sure enough, by the mid-1960s, the frequency of *HbS* was only 6.5% on Curacao but 18.5% on the adjacent mainland[57]—clear evidence of relaxed selection on the *HbS* allele.

The coevolutionary dynamics are more complicated than thus far described, however. In researching this example, my collaborator, archaeologist Mike O'Brien, and I were astonished to discover that yams may actually provide relief from the symptoms of sickle-cell anemia![58] Some foods, including horseradish, cassava, corn, sweet potatoes, and, yes, *yams*, contain cyanogenic glucosides; as natural plant compounds, these interact with bacteria in the large intestine and aid the body in producing a type of hemoglobin that can effectively carry oxygen through blood cells, leading to less pain.[59] For the Kwa-speakers to have chosen to cultivate a crop that by chance alleviates the symptoms of the disease that their planting inadvertently promoted, indirectly via another disease, would be too much of a coincidence. Far more plausible is the hypothesis that these agriculturists originally planted other crops and subsequently switched to planting yams once their medicinal properties had been discovered.[60]

Inspired by this example, Luke Rendell and Laurel Fogarty, postdoc and graduate student, respectively, at the University of St Andrews, carried out theoretical work that modeled the coevolution of agricultural practices and genes favored in the modified environmental conditions. The study revealed that even if costly to engage in and difficult to learn, the agricultural practice (slash-and-burn for planting crops) nonetheless spreads rapidly by creating conditions under which particular genetic variants are favored (e.g., *HbS* allele carriers), because the individuals that carry such genes are far more likely to be those who engage in the agricultural practice than those that do not.[61] In other words, the cultural practice can hitchhike to prevalence on the back of the selection it generates, through gene-culture coevolution. The study implies that slash-and-burn agriculture could have spread by increasing the incidence of malaria in adjacent regions, allowing the agriculturalists to move in on their neighbors' patch because they are better able to survive the diseased-riddled conditions that they themselves created.[62]

We can now start to see the full extent of the feedback from a single human cultural activity.[63] The initial crop planting generated selection for the *HbS* allele by the convoluted route described, eventually leading the agriculturalists to switch to the cultivation and consumption of yams. Crop growing not only modified selection on the *HbS* allele, but ultimately led to the development of medical treatments for malaria, pesticide treatments for mosquitoes, medical treatments for sickle-cell disease, and perhaps even to the spread of slash-and-burn agriculture to neighboring regions. These cultural activities, in turn, resulted in natural selection that favored alleles conferring resistance to pesticides in mosquitoes, and also further modified selection on the *HbS* allele. Causation is flowing through this ecosystem, from cultural practice, to evolutionary response, back to modified cultural practice, and from one species to the next, in perpetual cycles.

Anthropological evidence provides the clearest indication so far that gene-culture coevolution is a fact of human history. Undoubtedly, during the course of human evolution, genes and culture have shaped each other's characteristics in a multitude of ways. What such examples do not make transparent, however, is just how much gene-culture coevolution has occurred. The sheer scale of this feedback only becomes apparent through consideration of the genetic evidence for gene-culture coevolution. Only in the last few years has the attention of geneticists been drawn to the central role of culture in human evolutionary dynamics. A major catalyst for this was the development of methods for detecting statistical "signatures" in the human genome of recent, positive selection;[64] these are methods for identifying genes that have been favored by natural selection over the past 50,000 years or so.[65] Thus far, somewhere between a few hundred and a couple of thousand distinct regions in the human genome have been identified as subject to recent selection. Genetic variants showing signs of recent positive selection are not restricted to simple mutations in the protein-coding regions of the human genome, as in the lactose case, but also include chromosomal rearrangements, copy-number variants, and mutations in regulatory genes.[66] What is particularly interesting about this data is a large proportion of the variants that the geneticists identified as subject to recent selection appear to have been favored by human cultural practices.[67]

Recently, with evolutionary biologist John Odling-Smee at Oxford University and human geneticist Sean Myles at Acadia University in Canada, I collated genetic evidence for gene-culture coevolution. We compiled records of human genes, known to have been subject to recent selection, where the most likely source of selection is a human cultural practice.[68] In the vast majority of cases, whether cultural activities were what triggered the genetic response has yet to be conclusively proven. A lot of work still has to be done to complete the connections from the genes to their impact on human bodies, and to the source of the natural selection for the long list of human genes implicated as subject to recent selection in genome-wide scans. Nonetheless, the data is highly suggestive, and in a very large number of cases it is hard to imagine that culture did not play a crucial role in the evolutionary episode.

Some of the most compelling examples concern genetic responses to changes in human diet. Consider, for instance, the evolution of the human ability to eat starchy foods.[69] Starch consumption is a feature of agricultural societies, whereas most hunter-gatherers and some pastoralists consume much less starch. This behavioral variation raises the possibility that different selective pressures have acted on amylase (the enzyme responsible for breaking down the starch in our diets) across populations with variant dietary habits. Sure enough, humans from different populations have dissimilar numbers of copies of the salivary amylase gene (*AMY1*), and the number of copies correlates positively with the amount of the enzyme amylase in the carrier's saliva. Individuals from populations with high-starch diets have, on average, more *AMY1* copies than those with traditionally low-starch diets. Higher *AMY1* copy numbers and protein levels are thought to improve the digestion of starchy foods and may also provide protection against intestinal disease.

Humans have exposed themselves to a variety of novel foods through their cultural activities, including through the colonization of new habitats that contain different flora and fauna, the initial domestication of plants and animals, and the advent of full-blown agriculture. That such activities have been a major source of selection on the human genome is supported by strong genetic evidence.[70] Several genes related to the metabolism of proteins, carbohydrates, lipids, phosphates, and alcohol all show signals of recent selection, including genes involved in metabolizing mannose, sucrose, cholesterol, and fatty acids,[71] as well as other nutrition-related genes.[72] Several distinct cases of links between differences in diet and specific genetic variants, as observed for AMY1, have now been identified.[73] Evidence for diet-related selection on the thickness of human teeth enamel,[74] and on bitter-taste receptors on the tongue has also been forthcoming.[75] As novel food items have largely been introduced into human diets through cultural practices, the notion that a gene-culture coevolutionary process has shaped the biology of human digestion is hard to escape.[76]

Previous chapters described how through culturally informed access to nutritionally rich food sources, our ancestors were able to pay the energy costs of brain growth. Consistent with this, increases in hominin brain size coincide with advances in technology.[77] In turn, culturally transmitted food-processing methods, from chopping and grinding to cooking, externalized aspects of digestion and allowed nutrients to be garnered from otherwise inaccessible sources; this resulted in reductions in the size of the gut. In effect, control of fire, cooking, and other food-preparatory techniques created "predigested" food.[78] At the same time, the evolution of an expanding brain, a shrinking large intestine, and lengthening small intestine forced humans to eat nutritionally dense foods.[79] Omnivores like humans must seek out and consume new items to fuel their varied diet but also need to avoid novel toxic items. Humans are thought to have resolved this dilemma through the evolution of an inborn mechanism that generates "palate fatigue," a sensory-specific form of satiety that results in growing tired with eating the same thing again and again, thereby ensuring a variety of foods are consumed.[80] Cultural traditions help facilitate the requisite nutritional diversity by specifying which items found in nature are edible, and which (often entirely edible) foodstuffs are to be avoided ("food phobias").[81] How products are transformed into food, the flavors added to render food consumption enjoyable, and even the etiquette associated with eating are also socially learned traditions that vary across societies.[82]

Several other categories of human genes are overrepresented in lists of alleles subject to positive selection. One study reported that among 56 separate heat-shock genes—those activated by stressful temperatures—28 showed evidence of a recent selective sweep in at least one human population, probably in response to culture-facilitated dispersal and local adaptation.[83] Evolved responses to cope with disease is another such category, making up nearly 10% of recent selective events.[84] As the sickle-cell example illustrates, shifts from a nomadic hunter-gatherer lifestyle to a more sedentary, agrarian way of life were likely to have facilitated the spread of infectious agents and other diseases. This led to a rapid rise in the frequency of alleles that protect against disease, which was often one of the fastest genetic responses.[85] Genes involved in the human immune response are extremely well represented among those subject to recent selection.[86] These data confirm that cultural activities have played an active role in the evolution of human responses to diseases, by both inadvertently promoting conditions under which diseases, and resistance to them, become widespread; and (typically much later) devising treatments and practices that mitigate against them.

Genes associated with our external appearance provide some of the strongest signatures of recent local adaptation. For example, the lighter skin of non-African populations is the result of selection on a number of skin pigmentation genes.[87] Ancient DNA extracted from a 7,000-year-old skeleton from Spain reveals that the carrier possessed ancestral variants of several skin pigmentation genes, implying that the light skin color of modern Europeans has evolved quite recently—over the last few thousand years.[88] Various genes involved in skeletal development show signatures of recent local adaptation,[89] as do genes expressed in hair follicles, eye and hair color, and freckles.[90] The frequencies of these alleles typically vary across societies. Although some of this variation can be attributed to natural selection, many differential evolutionary responses could potentially be explained through a form of sexual selection,[91] in which population-specific, culturally learned mating preferences favor biological traits in the opposite sex.

Some years ago I developed a mathematical model that combines sexual selection and gene-culture coevolutionary theory to explore such interactions.[92] I found that even if human mating preferences are learned, socially transmitted, and culture specific, sexual selection will still result. Indeed, culturally generated sexual selection was found to be faster and more potent than its gene-based counterpart. Genetic and psychological data support this hypothesis. A recent study identified several genes that correlate with pair formation in humans, including genes involved in skin appearance, body shape, immunity, and behavior.[93] The study indicates that the subset of people from whom we will choose our partner can to a large extent be predicted by consideration of the genes that we share, although the learned mate preferences that we share are what is most likely to be causally relevant here. Little overlap was found between three populations of African, European American, and Mexican origins, and the study concluded that mate choice is population specific and socially learned. Indeed, experimental data shows that humans copy the mate choice decisions of others, which can lead to the social transmission of preferences for particular characteristics in the opposite sex.[94] Given the pervasiveness of cultural influences on human mating preferences, social transmission may exert a powerful influence on the selection of secondary sexual characteristics and other physical and personality traits.[95]

Some human genes subject to less recent selection have spread to fixation, and are now be carried by all humans. An interesting case is known as the sarcomeric myosin gene *MYH16*, which is expressed primarily in the human jawbone, and which underwent a deletion among our hominin ancestors, whereby a sizeable section of the gene was lost.[96] This deletion is thought to have resulted in a massive reduction in jaw muscle, with a timing that coincides with the appearance of cooking, over two million years ago. Unlike other apes and early hominins, humans and most other members of the genus *Homo* do not possess powerful chewing muscles. Seemingly, a cultural process (cooking) removed a constraint (the requirement for jaw muscle to chew raw meat), allowing genetic change to occur that would have been severely deleterious in its absence.

The preceding chapters implicated gene-culture coevolution in the evolution of large brains, enhanced learning capabilities, and many cognitive traits, including language. Evidence for genetic change associated with these characters is also forthcoming. Several genes known to be involved in brain growth and development show signs of recent natural selection,[97] as do genes expressed in the nervous system,[98] and those implicated in learning and cognition.[99] Other genetic changes involved in brain enlargement are more ancient, being shared with Neanderthals and other extinct hominins.[100] Genes expressed in neuronal signaling and energy production have been upregulated in the human neocortex (i.e., the genes increased their rate of production of a cellular component, such as a protein or RNA),[101] and the plasticity characteristic of early brain development has been extended in humans relative to other apes.[102] The human brain has experienced far more evolutionary changes than the chimpanzee brain over the last few million years, particularly in the prefrontal cortex,[103] a region thought to be responsible for decision making, planning, and problem solving.

Structural reorganization of the brain during evolution is as important as increases in brain volume.[104] Indeed, as we have seen, the two typically go together, because enlarged brains not only have more neurons but also have greater organizational complexity.[105] This reorganization includes changes in the proportions of different brain regions, the amounts of white and gray matter, the size and patterns of folding of the neocortex and cerebellum, hemispheric asymmetry, modularity, neurotransmitters activity,[106] and in a variety of other factors.[107] In many instances, the genome regions underlying these changes have been identified and are found to have been subject to recent selection, or to exhibit differences from the homologous regions of the chimpanzee genome.[108]

The hypothesis that language use generated selective feedback acting on the organization of human brain has received considerable attention.[109] Consistent with this, genes involved in language learning and production are among those subject to recent selection.[110] The best-known example is the *FOXP2* gene, in which mutations cause deficiencies in language skills.[111] Only four *FOXP2* mutations occur in the evolutionary trees of mice, macaques, orangutans, gorillas, chimpanzees, and humans, and two of these occur in the evolutionary lineage leading to humans, which is suggestive of positive selection.[112] The most common interpretation is that this selection introduced a change in the *FOXP2* gene that was a necessary step to the development of speech, although the gene may also have been favored for other reasons such as vocal learning or lung development.[113]

Preceding chapters also made the argument that the longevity of some primates, including great apes, has been extended through cultural practices that furnish individuals with the knowledge and skills to survive. Here too there is supportive genetic evidence. Vacuole-protein genes are essential in humans because they remove dangerous toxins that gradually accumulate in cells, but the same genes are not so essential in other animals, such as mice, where mutations are less problematic.

Recent increases in the longevity of humans have greatly amplified the importance of these genes because toxins can now build up over more time, to the point where they would be lethal if they were not removed by the vacuole-protein genes.[114] Cultural processes have thus rendered these "housekeeping" genes critical, leaving any mutation lethal.

Another way that gene-culture interactions can play a part in human evolution is when linguistic and cultural differences affect patterns of gene flow between human populations. For instance, in patrilocal societies the married couple resides with the husband's parents, and males typically marry females from other societies, whereas matrilocal societies exhibit the reverse pattern. These societal differences affect the spread of genes, with genetic variants carried by females, such as mitochondrial DNA, flowing into patrilocal societies from adjacent ones. This occurs, for instance, in the Gilaki and Mazandarani in southern regions of Iran.[115] Whereas genetic variants carried by males, such as Y chromosomes, flow into matrilocal societies, as witnessed in Polynesia.[116] Other examples of indirect effects of cultural traits on human genetic variation arise through differences in social systems.[117] A recent study found that differences in the social structure and cultural practices of South Amerindian populations can dramatically affect their rate of biological evolution, with Xavánte Indians experiencing a remarkable pace of evolution compared to their sister group, the Kayapó.[118]

A common misconception is that modern hygiene, medicine, and birth control have caused natural selection to stop working on human populations. However, natural selection is relentless, and could only stop if all individuals had exactly the same reproductive success, which is unlikely to ever happen.[119] Even in modern societies, extensive evidence for human evolution exists. For instance, at least in some populations, an earlier age for the birth of the first child is being selected for among both fathers and mothers. Likewise, there is some evidence of selection for a later age for the birth of the last child and later onset of menopause in women.[120] According to evolutionary biologist Stephen Stearns from Yale University and his colleagues,[121] the important role that culture plays in shaping that selective environment becomes clear in the contrast between the developed and the developing world: "In the developed world, it is primarily variation in fertility rather than mortality that shapes variation in lifetime reproductive success. In developing countries, the variation in mortality has a greater contribution to selection, particularly the variation in infant and child mortality that is associated with infectious disease and deficiencies in child nutrition."[122]

The studies that I have described only scratch the surface of the now extensive data from numerous academic fields that demonstrate genes and culture can, and do, coevolve. Theoretical models, such as the handedness, sexual selection, and lactose tolerance case studies, illustrate coevolutionary mechanisms, while genetic, anthropological and archaeological data establish that this feedback is not just a hypothetical possibility, but a fact of human evolution. The evolution of sickle-cell disease in Kwa-speakers, like countless other examples, shows that genetic and cultural change can occur on similar time scales, while analyses of the human genome strongly imply that gene-culture coevolutionary interactions are widely prevalent.

Eminent Harvard entomologist Edward Wilson, famous as one of the architects of "sociobiology,"[123] the modern approach to the study of animal behavior, once controversially claimed that "genes hold culture on a leash."[124] What Wilson meant was that genetic propensities shape behavior and the acquisition of cultural knowledge. Some truth can be found in this assertion; for instance, people are more likely to eat particular diets if they possess the genetic variants that allow them to metabolize those foods. However, Wilson's claim was contentious because it was understood to imply that human culture is constrained by our biology to be adaptive.[125] Irrespective of Wilson's intent, this reading would be difficult to defend. Human culture, like all aspects of human development and behavior, is flexible, open-ended, and capable of generating tremendous novelty, including the creation of new circumstances that impose selection on our genes. What Wilson failed to emphasize was that the gene-culture leash tugs both ways. Human behavior, culture, and technology are shaped by genes to some degree, but the architecture of the human genome has equally been profoundly shaped by our culture, as the aforementioned genetic data affirm.

Cultural knowledge expressed in human behavior, tools, and technology is amply manifest in our species' extraordinarily potent capacity to modify the circumstances of our lives.[126] Our ancestors didn't just evolve to be suited to their world; they shaped their world. The landscape of human evolution did not pre-exist us; we built it ourselves.

We constructed our niche. While all organisms engage in niche con-struction,[127] our species' capacity to control, regulate, and transform the environment is uniquely powerful, chiefly due to our extraordinary capacity for culture. That capability for environmental regulation may itself have been caught up in the runaway process that I have been emphasizing.[128] Theoretical analyses reveal coevolutionary dynamics in which cultural processes can spread rapidly by hitchhiking on the selection they themselves generate.[129] That humans, the species most reliant on culture, should have the most potent capability for niche con-struction may be no coincidence; autocatalytic and runaway effects may have fueled ever more powerful niche construction in our lineage, just as they fueled more efficient copying, bigger brains and more sophis-ticated communication.

Cultural niche construction did not just impose selection on our bodies, thereby shaping our physical appearance, skin color, suscepti-bility to disease, and ability to digest foodstuffs, but it also transformed the human mind, leaving our cognition specifically adapted for cultural life.[130] Among those genes subject to recent selection are numerous genes expressed in the human brain and nervous system, including those expressed in our learning, cooperation, and language.[131] These genetic data reinforce the message of comparative neural anatomy specified in previous chapters, which suggests that human evolution has been accompanied by expansions of regions of the brain linked to innovation, imitation, tool use, and language.[132]

As documented earlier in this book, social learning is widespread among animals and some species exhibit relatively stable behavioral traditions. I would be very surprised if such traditions had not triggered bouts of gene-culture coevolution in other animals too. Indeed, there are hints that this has happened. Both sponging dolphins and tool-using New Caledonian crows, for instance, are able to exploit different foods than other dolphins and crows by virtue of their innovative foraging skills, which are thought to be socially transmitted.[133] Likewise, genetic differentiation has followed cultural diversification in both sperm and killer whales.[134] How prevalent gene-culture coevolution is outside of hominins remains to be established, and it may still be found that its impact is substantial. The situation is much clearer for our own species, where gene-culture coevolution may even be the dominant form of evolution. Theoretical models consistently find that gene-culture dynamics are typically faster, stronger, and operate over a broader range of conditions than conventional evolutionary dynamics. By modifying selection pressures and increasing the intensity of selection, cultural processes can speed up evolution. Granted, in other circumstances, by providing an alternative means of responding to ecological and social challenges (for instance), cultural processes can also remove the requirement for an evolutionary response. However, the data imply that on average, human cultural activities have increased the pace of biological evolution, probably substantially so. Much as Allan Wilson envisaged decades ago, an autocatalytic process accelerated the evolution of cognition in our lineage. Our potent culture and the feedback it generated, switched on the evolutionary afterburners and allowed our cognition to race ahead of that of other species, thereby generating a major gulf between the intellectual capabilities of humans and other animals.

This picture of the evolution of the human mind is radically different from the portrayal advanced by evolutionary psychologists and many popular science writers. These authors have often claimed that those humans that walk the streets of today's urban metropolises are left struggling to cope with the modern world by the legacy of brains suited to ancestral primate or stone-age conditions.[135] This would imply that modern humans have manufactured environmental conditions fundamentally ill suited to their biology. While such "mismatch" does occur, as the aforementioned evolved preferences for salt, sugar, and fat testify, the extent of this incongruity is far more limited than these authors envisioned. Humans do not modify their environment randomly or haphazardly. Rather, like other animals who construct nests, mounds, webs, and dams that help them and their offspring survive, humans build structures and have other impacts on their world that broadly enhance their evolutionary fitness.[136] Animals also deplete resources and pollute environments, but this too increases fitness in the short term and is often tied to life-history strategies that take account of this activity—for instance, through dispersal or migration when resource levels are low or the environment becomes uninhabitable. The same holds when humans plant crops that, under some circumstances, have negative consequences (for instance, if the agriculture promotes disease), as those crops typically provide nutritional benefits that outweigh the costs. Niche construction does have diffuse and unintended negative impacts on fitness, but in spite of this it will typically increase the fitness of the constructors, at least in the short term.[137] This is hardly contentious; the fitness benefits of animal artifacts are well documented.[138] In this respect, humans are no different from other animals. We too construct our world to suit ourselves, leaving our behavior largely adaptive in spite of the radical transformations brought about in the environment.[139]

The changes that we impose on our world inevitably spill over to affect other species—often for the worse, but sometimes for the better. For instance, the use of fire by Australian Aborigines to clear vegetation and thereby help in finding prey, such as small mammals that live in burrows, has actually benefitted lizards who thrive in the resulting patchy mosaics.[140] Likewise, many commensals and inquilines thrive in human-constructed environments. Eventually, human habitat degradation, most obviously anthropogenic climate change, may well negatively influence human fitness. However, for the time being, other species bear the brunt of our activities, for whom the consequences are rarely so positive. Siberian tigers, golden lion tamarins, checkerspot butterflies, and countless other species are all endangered because of our habitat degradation, deforestation, urban development, and agriculture. In the meantime, the human species thrives and our numbers rise relentlessly, a clear sign that in the short term humanity benefits from our niche construction. Contrary to popularized narratives of environmental exploitation leading to cultural collapse, recent studies demonstrate that agricultural and other cultural practices shifted human carrying capacity upward.[141]

Far from being trapped in the past by an outdated biological legacy, humans are characterized by a remarkable plasticity. Our adaptiveness is reinforced by both cultural and biological evolution.[142] Through our culture, we are frequently able to counteract any mismatch between our biological adaptations and the world in which we find ourselves; for instance, clothing, fires, and air conditioning buffer extremes of temperature, while new agricultural practices and innovations alleviate food shortages. Other animals exhibit plasticity in their responses to novel conditions, but most do not possess anything approaching the remarkable flexibility and problem-solving capabilities conferred by human culture. On those occasions where humans fail to eradicate novel conditions through cultural activities, natural selection ensues and, as we have seen, culturally induced selection can be fast. In the malaria-stricken regions of the Kwa homeland, being a heterozygote for the sickle-cell allele is adaptive. Similarly, in dairying societies, genes expressed in high lactase activity pay fitness dividends. These genetic changes, induced by human cultural activities, have restored human adaptiveness.

Hence, in spite of extensive change, both to ourselves—in our anatomy, physiology, and cognition—and to our environments, which for many of us are very different from African forest or savanna, we humans nonetheless remain extremely well adapted to our surroundings. That is because we built those surroundings to benefit ourselves, and then adapted to life within them. Human minds and human environments have been engaged in a long-standing, intimate exchange of information, mediated by reciprocal bouts of niche construction and natural selection, leaving each beautifully fashioned in the other's image.

CHAPTER 10

THE DAWN OF CIVILIZATION

The pace of change experienced by the members of our evolutionary lineage has accelerated over recent times, and continues to accelerate.[1] Like all species on Earth, humans possess an evolutionary history that stretches back at least 3.6 billion years to the beginning of life. For the vast majority of the many millions of living species aside from us, that history of adaptive change has been written solely in terms of biological evolution, and for at least 99.9% of our history we too adapted to the world in this manner. However, for around the last two to four million years and perhaps longer, our history has also been crafted through gene-culture coevolution. The contribution of gene-culture dynamics to our adaptive evolution was almost certainly modest at first, but grew in influence over time as our cultural capacity was enhanced, and our control of the environment increased. Eventually, culture began to take over altogether, carrying us into an entirely new realm.

We are probably alone among the planet's residents in having experienced three ages of adaptive evolution. First, there was the age in which biological evolution dominated, in which we adapted to the circumstances of life in a manner no different from every other creature. Second came the age when gene-culture coevolution was in the ascendency. Through cultural activities, our ancestors set challenges to which they adapted biologically. In doing so, they released the brake that the relatively slow rate of independent environmental change imposes on other species. The results are higher rates of morphological evolution in humans compared to other mammals,[2] with human genetic evolution reported as accelerating more than a hundredfold over the last 40,000 years.[3] Now we live in the third age, where cultural evolution dominates. Cultural practices provide humanity with adaptive challenges, but these are then solved through further cultural activity, before biological evolution gets moving. Our culture hasn't stopped biological evolution—that would be impossible—but it has left it trailing in its wake.

Through cultural evolution, our species is utterly transforming the planet, and at breakneck speed. For hominins to evolve from a chimpanzee-like creature to *Homo sapiens* took approximately 6 million years, but in the last 10 to 12 thousand years of cultural evolution, humanity has been to the moon, split the atom, built cities, compiled encyclopedic knowledge, and composed symphonies. A frighteningly high proportion of other species are simply unable to cope with the express-train transformations of their environment that human beings have imposed; as a consequence, they will almost certainly go extinct before being able to adapt through natural selection. Our impact on the planet is now so devastating that scientists have marked it as a new geological epoch called the "Anthropocene."[4] We, on the other hand, adapt with no problem at all, because we uniquely possess a culture that enables it.

Culture provided our ancestors with food-procurement and survival tricks, such as how to gain access to nutrient-rich foods, and how to build a fire or make a cutting tool. As new inventions arose, hominin populations were able to exploit their environments more efficiently. Our forebears broadened their diet, for instance, to eat significant amounts of meat, which is a more concentrated source of protein, vitamins, minerals, and fatty acids than plant material. Later, our ancestors began to cook their foods, which greatly reduced the digestive burden and allowed further time and energy gains. Ultimately, an elevated rate of nutrient harvesting paid the substantial energy costs of growing and running a huge brain. The human brain uses up more energy than any other single organ; it weighs about 2% of our body weight, but burns about 20% of our energy. That extra energy has to come from somewhere, compelling hominin species to forage in a manner that yielded a bonanza of calories; the alternative was to go extinct. Only when our intelligence paid for itself by procuring more energy than it cost, could it become the direct target of positive selection.[5] Our brain growth, in turn, further augmented our capacities for innovation and social learning.[6]

Increasing efficiency in exploiting our environment not only fueled brain expansion, but population growth as well. Human numbers really took off after the advent of agriculture, and our domination of Earth's ecosystems rapidly followed.[7] With each new invention—the domestication of plants and animals, irrigation methods, plant and animal breeding programs, hybrid cereals, fertilizers, pesticides, biotechnology—food production increased, and the population expanded further. The global population increased by around five orders of magnitude, further fueling the adaptive evolution of our species at both genetic and cultural levels. Larger populations typically mean faster rates of biological evolution, due to increases in the numbers of novel mutations, and a stronger effect of natural selection relative to genetic drift.[8] At the same time, cultural evolution flourished through corresponding increases in the number of people who could devise new inventions, and because innovations and knowledge were less likely to be lost through chance events.[9]

With each increment in technology, the human population became more and more dependent on socially transmitted knowledge to acquire the skills necessary for survival. In turn, as the social learning strategies tournament teaches us, with increasing reliance on social learning, characteristic features of modern culture automatically surface, including a larger cultural repertoire; the retention of knowledge for long periods of time; some semblance of conformity in behavior; and fads, fashions and rapid technological change.[10] The remarkable success of our species, our enhanced capacity to adapt, our astonishing diversity, and the bewildering amounts of information that we have generated, all follow directly from our heavy reliance on social learning.[11] Insights from mathematical analyses exploring the importance of transmission fidelity, described in chapter 7, are also relevant here. These studies suggest that with the evolution of more accurate transmission mechanisms (either as legacies of our biological heritage such as our capacities for teaching and language, or as products of cultural evolution such as writing and record keeping), more cultural knowledge would be retained, and lead to a more extensive cultural repertoire.[12] With each increment in transmission fidelity, knowledge would be preserved more effectively and for longer periods of time, thereby generating opportunities for conceptual lineages to cross-fertilize and to further stimulate cultural accumulation. Thus, drawing solely on established empirical and theoretical findings, not just one, but a whole a series of interwoven feedback mechanisms can be envisaged, with each reinforcing and invigorating the others to generate runaway niche construction and adaptation in our lineage and in our lineage alone.

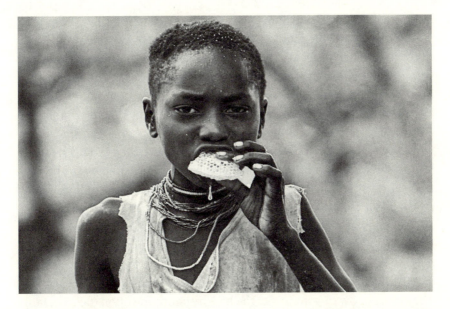

FIGURE 8. Honey is one of the most energy dense foods in nature, and is thought to have been an important food source for our ancestors, and for many foraging populations today. Surprisingly extensive cultural knowledge is required to harvest honey safely. By permission of Joanna Eede/ Survival.

Yet if this feedback was operating so relentlessly, why did the members of our genus spend such a long time as hunter-gatherers? And why are there still so many small-scale preindustrial societies around the globe that have not devised wheels, arches, or metal extraction, let alone quantum mechanics or gene-editing technology? The answer to these questions is that the life of a hunter-gatherer imposes tough constraints on the growth of cultural knowledge.

Hunter-gatherer societies are no less dependent on socially transmitted knowledge than postindustrial societies. A simple illustration suffices to make this point. Consider the seemingly uncomplicated foraging challenge of harvesting honey, a behavior common to many contemporary hunter-gatherer communities (figure 8).[13] Honey is one of the most energy dense foods in nature, and is thought to have been an important food source for the earliest members of our genus.[14] Among the Hadza people of Tanzania, honey comprises approximately 15% of their diet.[15] This valued commodity is shared widely outside of the household, and is the primary weaning food for children.[16] You might well imagine that you could easily pick up the ostensibly uncomplicated skill of gathering honey without help from others. Of course, you almost certainly already bring to the exercise the prior knowledge that honey is nutritious, produced by bees, found in their nests, and more; this is knowledge that you acquired socially. Yet even thus informed, there is a great deal of additional knowledge that the novice honey harvester would require in order to succeed. For instance, they would need to know which bees to target because there are many species, and while some are stingless, others, such as the African killer bees, will viciously attack humans and repeated stings frequently lead to death. They also need to know what the nests of the favored bees look like, where they are typically located, and in which species of trees. The honey harvesters need to know how to climb the trees safely—for instance, they might hammer pointed sticks into the trunk to make a ladder, as the Hadza do. They must know how to reveal the hive by chopping open the tree, how to lull the bees into inactivity with smoke, as well as how to build the fire to make the smoke. They need to know how to remove the nest, or a part of it, from the tree without getting stung, and how to extract the honey. Finally, if they are members of the Hadza, they need to know how to speak the song of a bird known as the honeyguide.[17] A remarkably mutualistic relationship exists between the honeyguide and humans; in order to obtain beeswax and larvae, this bird leads people to the nests of wild bees. The Hadza engage in a "conversation" with the honeyguide, which comprises a series of chatters and whistles, until the bird leads the gatherer to the hive.[18]

What is relevant here is that all of this honey-harvesting knowledge is acquired socially. Communities such as the Hadza, the Aché of Paraguay, or the Machiguenga of Peru are only able to forge an existence because they draw on a very long history of accumulated adaptive know-how.[19] Honey gathering is no different from countless other essential skills, such as trapping animals; skinning and butchering a deer; making a fire and keeping it lit; knowing which fruits and fungi are safe to eat and which are not; making a spear, bow, and arrow; and processing plant foods to remove toxins. Knowledge of how to carry out such tasks is the lifeblood of these societies and must be passed down through generations to ensure survival.[20]

In hunting and gathering societies of today, plant products supply the bulk of the calories, while fish and game provide the major share of the protein.[21] However, humans are not well adapted biologically to extract nutrients from many plant materials (we cannot digest cellulose like ruminants, or detoxify the numerous chemicals that plants produce as defenses), and this leaves our gathering largely restricted to fruits, seeds, and tubers.[22] Food-processing methods allow human populations to broaden their diets to some degree, but such methods also require a great deal of cultural knowledge.[23]

Acorns, for instance, can be consumed if first dried, their shells removed, mashed, soaked in water, and rinsed repeatedly to remove the tannin. The resultant mash can then be dried again to make a kind of flour used for breads, a method deployed by some North American hunter-gatherer communities. The leaching of the tannin is critical, because tannin both makes the food taste bitter and binds its proteins to inhibit digestion; acorns can lead to sickness if consumed unprocessed.[24] Techniques such as these both improve yields and broaden diets, but such complex procedures are a relatively recent advance in human evolution. Most hunter-gatherer populations throughout history had to craft a difficult living on slim pickings.

As a consequence, a typical human hunting and gathering band must constantly be on the move to gather more food elsewhere as they rapidly exhaust all of the easy returns in their locality.[25] Often, they will move as frequently as every few days. The rate at which they move depends critically on two factors: the depth of the population's accumulated cultural knowledge of how to extract resources from the environment, and the general productivity of the region. Hunter-gatherers will, of course, forage some distance from their home base, but there is a limit beyond which moving camp is easier than to forage at great distances and return.

Early *Homo* probably did not even have home bases. Nomadic hunting and gathering, perhaps following seasonally available wild plants and game, is generally thought to be the oldest human subsistence method. Human populations with a restricted repertoire of food-procurement techniques, and communities in regions of low productivity where an efficient strategy for exploiting scarce resources is required, are well suited to this nomadic lifestyle. Even many contemporary pastoralist societies that keep herds of cattle or goats must continuously drive them in patterns that avoid depleting pastures beyond their ability to recover.[26]

Constant movement imposes severe constraints on the size and sophistication of the technological tool kits exhibited by a society.[27] A band of hunter-gatherers cannot afford to possess extensive equipment when all of its members' belongings must be carried from place to place every few days. Everything that the group possesses must be light and easily packed. A nomadic lifestyle does not lend itself to making equipment that will not be required for weeks, or in different seasons, because that paraphernalia would have to be transported. Once a group has to move all the time, long-term food storage also becomes practically impossible.

In addition, movement on foot effectively limits each woman to one dependent child at a time, because she cannot gather food while holding multiple young. Hunter-gatherer communities will space out the births of children deliberately so that they only have to carry one at a time. For instance, the !Kung San people who live in the Kalahari desert will typically raise children about four years apart in age.[28] This puts a severe constraint on population sizes.

Discounting the division of labor between the sexes, in hunter-gather societies all individuals essentially have the same job of foraging.[29] As a consequence, no one can accumulate wealth or resources or have a higher status than anyone else, and there is little social structure. Specialized, permanent roles are normally absent.[30] While some individuals are recognized as a Shaman, or as various types of toolmakers, none are able to earn a living in those ways; rather, all able-bodied individuals must hunt or gather. Only the very richest hunting and gathering societies possess distinct occupations, including chief, canoe builders, arrow makers, and so forth.[31]

This helps us understand why hunter-gather technology was only slowly changing for such a long time, and also why, even today, many small-scale societies possess limited technology.[32] Hunter-gatherers are effectively trapped in a vicious cycle that severely constrains their rate of cultural evolution. Unless the group devises clever means of exploiting their environments more efficiently, or just happens to live in an extremely rich location, they must constantly be on the move in order to garner sufficient resources merely to exist. Any new innovations would also have to meet these constraints; a new tool or device, however fantastic, is of little value if it cannot be easily transported. In any case, there are likely to be few new inventions, both because there are few individuals to come up with them, and because people would have little time to innovate.[33] Compelled to live in small societies with limited technology, few opportunities arise for people to combine different cultural elements into new and more complex inventions. As no wealth or resources can accumulate, and everyone (even children) is pressed to hunt or forage, the society is naturally egalitarian, with little social structure. These factors combine to generate a comparatively slow pace of change in cultural repertoires, at least relative to postagricultural societies.[34]

The flip side of these constraints, however, is that once a population does pass the critical threshold when the spread of novel technology allows it to exploit its local environment with sufficient efficiency, then lots of consequences immediately follow. First, the population does not need to constantly be on the move. Second, they can have a stable home base, or even lead an entirely sedentary existence. Third, they can now store food and equipment, and thereby accumulate technology. Fourth, expanding tool kits provide more opportunity for combining ideas or borrowing elements from one domain and applying them to another. Fifth, being more sedentary allows populations to reduce their birth spacing,[35] and permits an increase in foraging efficiency to raise the carrying capacity; consequently, the population grows. Sixth, with more people, there are more potential innovators, each of whom may have more time, more resources, and more uses for the resulting innovations. Seventh, more innovations potentially allow the population to exploit new resources in their environment, or to exploit existing resources with greater efficiency, further increasing the carrying capacity and allowing the population to grow some more. Finally, with the opportunity for the accumulation of wealth, social structure and division of labor is now possible, leading to specialization and efficiency gains. With the advent of a non-nomadic lifestyle, cultural and technological evolution can really take off. That is the reason why cultural evolution appears to accelerate with, or after, the appearance of agriculture and the associated domestication of plants and animals. Only through agriculture could most populations throughout human history throw off the shackles that the requirement for constant movement places on the pace of cultural change.

Considerable work within the field of archaeology has focused on the origins of agriculture, particularly the questions of *what* species were domesticated, *when*, and *where*.[36] Researchers have established that agriculture did not originate only once, but rather there were eight or more locations of early plant domestication.[37] This naturally raises the question of what factors contributed to the origins of agriculture. Classical arguments have mainly emphasized environmental change as the trigger.[38] Certainly, it is difficult to imagine how agriculture could have been cost effective during an ice age, so we should not be surprised that it first appears shortly after the end of the last ice age, around 11,500 years ago.[39] However, researchers also have assumed that it would only have been economical for our ancestors to turn to intensive agriculture when the availability of wild resources was curtailed.[40] Yet, Bruce Smith at the Smithsonian Institution, a leading authority on the topic, has recently noted a frequent context of a rich resource zone for domestication of plants and animals worldwide.[41]

These *prima facie* conflicting conclusions reflect the common assumption that some external condition or circumstance was what triggered the initial tentative steps toward plant and animal domestication and what rendered the transition to complete reliance on agriculture economical.[42] Together with my archaeologist collaborator Michael O'Brien of the University of Missouri, I reached the conclusion a few years ago that instead of (or in addition to, perhaps) searching for an external "prime mover" for the origins of domestication and agriculture, researchers would be well advised to focus greater attention on the co-evolutionary interactions between humans and domesticated plants and animals, as well as the cultural niche-constructing practices that underpinned them.[43] An emphasis on external environmental factors as the cause of the origins of agriculture is not wrong. During the last glacial period, the climate was not only cold, but highly variable, and plant and animal communities were forced to shift their ranges frequently.[44] Hominin groups would have been compelled to find a favorable region that contained suitable forage and game, but then move locations repeatedly as they chased their food across the landscape.[45] Such unstable conditions mitigate against a sedentary existence. Conversely, in the Holocene, the period since the end of the last ice age, climates have not only been warmer, but also wetter and more stable, with higher carbon dioxide concentrations in the atmosphere; these are conditions that favor a much higher reliance on plant resources.[46] Of the 2 million years that our genus *Homo* has ranged the Earth, probably the last 10,000 or so have been the most conducive to agriculture.

Certainly, the climate had to be suitable for domesticating plants before agriculture could arise, but I tend to regard such external factors as necessary preconditions rather than the direct cause. At best, climate change is only part of the causal story. After all, Earth has experienced many warm, wet periods of long duration in the past, and yet no species devised agriculture.[47] Nor do I envisage that, had the current interglacial occurred, say, 500,000 years ago instead of in the present, hominins would have devised agriculture then. The right species, with the requisite knowledge and abilities, had to be in the right place, at the right time and, crucially, had to behave in the right way. To understand the origins of agriculture we need to consider how humans constructed a niche in which domesticating plants and animals was economical.

Animals and plants coevolve as natural selection favors characters in each that have an impact on the evolutionary fitness of the other. In this instance, as the animal population is human, both genetic and cultural evolution are involved.[48] Agricultural practices are examples of cultural niche construction that, as described in the previous chapter, can trigger evolutionary episodes in both the domesticates and, via selective feedback, in the human populations too.[49] Cultural niche-constructing processes that contribute to plant domestication include selective collecting, transporting, storing, and planting of seeds; setting fire to grasslands and forest, either intentionally or accidentally; cutting down trees; tilling; weeding and the selective culling of competing species; irrigation; and creating organically rich dump heaps.[50] The skills and information that underlay these processes were passed from one generation to the next through a combination of teaching, imitation, stories, myths, and ritual,[51] with the knowledge base regularly accumulating and being updated. Over time, these agricultural practices had an impact on the plants, which underwent a series of dramatic changes, such as major increases in size of the plant or its seeds, faster seed germination, simultaneous ripening of the seed crop, and so forth. The changes benefited both species by increasing the fitness of the plant community and elevating its yield. Sowing seeds in prepared substrates, for example, both induces changes in germination and dispersal mechanisms through inadvertent artificial selection, and helps the tended plants by increasing their likelihood of being included in next year's seed stock.[52] The increased yield, in turn, encouraged humans to perpetuate the practices that maintained or increased plant productivity, thereby triggering natural selection that modified human digestive enzymes. However, the methods of sowing selected seeds and harvesting plants inadvertently imposed selection on the crops that eventually left many inviable when in open competition with wild counterparts, and hence utterly dependent on humans.[53]

The same reasoning applies to animals, where domestication again selected for increased yields of animal products, such as milk, but also a variety of other traits, including lowered reactivity to environmental stimuli and a dependence on humans for survival and reproduction.[54] The protection provided by corrals and pens, and selection of animals that were easy to manage, again modified the impact of natural selection on animal breeds. When removed from anthropogenic settings, that selection left the animals concerned much more vulnerable to predation.[55] Hence, those plants and animals tended by humans became increasingly dependent on their keepers. Through this reinforcement of the mutually beneficial relationship between humans and the plants and animals we domesticated, a codependence or coadaptation evolved.

In fact, the initial domestication of plants and animals typically occurred well before the emergence of full-blown agriculture,[56] which can be thought of as the near total reliance of human societies upon domesticated plants or animals.[57] In some areas of the world, initial domestication and agriculture were separated in time by thousands of years.[58] During that transitional window, small-scale societies had low-level food production economies that included domesticates, but were also strongly reliant on wild species of plants and animals.[59] A key factor determining if domestication occurred and transitioned to support stable agricultural societies, was whether a suitable plant species was locally available.[60] The first domesticated crops were generally annuals with large seeds or fruits. These included pulses such as peas, and cereals such as rye, wheat, barley, and maize.[61] An important advance was the appearance of a new form of wheat (with large heads of seeds) in the region surrounding the eastern Mediterranean, close to 9600 BCE.[62] Prior to that, wheat was nothing like we know it today, but rather resembled a wild grass. Wild wheat was, either naturally or through human action, crossed with other types of grass to produce a fertile polyploid hybrid,[63] which possessed a fat and easily harvested ear, on account of a quadruple dose of chromosomes. However, this particular plant had lost the ability to propagate its swollen seeds in the wind, leaving the it reliant on humans to gather the ears and scatter the seeds for next year's harvest.[64] Barley appeared in the Fertile Crescent around the same time.[65] Maize was first farmed in southern Mexico, around 9,000 years ago, and subsequently spread throughout the Americas,[66] as did squash, beans, and cassava. By 10,000 BCE the bottle gourd,[67] which was used as a container, also appears to have been domesticated. Use of this plant had spread to the Americas from Asia by 8000 BCE, most likely due to the migration of people.[68] The first use of millet, one of the earliest crops to be farmed in Asia, is dated to between 10,300 and 8,700 years ago in China, with the success of this species attributed to an excellent resistance to drought.[69] Rice too had been domesticated in China by 9,000 years ago, and spread to India around 5,000 years ago.[70]

Plant domestication might reasonably be thought to have contributed positively to the health of the population by reducing the risks of starvation. In fact, that could not be further from the truth.[71] Certainly health was dramatically affected by the onset of agriculture, at least in many parts of the world,[72] but not generally for the better. Initially, and temporarily, the transition of many human societies from a lifestyle of hunting and gathering to one of agriculture and settlement dramatically narrowed diets by decreasing the variety of foods consumed. A shift to intensive agriculture thereby first created a decline in human nutrition, with resulting impacts on human health.[73] The health of many Europeans began to decline markedly about 3,000 years ago, after agriculture became widely adopted in Europe and during the rise of the ancient Greek and Roman societies.[74] The excess of genes expressed in immunity and resistance to disease that have been subject to recent selective sweeps, as described in the previous chapter, is a signature of this fallout from the widespread implementation of agricultural practices and the consequent increase in density and exposure to animal diseases.[75] Similarly, researchers document from these periods a growing number of skeletal lesions from leprosy and tuberculosis—zoonotic diseases caused by living close to livestock and other humans in settlements where waste accumulated.[76] The frequency of dental cavities and enamel problems also increased as people switched to a grain-based diet with fewer nutrients and more sugars.[77] The net result was that Europeans actually became shorter for a time, with males shrinking an average of 7 cm between about 2,300 years ago and 400 years ago; this was a sure sign that children who were not members of elite families were eating less nutritious food or suffering from disease.[78] Agricultural innovations may have repeatedly increased yields, but populations quickly responded by growing in numbers, leaving human misery much as it was before.

Yet anthropologists have long held the view that the development of large-scale, socially structured societies with extensive division of labor would not have been possible without the extensive food base that only intensive agriculture could provide.[79] Such reasoning suggests a positive connection between agriculture and evolutionary fitness, and that connection may have motivated the agricultural innovation of our forebears. Presumably, agriculture was attractive to our ancestors because they believed that control of food production offered a means both to increase the volume of food, and to reduce variation in food availability.

In the same way that animals store food to cope with harsh times, one might also expect humans to have engaged in potentially intensive forms of food production when the alternative, in the form of readily caught or easily gathered wild food, was not freely available. On this reasoning, agriculture might be expected to have evolved when the availability or productivity of wild resources was low. However, other factors are being overlooked here; for instance, poor environments often do not provide good conditions for growing crops,[80] and individuals living under such circumstances are unlikely to have had the time to invest in innovation. Moreover, despite the benefits that agriculture offers, little evidence supports the suggestion that its appearance greatly reduced the risk of resource failure and starvation.[81] Agricultural peoples may have developed a more advanced knowledge of food storage and transportation than hunter-gatherers, and they may not have faced some of the severe fluctuations in natural resources that nonagricultural groups did; however, the strategies that sedentary populations used to reduce or eliminate food crises generated costs as well as benefits.[82] In fact, the advantages may often have been outweighed by such factors as the greater vulnerability of domesticated crops to climate fluctuations or other natural hazards—a vulnerability that could then be exacerbated by the specialized nature or narrow focus of many agricultural systems.[83] Such counterintuitive observations beg the question of why, when it brought high labor costs, high rates of failure and, in some cases, no clear economic incentive, numerous communities adopted systematic agriculture. What circumstances favored the appearance of agriculture, and how did it manage to flourish in spite of the problems generated?

Almost certainly, rich, rather than poor, environments favored agriculture, although the associated technology often subsequently spread to poorer regions.[84] In a rich environment, populations could afford to be more sedentary and thereby support a larger repertoire of technologies. Such environments promoted human population growth, which was frequently followed by resource depression and the reduction in availability of desirable foods due to overexploitation. In turn, the increased efficiency with which the population exploited the natural world rendered food storage beneficial, and raised the economic value of investing in agriculture.

Humans constructed the agricultural niche. Environments are not fixed as rich or poor; they are dynamic variables, able to change as a result of the activity of potent niche constructors like humans.[85] The wealth of resources that a particular locale affords depends critically on the cultural sophistication of the populations that live there, the depth of their knowledge bases, and ultimately on what they do to their environments. We are champions of niche construction, largely because human manipulation of the environment draws on a long legacy of culturally acquired and transmitted information through which we can modify our world and increase its carrying capacity.[86]

As to the reason why agricultural societies flourished in spite of all the negative consequences, I suspect that there are two major factors. The first really comes down to natural selection. In simple terms, agricultural societies outbred hunter-gatherer communities. Agriculture may well have initially had a negative impact on human health but, at least in some societies, it afforded a gigantic leap in the environments' carrying capacities.[87] Through farming, humans were able to increase the productivity of their local environments by several orders of magnitude, allowing the support of far larger communities. Agricultural societies took a hit with viability selection (that is, selection acting on the ability to survive) as survival rates reduced marginally, but they were more than compensated by fecundity selection (selection acting on the ability to reproduce) as people had more surviving children.

Prior to the advent of agriculture, the world's population is thought to have stabilized at around a million, whereas it surpassed 60 million by the time of the Roman Empire.[88] Genetic data reinforce this view. A recent analysis of human mtDNA has revealed strong evidence for distinct recent demographic expansions, with timings corresponding neatly to the origins of agriculture derived from archaeological evidence.[89] With the industrial revolution, like the agricultural revolution, life expectancy again increased dramatically, which reinforces the conclusion that population growth is genuinely linked to societal development.[90] Labor-saving farm machinery allowed the bulk of the workforce to be employed elsewhere and produce other goods, including other labor-saving devices, medical innovations, and improvements in sanitation and living conditions. The genetic data demonstrate how advances in technology and their spread can increase human reproductive success.[91]

Agriculture led to more births than other forms of subsistence did, primarily because it was accompanied by a settled, stationary existence; this removed the major constraint on the birth rate, which had been the need to constantly be on the move.[92] No longer forced to impose severe birth spacing, birth rates increased, slowly at first and then more rapidly. As populations grew, the earliest cities appeared around 6,000 years ago.[93] The agricultural niches of many of these farming societies expanded, evolving into larger and more complex social systems characterized by urban centers and class inequalities. The relentless spread of agriculture into new territories occurred through a combination of population expansion and the diffusion of agricultural innovations.[94] As technology became more sophisticated, farming became a plausible form of subsistence in regions that hitherto had proven unproductive. The agricultural niche grew inexorably, extending beyond its original zones through mechanisms such as conquest and trade. Indeed, with this demographic expansion the very idea of leading an agricultural existence spread through a process of selection acting on cultural groups, in which agricultural centers swelled to the point where they gave rise to daughter communities that embraced the same way of life.[95]

In addition, from its inception, agriculture triggered a raft of further innovations that dramatically changed human society. This is the second major factor that explains the remarkable success of agricultural societies. Some inventions are so momentous that they spawn an array of secondary products or behaviors, with each new invention quickening the pace of cultural evolution. Such discoveries are sometimes described as "key innovations,"[96] and agriculture certainly falls into this bracket. Agriculture created a world in which spades, sickles, ploughs, draught animals, harnesses, looms, storage containers, irrigation methods, building materials, and countless other innovations became of value. Agriculture also had a tremendous impact on social structure, creating circumstances in which an organized workforce, division of labor, major wealth differentials, and a much more stratified society flourished.[97] Farming and urban living created new niches for a curious assortment of other creatures, including cats, dogs, rats, mice, houseflies, pigeons, bedbugs, rhododendrons, weeds, and microbes, with feedbacks that shaped the cultural practices that followed.

With division of labor came enormous gains in efficiency and productivity. In the *Wealth of Nations*, economist Adam Smith used the example of a pin to illustrate this point: "One man draws the wire, another straightens it, a third cuts it, a fourth points it."[98] As many as seventeen people might work on a single pin, with a collective output of 4,800 pins per day. Without the division of labor, each worker would produce less than twenty pins. (The industrial revolution increased productivity further. The earliest pin-making machines made 72,000 pins per day.[99])

Agriculture is regarded as a key innovation because it generated so many opportunities for the creation and spread of further inventions. The plough is a case in point.[100] When agriculture was first developed in rich zones such as the Nile Delta and Fertile Crescent, simple handheld digging sticks were used to create holes in which to plant seeds. Known historically as "the cradle of civilization," the Fertile Crescent is regarded as the (first) birthplace of agriculture, as well as of urbanization, writing, and organized religion. The name derives from the fertility of the land in the vicinity of the Nile, Euphrates, and Tigris rivers, whose annual flood waters regularly rejuvenated the soil and promoted vigorous plant growth—excellent conditions for cultivating crops. That agriculture was first developed in fertile regions that were rich in natural biodiversity comes as no surprise, because the regular flooding meant the land could be farmed without turning it over.[101] The earliest farmers scratched at the earth with a hoe to form a tilth in which seeds were sown. This traditional tillage method is still practiced in some tropical or subtropical regions today. However, as soon as a society's struggle for survival became closely linked to the success of farmed crops, the need to cultivate the land necessary to sustain a burgeoning community generated a demand for greater farming efficiency. The more the soil was tilled, the better the germination of the seeds and higher the crop quality. In the struggle to become more efficient, handheld hoes eventually evolved into simple ploughs.

The domestication of cattle in Mesopotamia and the Indus valley civilizations, perhaps as early as 6000 BCE, provided humanity with the draft power necessary to develop the larger, animal-drawn scratch ploughs. This invention was used by the Egyptians over 4,000 years ago, as can be seen in the many paintings, ceramics, and stone carvings in which primitive ploughs are pulled by oxen (figure 9). Animals enabled the land to be tilled more easily and faster, and allowed greater amounts of food to be produced. In ancient Egypt, the fields were ploughed and sown with food crops, as well as with flax for linen, and other products. The Egyptians made considerable advances in the plough's design, including innovations that stabilized the blade, and lifted and turned the soil. These advances, combined with the invention of complex irrigation systems, meant that the Egyptians could grow many crop varieties in an arid climate.[102] The Greeks developed Egyptian ploughs further, by fitting them with wheels, which provided far greater control and maneuverability, and with iron cutting blades.[103] Evidence of wheeled vehicles appears from around 5,500 years ago, virtually at the same time in Mesopotamia, Russia, and Central Europe, so the question of which culture originally invented the wheel is still unresolved.[104] The ancient Greeks were almost certainly not the source of this innovation, but rather adopted the wheel from elsewhere, as they did the plough. The Greeks

FIGURE 9. Agriculture in ancient Egypt. This painting by Charles Wilkinson, held in the Metropolitan Museum, New York City, is a 1922 modern reproduction of a tomb painting dated to 1299–1213 BCE.

nonetheless showed considerable ingenuity in integrating wheel and plough into a powerful new combination.

Terracing and irrigation are also major consequences of agriculture. High up in the Peruvian Andes, 2,500 meters above sea level, sits the spectacular city of Machu Picchu. Built by the Incas at the height of their powers, Machu Picchu was abandoned and virtually forgotten at the time of the Spanish conquest until it was famously rediscovered in 1911 by Hiram Bingham, an archaeologist from Yale University. What makes Machu Picchu such a dramatic setting is not just its altitude, but the incredible, extensive mass of terraces that surround it, cut like giant steps into the mountainside. Inca terracing covers huge distances, and still shapes the Andean landscape today. Like any city, Machu Picchu required a strong agricultural foundation, but suitable land for growing crops is not easy to find on a mountain. Terraces were the ingenious solution the Incas adopted. The terraces allowed them to drain and siphon rainwater, and stabilize the soil, thereby allowing maize and potatoes to be cultivated. Each terrace was multilayered, with topsoil above the subsoil, sand, and finally stone chip layers (figure 10). The Incas

FIGURE 10. Peruvian terraces spread over vast distances, and are thought to be thousands of years old. They were the solution that the Incas and populations that preceded them adopted in response to the challenge of growing crops on steep mountainsides. By permission of Ros Odling-Smee.

could then retain rainfall to water their crops rather than losing it as runoff, while sophisticated channeling systems, complete with canals and aqueducts, provided irrigation for the artificial fields.

The Incas, like the Egyptians and all of the world's major, large-scale, and complex societies were critically dependent on irrigation technology that took water from natural sources and diverted it via artificial channels to the crops. These irrigation systems required a huge investment of resources and labor. Early cities were built on this technology. In Mesopotamia, for instance, there was such low rainfall that growing crops was only possible with water drawn from the Tigris and Euphrates. The earliest city-states of Mesopotamia lay in the lowest, most water-rich areas, and each city-state devised its own irrigation systems. In fact, the cities were essentially local irrigation schemes around which administrative, marketing, and defensive centers rose.[105] The hot and dry Mediterranean regions were the same, where irrigation was indispensable to the ancient Greeks and Romans.[106]

The central buildings of Machu Picchu are all built from blocks of stone, fit together without mortar, but nonetheless cut with such precision that they made remarkably stable drystone walls.[107] Machu Picchu's 200 buildings are arranged around a huge central square, with numerous stone stairways set in the walls for access to the different levels.[108] The buildings include granaries and warehouses, army barracks, and living quarters. Roads and bridges extend from this center, along which goods and materials were transported, as well as orders dispatched from a central authority known as the supreme Inca. Ancient numerical records were produced through an intricate series of knots on strings called *quipus*; their subject matter was a numerical record of the harvests.[109]

The Incas worshipped the sun, and the fact that the sun god, called *Inti*, was also the god of agriculture clearly indicates the importance of agriculture to their society. People were well aware that sunshine was necessary for the production of crops like potatoes, maize, and other grains.[110] The same vital importance of agriculture was also apparent to the ancient Egyptians, and led them to make many offerings to *Renenutet*, the god of the harvest and nourishment. According to hieroglyphic writings on the walls of the pyramids, Renenutet was the goddess of plenty, and good fortune. She was depicted as a woman with the head of a cobra. Snakes were often seen in the fields around harvest time, hunting the rodents that threatened the crops. Renenutet was thought to protect the harvest, and hence was given the epithets of "Lady of Fertile Fields" and "Lady of Granaries."[111] Much the same is found in Mesopotamia, where *Ashnan*, the goddess of grain, and her brother *Lahar*, the Sumerian goddess of cattle, were thought to provide people with food.[112]

In Ancient Greece, *Demeter* was one of the twelve Olympians, or major deities, and was the goddess of the harvest who presided over grains and was responsible for the fertility of the earth.[113] According to Greek legend, Demeter's greatest gift to humankind was agriculture. The Romans recognized over 30 Roman agricultural deities, included *Vervactor* ("He who ploughs"), *Subruncinator* ("He who weeds"), and *Messor* ("He who reaps"). The most important was *Ceres*, the Roman counterpart of Demeter, who was ranked among the *Dii Consentes*, Rome's equivalent to the Twelve Olympians.[114] Ceres was credited with the discovery of wheat; the invention of ploughing; the yoking of oxen; and the sowing, protection, and nourishing of seeds. Like Demeter, Ceres is thought to have gifted agriculture to humankind. Before this, it was said, humans had subsisted on acorns, and wandered without settlement or laws.[115] In all these populations, countless rituals, ceremonies, dances, music, and theatrical performances grew in efforts to please the gods who were thought to hold peoples' lives in the balance.[116]

With agriculture, then, came so much more. Terraces, aqueducts, bricks, complex buildings, road networks, bridges, stairs, warehouses, food stores, armies, record keeping, hierarchical social structures, cities, heads of states, and even gods all followed once the agricultural Rubicon was crossed. From one innovation sprang countless others— each, in turn, the source of further creativity. This is the nature of cultural evolution. With every new invention the diversity of cultural life ratchets up, and the opportunities for cross-pollination of ideas grow geometrically. Innovations not only spawn better, faster, and cheaper variants, but generate a cascade of new possibilities that deliver entirely new kinds of functionality.[117]

Agriculture was not the only key innovation. The wheel, which the Greeks fused with ploughs, was originally devised for pulling heavy loads on a sledge-like structure. From this simple beginning, it was converted into a potter's wheel (in Mesopotamia), a cart (also in Mesopotamia), a chariot (in Greece), an instrument for grinding wheat (harnessing animal, wind, and water power, in diverse regions), a pulley mechanism to draw water (in Sumer and Assyria), a spinning wheel (in Asia), and innumerable other uses.[118] Today, it is extremely difficult to imagine how virtually any mechanized instrument would have been possible without the seemingly simple invention of an object moving around an axis. Such technology ranges from gears, to bicycles, cars, jet engines, and computer drives.

The Greek chariots were pulled by horses, but they did not develop horse domestication; that was first accomplished by the Scythians from Central Asia. It is said that when the Greeks first saw the Scythians on horseback they believed the horse and rider were a single animal. Allegedly, that is how the legend of the centaur arose.[119] Some of the earliest evidence of horse domestication comes from the Botai people of northern Kazakhstan, a community of foragers who adopted horseback riding to hunt wild horses, around 3500–3000 BCE.[120] Horseback riding transformed warfare by allowing mobile hordes to sweep into battle at great speed, to which numerous armies from the masses of Attila the Hun to Genghis Khan attest.[121] However, horses have also been used as working animals on farms, and in factories and mines, as well as for food. The ancient Greeks and Romans may be renowned for their chariots, but they adopted this invention from other societies. The critical innovation that allowed the construction of the light, horse-drawn chariots for which the Greeks and Romans are famed was the spoked wheel, which dates to around 2000 BCE in western Siberia.[122] Chariots were initially devised for warfare, but long after they had been superseded in this role, they were used for travel, hunting, games, races, and in processions.

In this richly interwoven manner, cultural evolution proceeds. Novel discoveries are copied, replicating first within the society, and later adopted by other communities through diffusion, conquest, or trade. These innovations are then adapted to other purposes, frequently by uniting the idea with pre-existing elements into powerful new complexes. So the great inventions of the ages in manufacturing, engineering, architecture, the sciences, and the arts express an intricate history of human endeavor, forged in constant debt to innovation and social transmission.

Agriculture precipitated a transformation in society because it not only led to directly related technologies such as spades, sickles, and ploughs, but to the idea of domesticating animals, the wheel, new means of transporting both people and goods, and to city-states; surely no early farmer could ever have imagined such profound ramifications in their wildest dreams. The same goes for the wheel, the arch, money, writing, the printing press, the internal combustion engine, the computer, and the internet. Each successful innovation propagates through repeated bouts of social learning and refinement, giving rise to a barrage of new ideas and devices in a never-ending hotchpotch of diversification and adaptation where the pace of cultural change ceaselessly accelerates.[123] In the process, cultural elements can accumulate a stunning complexity of subcomponents, which operate in exquisite coherence and synchrony, as well as a mind-blowing diversity.

Prior to the advent of agriculture, each population would have possessed at most a few hundred types of artifacts, while today the inhabitants of New York are able to choose between 100 billion barcoded items.[124] Now innovation is in the hands of the professionals.

New businesses have arisen that are entirely dedicated to devising and promoting a relentless stream of new products. Innovation is the preserve of smart, suited professionals in industry and state-sponsored research and development laboratories, while every major university is surrounded by a burgeoning science park desperate to translate scientific and engineering insights into patents, or new spin-off companies. Invention in other animals may be driven, in part, by necessity, but human innovation relates far more to *wants* than *needs*.[125] Modern medicine clearly has improved survival rates and quality of life—yet in spite of that, comparatively few technological innovations affect our biological fitness. Now we live in a world in which "fitness" has been superseded by "virality" and what is trending on Facebook,[126] where advertising companies devote vast amounts of money to convince consumers that the latest new gadget is critical to their identity or social standing.

Not all inventions prove a success, of course. For every innovation that triumphs there are multiple failures, ranging from the spectacular to the comical. We have all witnessed such cases. New Coke, for instance, was a huge commercial flop, and older readers may remember Sony's Betamax, the also-ran to VHS in videocassettes. The latter illustrates how innovations can fail for reasons other than being a bad product. Betamax was simply outcompeted in a conformist market in which the dominant product would inevitably win. More recently, HD DVDs (digital optical discs for storing data and playing back video) went the same way in losing out to Blu-ray. Another spectacular disaster was the Sinclair C5, the first commercially available electric car that was subject to a glitzy launch in 1985 by a UK computer manufacturer. The vehicle's limitations (a short range, a maximum speed of only 15 miles per hour, a battery that ran down quickly, and being open to the elements), combined with some serious safety concerns, left the product devastatingly unpopular and the company went into receivership. Even more disastrous cases include asbestos, DDT, and thalidomide, which were initially commercial successes, but hidden negative ramifications emerged years later with tragic consequences.

In some instances, how anyone might have conceived that a particular invention could ever have been a success is hard to image. What possible market could there have been for the musical flamethrower, or the cheese-flavored cigarette?[127] In the nineteenth century a bicycle manufacturer produced a one-pedaled bicycle, designed to allow Victorian women to ride side-saddle; needless to say, it was not a commercial success.[128] More recent and slightly comical disasters include cutlericks, an extremely bizarre integration of cutlery and chop sticks, and phone fingers, a latex "glove" for your index finger, to ensure that your smart phone doesn't get covered in fingerprints.[129]

Such cases bring home how nothing is inevitable about the adoption of an invention, even for a good product. All new ideas and devices must battle it out in a tough market, in which success depends not just on the product's merits, but also its appeal, the learning strategies involved, the nature of the competition, the level of exposure, the extent of promotion, the size of the target population, and multiple other factors, including a major dose of chance. Often, the potential of a new invention is not immediately appreciated. In 1943, Thomas Watson, chairman of IBM, is reputed to have claimed, "I think there is a world market for maybe five computers."[130] However, stochastically and over time, the qualities of beneficial variants will give them a statistical edge that, while not guaranteeing success, stacks the deck in their favor. As for biological evolution, where cultural variation, differential "fitness" (i.e., different propensities to spread), and inheritance exist, cultural evolution will ensue, and cultures will adapt, diversify, and accumulate complexities.

Agriculture also changed the face of human adaptation in another manner, which resulted from its demographic impact. As we have seen, agriculture caused population growth, and those greater numbers strongly affected the pace of cultural evolution. Evolutionary biology teaches us that in small populations, chance factors dominate and the dynamics of evolving populations are dictated by drift, but as populations get larger, natural selection starts to become more important and advantageous mutations become more likely to propagate.[131] The same has been proven for cultural evolution, where again selective processes become increasingly important as population sizes increase.[132] This means in the larger societies that rich agricultural yields supported, beneficial innovations would be more likely to spread and to be retained.[133]

Cultures can be thought of as comprising a collective memory that provides a knowledge-base for the community.[134] The efficiency of these informational repositories depend on the size and structure of the populations, which means that demographic processes can both help and hinder the buildup and retention of cultural memory.[135] Formal theory suggests that below certain population density thresholds not only is it difficult for new innovations to accumulate, but adaptive cultural knowledge may actually be lost.[136] Why this should be is easily understood. If you have a good idea that is not shared through social learning, then it dies with you; while conversely, if enough people are around for others to take your idea onboard, the idea may well outlive you. Experiments have demonstrated that groups of individuals are better able to solve complex problems than individuals alone, and that the larger and better connected the population, the more complex the innovations that can be devised.[137] Most truly tough challenges are not solved by a single genius, but by groups of individuals working together.[138] Larger or denser populations not only generate more innovations, and more complex innovations, but are capable of preserving them for longer, leading to bigger cultural repertoires.[139] This relationship between cultural repertoire and group size was demonstrated through a theoretical analysis by Harvard University anthropologist, Joseph Henrich, using the example of Tasmanian culture.[140] The separation of Tasmania from mainland Australia, due to rising sea levels around 10,000 years ago, provides one of the most famous examples of the loss of cultural memory. When no longer in contact with Aboriginal populations on the mainland, the small Tasmania community simply forgot numerous skills and technologies, including how to manufacture cold-weather clothing, fishing nets, spear-throwers, and boomerangs.[141] More generally, anthropologists and archaeologists have found a positive relationship between the size of the population and the diversity of their tool kits (i.e., their technology). For instance, among Oceanic populations at the time of early European contact, larger populations had broader tool kits.[142]

One negative ramification of a long history of selection for high-fidelity copying mechanisms is that humans sometimes retain inappropriate or outdated knowledge.[143] A famous example is the Norse attempt to colonize Greenland, beginning around 1000 AD. The Greenland colony ultimately failed because the colonists persisted in trying to raise cattle for food, a socially transmitted norm that had been adaptive in their Scandinavian homeland, but which was extremely ineffective in the harsher environment of Greenland. The colony failed to shed reliance on their old cultural knowledge, and the Norse starved. Such examples are, however, more the exception than the rule. By and large, cultural knowledge is broadly adaptive.[144] One reason why human global population growth is unerring is that our populations die out with far less frequency than do those of most other species. We survive, in challenging circumstances, precisely because our culture typically confers an adaptive plasticity, rather than inflexibility.[145] That plasticity draws on a huge cultural knowledge base, which retains good ideas and effective solutions to past problems over long periods of time, including those that arise infrequently.

Language plays a central role in helping human societies to share and retain knowledge, and many traditional societies possess oral traditions and rituals that preserved historical information with remarkably veracity. A compelling illustration of the potency of verbal tradition is provided by an account of the history of the Māori.[146] This people's oral history depicts the tale of their arrival in New Zealand after traveling from their homeland *Hawaiki* in large, ocean-going canoes called *waka*. According to legend, the first inhabitants to come to the Bay of Plenty arrived in AD 1290, when *Takitimu* travelled across the ocean and landed his canoes at Mount Manganui on the North Island. Consistent with this account, archaeologists have dated the original Māori settlers to this region at some time between AD 1250 and 1300. More than 400 years passed before the community was encountered by the English explorer, Captain James Cook. In 1769, Cook sailed around New Zealand and stopped at the Bay of Plenty to try to communicate with the locals. The written account of his voyage describes how a Polynesian member of the crew called Tupia was able to communicate perfectly well with the Māori, despite the fact that he had lived his entire life on Tahiti, an island some 2500 miles northeast of New Zealand.[147] Back in Tahiti, Tupia had been a priest and was therefore steeped in the oral tradition of his community. Cook's crew described him as a polished orator who could recite the history of his people without error or omission, a feat that took several weeks; he could also name his ancestors for the last 100 generations.[148]

The account of Cook's voyage notes a remarkable similarity between the Māori language and that of Tahitians in French Polynesia, and lists large numbers of identical words. The report also notes extraordinary parallels in their oral histories—for instance, both claimed to originate from the same homeland to the northwest. Evidence from archaeology, linguistics, and physical anthropology supports these accounts perfectly, and indicates that the first settlers in New Zealand came from east Polynesia and became the Māori. Language evolution studies and mitochondrial DNA evidence suggest that most Pacific populations originated from Taiwanese aborigines around 5,200 years ago, and moved down through Southeast Asia and Indonesia;[149] that is, they traveled from a land to the northwest, exactly as the shared elements of both traditions specified.

Such examples are impressive, but spoken tradition also has limited fidelity and stability, even with ingenious inventions such as the aboriginal use of song to maintain and transmit geographic knowledge. However, the human knowledge base has been further enhanced by the creation, again largely through cultural means, of systems of memory that are external to our brains, such as written records, architecture, painting, or the Incan quipus. This external storage, which includes all aspects of material culture, has been crucial in the explosion of cumulative culture in the last 10,000 years, because it massively increases the pool of knowledge available to a population and encompasses the recorded memories of individuals that lived in the past.[150] Even for those other species that possess some rudiments of cultural transmission, animal memory relies largely on information residing in brains, and there are concrete limits to both the amount and duration of information any living population can carry in that manner. Externalized and persistent social information appears to be a very rare thing in nonhumans. Some ants leave pheromone trails to food sources, for instance, but these trails are highly ephemeral and hardly ever persist longer than 24 hours.[151] In contrast, modern human societies are hugely reliant on a collective, externalized, and persistent memory that now encompasses books, records, libraries, computer databanks, and the World Wide Web. These resources are a characteristic of entire societies, or even global knowledge, rather than any one individual.[152] The biggest library in the world is the British Library in London, which holds about 170 million books. A typical book contains around a megabyte of information, which would imply that the world's biggest library contains approximately 170 terabytes (170,000 gigabytes) of information. One recent estimate of the amount of information now stored on the internet is 1,200,000 terabytes.[153] Externalized cultural memory stores far outstrip the capability of any one individual's memory to contain information, or even any population's collective knowledge, by several orders of magnitude. Knowledge is becoming more difficult to lose, and cultural evolution has ratcheted again, from a process encompassing knowledge gains and losses, to a process greatly biased toward gains. Our major problem now is that we are so swamped with information that we cannot effectively filter it to locate what is immediately relevant.

With agriculture, then, came not just an upsurge in innovation, but a population explosion and an information overload. To get off the ground, agriculture required a favorable climate and soil, and a location rich in biodiversity that included plants and animals suitable for domestication. However, agriculture is not something that just happened to people when a suitable climate transpired. More than anything, agriculture required populations to draw on their guile and ingenuity and seize the opportunity by actively creating the circumstances that rendered it economical.[154] While innumerable positive consequences followed, the agricultural revolution has nonetheless proven a double-edged sword. Return to the Fertile Crescent today, and you will find that many of the famously rich plant resources of the region are being eroded through degradation of natural habitats, intensification of the cultivation of arable lands, expansion of cultivation into marginal areas, and overexploitation of natural pastures and grazing lands.[155] Now a grave and ever-growing danger is that the same inherent biodiversity that originally rendered this region the birthplace of agriculture, will continue to be destroyed by its relentlessly unfurling spawn. The same story is written the world over. Unless a credible approach to the management of ecosystems that is based on sustainable agriculture and development can be implemented on a global scale, no zones rich in biodiversity will be left on our planet.

Humans trail in their evolutionary wake a disconcertingly long history of biodiversity destruction.[156] In the period between 50 thousand and 10 thousand years ago, at least 101 out of 150 genera of Earth's megafauna (including woolly mammoths, giant sloths, and saber-toothed cats) went extinct, and archaeologists are increasingly confident that humans played important roles in these events.[157] By 12 thousand years ago our species had dispersed from Africa to the far corners of Eurasia, Australia, and the Americas, and the concomitant increase in global human populations is clearly associated with species extinctions and habitat modification.[158] Early colonists of New Guinea, Borneo, Australia, and the Americas burned and disturbed the pre-existing forests to promote the growth of useful plants and enhance hunting opportunities. The inexorable spread of agriculture from a small number of centers of early domestication to encompass large swaths of the planet had unprecedented impacts on species distributions, including the relentless promotion of domesticated crops and animals, commensals, weeds, and microbes carried by humans. When Polynesian people expanded across the Pacific, for instance, they carried with them taro, yams, banana, pigs, chickens, dogs, and numerous unintentionally transported species, including rats; thousands of endemic species are now known to have gone extinct in the face of this invasion, including two-thirds of nonpasserine birds.[159] Far from being a product of postindustrial societies, the Anthropocene began some tens of thousands of years ago, and was gradually ramped up through increasingly intensive cultural niche construction until virtually no "pristine" natural environments now remain.

Agriculture produced a food bonanza for many, and the global industrialization of food production has resulted in unprecedented diversity of diet in many postindustrial societies. However, now an abundance of inexpensive, high-density and high-caloric foods laden with sugar and fats is made easily available to many of us through means (such as driving to the supermarket) that require little expenditure of energy. The result is increasing levels of obesity, and a surge in heart disease, diabetes, high blood pressure, and cancer.

Agriculture gave rise to complex society, but with that shift arose vast inequalities. Today, the 80 richest people on the planet have a combined wealth comparable to the poorest 50% of the world's population.[160]

Billionaires' businesses employ a great deal of people, giving rise to some higher household incomes. However, any trickle-down effect of this wealth is insufficient to prevent the lack of basic services in many of the poorest countries, and on a global scale the gap between rich and poor continues to widen.[161]

Complex societies first began to appear when humans accrued sufficient socially transmitted acumen to increase the productivity of the land and lead a settled, rather than nomadic, existence. A uniquely potent innovativeness and enhanced capability for high-fidelity social learning has allowed our species to devise means of exploiting our environments with extraordinary efficiency. We have increased carrying capacities by orders of magnitude and our numbers have soared. Through our cultural niche construction, we have transformed and regulated the circumstances of our lives to a degree unprecedented in the natural world, and that transformation accelerated in the aftermath of those early experiments in plant domestication. Agriculture is where the modern era began; it was the first key innovation in an ever-multiplying cascade of novel products and ideas that in developed countries led inexorably to today's "innovation society." However, the ideology of this society is that innovations solve problems, but that is only half of the story. Innovations construct new niches, just as organisms do, and every "solution" has the potential to generate many new "problems."[162] When our ancestors first devised agriculture, they opened up a Pandora's box, and let lose the evil of the Anthropocene.

CHAPTER 11

FOUNDATIONS OF COOPERATION

You take a flight from New York to London. Thousands and perhaps millions of people, including the pilot and flight attendants, air traffic controllers, airport workers, baggage handlers, travel agents, and bank workers cooperated to get you there. No one stole your luggage, no one ate your in-flight food, and no one tried to sit in your seat. In fact, the hundreds of people on the airplane, despite being mainly strangers, behaved in an entirely civilized and respectful manner throughout. Cooperation that comprises so many unrelated individuals performing such diverse roles in such a coordinated manner is unprecedented in the natural world.

The previous chapter discussed the emergence of complex human societies, but arguably the most striking feature of early agricultural communities was the rapid emergence of what can only have been large-scale cooperative enterprises. Terraces cannot be carved out of the mountain, extensive crops cannot be harvested, granaries cannot be built, and city-states cannot function efficiently without an extraordinary level of cooperation among community members. Hunter-gatherers also coordinate their actions in cooperative endeavors such as group hunting and foraging, as well as through sharing food, labor, and childcare; they also gather together as a community when hostility or disputes with other societies arise. How is this scale of cooperation to be explained?

Kin selection, which occurs when individuals help their relatives, could certainly account for some aspects of the cooperation witnessed in human societies, but it would not account for larger-scale projects among unrelated or distantly related individuals. Our hominin ancestors are thought to have aggregated into small bands that were often kin based. However, modern hunter-gatherer societies are known to have many interactions with nonrelatives, even in societies numbering only a few hundred individuals.[1] Such cooperative activities are regulated by systems of norms and institutions. In agricultural societies also, cooperative arrangements have to be negotiated and maintained among thousands of often unrelated individuals.

Coercion certainly occurred in some instances, as when powerful leaders and groups forced the weak to do their hard labor. By 4000 BCE, the Sumerians were building cities of over 10,000 people, using slave labor captured from the hill country,[2] while the Old Testament book of Exodus reports, "the Egyptians enslaved the children of Israel with backbreaking labor." However, this cannot be the whole story. In order to get slaves, societies would need to engage in impressive acts of group cooperation by, for instance, putting together a functioning army that operated in an organized fashion. Moreover, historical research has established that some of the major architectural projects of antiquity did not get built by force. The pyramids, for instance, are now thought to have been assembled by paid laborers.[3] The builders came from poor families from the north and south of Egypt and were respected for their work—indeed, those who died during construction were bestowed the honor of being buried in the tombs near the sacred pyramids of their pharaohs. Ancient societies, just like today's modern ones and including contemporary hunter-gatherers, were built upon massive levels of cooperation in which most individuals participated entirely voluntarily. How was this possible? Conventional gene-based evolutionary explanations alone cannot account for all aspects of human cooperation.[4] The answer is multifaceted, and draws on some surprising and rarely appreciated connections between cooperation and social learning.

One important step in getting large-scale cooperation off the ground was the evolution of widespread teaching. Few people think of teaching as an act of cooperation but, in fact, that is precisely what it is. Cooperation is defined as behavior that provides a benefit to another individual (the "recipient") and that was favored by natural selection because of this positive effect.[5] In the case of teaching, the "benefit" provided is "useful knowledge." Theoretical work described in chapter 7 suggested that a generalized form of teaching coevolved in humans with our cumulative culture. The analysis strongly suggested that teaching is widespread in humans because cumulative cultural knowledge builds up over time, leaving otherwise difficult-to-learn, but highly advantageous information (such as how to make a functional stone tool, or gather honey) available in the population for tutors to impart to their pupils.[6] The claim is frequently made that human cooperation is unique,[7] but in what respects, and whether this claim is justified at all remains contested.[8] Our analysis of the evolution of teaching provided one possible answer to this conundrum. Clearly, "complex and unique mechanisms to enforce cooperation have arisen in humans, such as contracts, laws, justice, trade and social norms,"[9] and all of these mechanisms require teaching in order to operate. Much human cooperation apparently also requires a capacity for language, but as we saw in chapter 8, there are good reasons for thinking the origin of language is tied with teaching and cumulative culture. Language may initially have evolved to enhance the efficiency and accuracy of teaching and, even if it originated for some other reason, it could certainly have been applied in this context. Human cooperation may, therefore, be unusually extensive as a result of cumulative culture and may be uniquely reliant on important mechanisms that are either rare or entirely absent in other species—specifically, teaching and language.[10]

Two of the more intuitive findings to emerge from our investigation of the evolution of teaching were that the likelihood of teaching increases with the degree of relatedness between tutor and pupil, and decreases with the costs of teaching.[11] Empirical data from human populations support these conclusions.[12] Accordingly, we can envisage the kind of widespread, generalized teaching observed in humans to have initially appeared among close relatives; for instance, parents teaching foraging skills to offspring, or siblings helping each other learn toolmaking. Yet, without some rudimentary form of language, the teaching of more distant relatives would rarely have been economical, because the inclusive fitness benefits would be modest. By simultaneously reducing the costs and incrementing the accuracy of teaching, the advent of language would have allowed instruction to spread to more distantly related individuals, including small groups of kin such as hominin hunter-gatherer bands. This teaching may have sufficed to allow individuals in small groups to teach each other specific roles in coordinated tasks, such as hunting antelope or driving off predators, thereby broadening the scale of kin-based cooperation.

As documented earlier in this book, many animals that are proficient at social learning (primates or cetaceans) exhibit behavioral traditions, which are effectively society-specific conventions for, say, singing the local song or hunting prey. Our hominin ancestors also will have had group-specific habits maintained as social conventions. Some traditional behaviors could be picked up through observational learning, while other, more challenging habits would have spread through teaching among relatives. Most animal teaching merely creates enhanced opportunities for learning in the pupil,[13] as when meerkats provide pups with disabled scorpions.[14] However, in rare cases, animals show hints of "coaching," in which the response of the tutor functions to encourage or discourage the pupil's behavior;[15] an example is the maternal display of a mother hen who pecks and scratches the ground intensively to distract her chicks from unpalatable food.[16] Such error correction is, of course, a feature of contemporary human teaching. It follows that at some juncture in our history, our ancestors began systematically to correct the behavior of the individuals they taught; in the process, they shifted their society away from reliance on mere conventions and toward governance through norms.[17] People stopped illustrating *a* way to behave and began insisting on *the* way to behave. Eventually, each society was characterized by a particular set of norms that dictated how individuals *should* behave (e.g., how they should build a fire, how they should catch a turtle, how they should till the soil), each of which was propagated through verbal instruction. Norms specify rules of social interaction too, including specification of how people should respond to norm violation.[18] With the advent of norms, hominin social life effectively became transformed from simply living in groups to identifying with the group, abiding by its rules, and privileging in-group members.[19] Norms facilitated group coordination and thereby substantially enhanced the society's capacity for cooperative endeavor. In order to resolve conflicts or prevent social problems from arising in the future, institutionalizing the norm explicitly as a "rule of law" that all members of the society must abide by, and agreeing on sanctions for violators, would sometimes have been necessary.[20]

From that point onward, our ancestors lived in a society structured by cooperatively created and enforced conventions and norms for how to behave, many of which evolved into rule-governed social institutions.[21] These norms are far from obvious, and young members of a community must typically be taught both the nature of the norms and the need to conform to them. This much is clear in contemporary human populations where, say, how a check or bank transfer works; when, how, and why citizens should pay their taxes; or the rules of driving, are not transparent. All human societies possess such norms and laws, which children generally learn well, largely because adults actively instruct them, or else devise organized learning environments for that purpose.[22] Human cooperation extends far beyond individuals helping their kin or responding reciprocally to those that help them. The scale of human cooperation is unprecedented in large part because it is uniquely built upon socially learned and transmitted norms. These do not just specify how to behave but also set guidelines for rewarding good behavior and punishing the bad, as brought to prominence by the experimental work of leading economist Ernst Fehr.[23]

Human social life is far more cooperatively structured than that of any other primate,[24] partly as a consequence of our norms. Large-scale cooperation commonly requires coordinating the actions of many individuals. Such organized action among pairs and groups of contemporary humans typically involves shared intentionality and goals; and joint attention, perspective-taking, and commitment.[25] Here again, teaching is relevant. Teaching another individual how to prepare food, build a fire, or make a tool requires both joint attention and joint commitment, as well as shared intentions and goals. Effective teaching would also benefit from the tutor taking the perspective of the pupil. The cognitive abilities necessary for coordinating action may initially have evolved through teaching, and subsequently might have been applied to facilitate diverse, large-scale cooperative enterprises. In this manner, language, teaching and conformity have become central to human behavior, in a way that they have not for other animals.

Research into the evolution of cooperation has generated a huge literature focused on the problem of how cooperative societies prevent individuals from cheating. However, it largely neglects a second and equally challenging problem related to coordination: How do you induce individuals to work together to generate a profitable output?[26] Complex coordinated action among multiple individuals is very difficult without social learning, language, and teaching.[27] Cooperative foraging, hunting, and defense is seen in other animals (lions hunting, for instance, or the circular defensive maneuvers of musk oxen against wolves), but in such societies individuals rarely take up a variety of distinct and coherently integrated roles. That would require some means of coordinating the behavior of the collective, and such mechanisms are generally not present. However, through our language, teaching, and the inadvertent construction of learning environments for others, humans can solve the coordination problem; they can assign distinct roles to individuals, and ensure each is trained.[28] Indeed, experimental studies have demonstrated that language is an important mechanism through which humans resolve coordination tasks.[29]

Training through apprenticeships may also have played vital roles in organizing group coordination.[30] In early agricultural societies, the pressure to generate sufficient food to feed the mushrooming population demanded division of labor and occupational specializations. For society to function efficiently, relevant skills and expertise would need to be passed among unrelated individuals, but these skills would frequently be far too complex to pick up simply through imitation. They require the cooperation of the would-be tutor to instruct the pupil. Much complex human knowledge associated with manufacturing tools or goods, for instance, or the extensive knowledge associated with a particular profession can only reliably be passed on through a long period of apprenticeship. This requires the establishment of a long-standing cooperative relationship between a tutor and a pupil.[31] Here, large-scale cooperation took another step forward, as experts in a variety of trades appeared. Unlike their predecessors, who benefited from teaching only indirectly through helping their relatives, these experts began tutoring unrelated individuals in exchange for the direct advantages of resources such as food, clothing, or protection. Various classes of professional teachers emerged; most commonly, priests taught children of the wealthy the skills and knowledge necessary for their future roles, but early agricultural societies also had teachers of military combat, farming methods, and even dance.[32]

There are further grounds for thinking that the uniquely human capacity for cultural learning, teaching, and language substantially increased the scope of cooperation in our species.[33] Teaching may have originally evolved as cooperation among close kin; but with language, teaching could be extended to support other established cooperative processes, such as mutualistic exchanges, indirect reciprocity, and group selection.[34]

Some human cooperation, both among unrelated individuals and between groups of individuals, can be understood as mutualism. Chimpanzees, orangutans, macaques, and capuchins all exhibit population-level differences in their behavioral repertoires, but, migration and dispersal aside, there is no evidence of exchange of valued resources between communities. However, among those hominin ancestors capable of nascent language, such population-level diversity in material culture would have created the opportunity for trade in extracted and constructed resources. Mutualistic relationships, where two organisms provide resources that help each other, are widespread in animals. For instance, a bird called the oxpecker will commonly take a ride on the back of rhinoceroses, zebra, or cattle, and eat the ticks and other parasites that live on their skin.[35] The arrangement is mutually beneficial, because the birds find food and the mammals receive effective pest control. However, the evolution of this type of exchange is comparatively easy to comprehend, because the mammals do not want the parasites desired by the bird; to the contrary, they want rid of them. Equally common are cases where animals provide the same service to each other, such as horses or baboons grooming each other. Here again, the evolution of this behavior is no mystery, as the two services are equivalent, and hence the trade is easy to recognize as a fair exchange. Conversely, evolution of the kind of trade or barter commonly observed in humans, like swapping a tool for food, is more challenging to comprehend; here both parties value both of the commodities but, as the items are different, the exchange currency requires negotiation. This form of mutualism appears to be rare or absent among other animals.[36] The only documented example of which I am aware is the putative trade of meat for sex among some chimpanzees that hunt red colobus monkeys,[37] but this claim is contested.[38] A reasonable conclusion would be that the trade of distinct, desired commodities is exceedingly rare in other animals.[39] That this phenomenon may be unique to humans is not surprising, as it seemingly requires some capacity to agree on a rate of exchange, something that would be very difficult without at least protolanguage (or perhaps the flexible use of shared gesture). With the evolution of language, trade becomes a possibility; with trade comes negotiation, and selection for still more developed communication. Particularly in environments that vary from one location to the next, culturally transmitted population-specific diversity among linguistically capable hominins creates the opportunity for the mutualistic exchange of goods.

Trade exploits a division of labor in which valuable goods or services, available to some individuals or societies but not to others, are exchanged—this variation in the availability of resources is what makes the contract economical.[40] The emergence of large-scale, stratified societies, and the associated development of distinct professions, would immediately have created extensive opportunities for trade; indeed, such division of labor almost certainly could not have arisen without trade. The more division of labor exists within and between societies, the more opportunities for trade arise. Eventually, a threshold is passed whereby it becomes convenient to facilitate such exchange with a common currency, and the institution of money arises. As early as 9000 BCE, both grain and cattle were used as money or as barter.[41] The Hebrew shekel is thought to have originated as a weight of barley, corresponding to about 180 grains. That such agreements might arise without language is difficult to imagine. In this manner, norms, institutions, and laws stabilized cooperative interactions between nonrelatives on huge scales,[42] while language helped to ensure that these rules or agreements were specified in detail and widely known.[43]

I suspect that the advent of teaching through language was a game changer for our species, because it hugely enhanced the scale and mechanisms of cooperation.[44] For instance, that individuals will tend to help those who help them in return is well established,[45] but whether individuals will tend to help those who help others,[46] and how prevalent this phenomenon (known as "indirect reciprocity") is in nature, remains far more contentious. Theoretical models have demonstrated that indirect reciprocity can lead to cooperation, and explain why it might pay individuals to develop a reputation for this.[47] However, as Martin Nowak, the Harvard evolutionary biologist who has pioneered research into this mechanism stated: "Language is intimately linked with cooperation. For the mechanism of indirect reciprocity to work efficiently it needs gossip, from names to deeds and times and places, too."[48] In addition, verbally taught social norms allow humans to institutionalize the punishment of noncooperative individuals—for instance, through policing or socially sanctioned retaliation. Theory and experimentation show institutionalized punishment is a more effective means of preserving cooperation than individual-level retaliation.[49] Such cooperation nonetheless allows the possibility of cheating over cooperative endeavors. At least in humans, subtle cheats can operate by gaining control of communication networks to ensure that messages sent maximize their returns, or those to their group. This form of cheating might again have selected for more competent, skillful communicators. At the dawn of language, as the repertoire of local population-specific symbols expanded, it would become increasingly difficult to understand the protolinguistic symbols of other societies without effort, which further reinforces the benefits of biasing learning toward the local population. With time, local protolinguistic variants would start to resemble dialects, and may define and signify a community.

Interpopulation cultural diversity sets a premium on recognizing and preferentially learning from the members of one's own group who have useful local knowledge, rather than from outsiders who come from other groups. Theoretical analyses suggest that conforming to local traditions is favored in such circumstances, with several important consequences, including the evolution of "ethnic markers" that symbolize group membership, increased cooperation within groups, and the potential for greater conflict between groups.[50] Languages or dialects can function effectively as ethnic markers and promote local learning and other parochial tendencies.[51] In turn, imitation, teaching, languages, and local conventions all act to ensure that local differences in behavior among groups are maintained in the face of the dispersal of individuals. This allows an unusually stable form of group selection to arise, known as "cultural group selection," which has shaped human history.[52]

Anthropologists Rob Boyd and Peter Richerson first brought to prominence the hypothesis that group selection works at the cultural level through the selection of cultural traits,[53] such as a population's reliance on agriculture. Those groups that possessed more effective or efficient traditions, norms and institutions fared better in competition with other groups. For instance, (1) societies with an organized army are more likely to win conflicts than those without, (2) city-states with division of labor and occupational specializations would tend to out-compete those without these innovations, (3) agricultural communities that have devised irrigation systems would flourish more readily than others, and (4) societies with religious doctrines that stabilize within-group cooperative activities will thrive at the expense of those with no gods to help ensure compliance. The net result is the spread of military technology, division of labor, irrigation, religious doctrines, and many other cooperative endeavors. By these means, societies are able to resolve many collective action problems.[54]

You might be wondering why, in this instance, I envisage that natural selection will have operated on groups of individuals, rather than the more conventional selection of individuals. If individuals in agricultural societies end up having more offspring than those in nonagricultural societies, then, provided offspring tend to adopt the means of subsistence of their parents, agriculture will increase in frequency. What is missing from this line of reasoning is the recognition that most of the fitness benefits associated with agriculture derive from group-level activities. A lone farmer scratching a living solely through his own efforts, would not typically produce any more food or raise any more offspring than a hunter-gatherer. Only when practiced by groups of people that engage in shared enterprises to produce resources that benefit the collective (known as "public goods") does agriculture start to become highly productive. Several hundred people are typically required to construct a decent irrigation system. A similar number are required to build a corral suitable for catching antelope or horses. Some fish weirs, which are traps built of stone, nets, or wooden fencing for catching fish, are several hundred meters long and would have required a large group of individuals to manufacture. Burning the land, sowing the seeds, harvesting the crops have all traditionally been activities in which entire communities engaged. Such activities are infeasible for an individual farmer, but a group of agriculturalists working together could yield substantial dividends and those that cooperate in this manner generally outcompete those that do not. The end result is the propagation of cooperative practices.

Group-level cooperative enterprises are far from restricted to agricultural communities. The Nuer and the Dinka are two African cattle herding societies that live in Sudan. The two groups have a long history of conflict, with the Nuer dramatically expanding their territory at the expense of the Dinka throughout the nineteenth century. The success of the Nuer in warfare is attributed to their social structure, which allowed the Nuer to call on larger war parties than the Dinka. As a consequence, the Nuer's beliefs and practices spread.[55] More generally, many small-scale societies engage in cooperative hunting and foraging, and form war parties when conflicts arise,[56] and cultural group selection likely operated on these societies to propagate their cultural traditions.[57]

Boyd and Richerson place emphasis on accurate social learning, a "when in Rome do as the Romans do" conformity in which individuals adopt the behavior of the majority, as well as on the norms and institutions that regulate social behavior.[58] The latter include the institutionalized or socially sanctioned punishment of noncooperators.[59] The significance of conformist transmission is that it minimizes behavioral differences within groups, while maintaining differences between groups. Moreover, as the payoff to any behavioral strategy depends on the local frequency of its use, then even subpopulations in identical environments may end up performing quite different behavior.[60] In other words, cultural processes generate plenty of variation among human groups for natural selection to act upon. Extensive data now demonstrate that the differences between human societies result far more from cultural rather than genetic variation.[61]

Several factors make Boyd and Richerson's idea highly plausible. First, in cultural inheritance, unlike genetic inheritance, a descendant can learn from individuals other than their biological parents, allowing them to be sensitive to the most frequent cultural traits in their society and to conform to the dominant local behavior. This helps preserve cultural differences. Second, because a threatened or defeated people may switch to the traits of a new conquering culture, either voluntarily or under duress, migration will not weaken the cultural differences between groups (unlike gene flow in genetic group selection). Thus the movement of people between groups does not typically greatly erode group-level variation. Third, symbolic group marker systems, such as rituals, dances, songs, languages, dress, and flags, make it considerably easier for cultures to maintain their identities and to resist imported cultural traits from immigrants, than it is for local gene pools to maintain their identity by resisting gene flow. Fourth, institutionalized punishment (e.g., by a police force) or socially sanctioned retaliation (e.g., beating of deserters during warfare) can stabilize cooperative norms within the society, as both data and theory strongly attest.[62] Collectively, this leaves cultural group selection substantially more likely to be operational than genetic group selection, and indeed, the former now commands extensive empirical support.[63]

Cultural processes, including cultural group selection, can feed back through a gene-culture coevolutionary dynamic to affect genetic evolution, and thereby influence our evolved cognition. This interaction is thought to favor evolved psychological predispositions for cultural life, which anthropologists Maciej Chudek and Joseph Henrich label a "norm psychology."[64] Norm psychology refers to "a suite of psychological adaptations for inferring, encoding in memory, adhering to, enforcing and redressing violations of the shared behavioral standards of one's community."[65] Considerable theoretical evidence supports the suggestion that once a species becomes heavily reliant on social learning and culture, it will probably evolve a specialized norm psychology.[66] Theoretical analyses suggest that humans should be particularly adept at recognizing, representing, and adopting the local norms of their society, as well as notice, condemn, and punish violations of those norms.[67]

For instance, moral norms could plausibly have generated natural selection acting on human genes to favor cooperative tendencies. Individuals who are more inclined to conform to norms would find it easier to enter larger norm-bound societies and to abide by the rules, than individuals lacking this tendency. These more "docile" individuals would be at an advantage, to the extent that they would be better placed to benefit from the society's technologies and less vulnerable to exclusion or punishment.[68] In turn, a population of more docile individuals could then permit the cultural evolution of more sophisticated and effective norms, and allow groups to maintain more reliable cooperation. A similar mechanism could have favored a tendency of individuals to feel shame or guilt when they violate a social norm.[69] While such arguments are speculative, the fact that we would struggle to imagine any other primate that could live in such a norm-governed cooperative manner, combined with the observation that artificial selection in many domesticated animals has favored docility, suggests that selection for such tendencies in humans is credible.[70]

In addition, the extensive human history of coordinated group behavior and the solving of collective action problems through social learning, teaching, and language have seemingly also fed back to shape human psychology in ways that leave us uniquely able to understand and share the goals and intentions of others.[71] We humans appear unusually inclined and able to share experiences with others, via joint attention, cooperative communication, and, of course, teaching.[72] Humans have not just evolved advanced levels of individual cognition, but also extensive skills and motivations for shared cognition (where knowledge is constructed through dialogue between individuals). The typical pattern of social foraging among other primates, as was seen in the cumulative culture problem-solving task described in chapter 8,[73] is strong competition for food, low tolerance for food sharing, and almost no food offering at all.[74] Our ancestors somehow broke away from this pattern to become cooperative foragers.[75] I strongly suspect that one ramification of the cultural drive process described in chapter 6 was that in the quest for efficient and accurate information transmission, changes in human temperament that left us more socially tolerant (together with a generalized capacity for teaching) were favored by selection.[76] Social tolerance may also have been favored by cultural group selection, or by other gene-culture coevolutionary processes.

Gene-culture coevolution could equally be implicated in the evolution of the enhanced human capacity for imitation, as well as the social bonding that imitation generates and the activities of caretakers that elicit childhood imitation. For an adult to be imitated by a child may be appealing or advantageous for a number of reasons. Being imitated may be flattering to the imitated party to the extent that it implies that they, their values, or choices have been noticed or chosen. Second, the imitated individual may wish to encourage the learning of life skills by the child, either because they inherently want the child to develop and mature, or because they want to be relieved of their child-caring responsibilities as soon as possible. Third, by engaging in similar activities, children who imitate may be less disruptive, or even helpful, than children who do not (consider children helping with foraging activities). That might explain why parents and helpers commonly encourage imitation in the children they look after, often showering them with praise and positive reinforcement and sometimes even imitating the child's actions and sounds. Theory and data suggest these responses strengthen imitative competence, and elicit further imitation from the child.[77] Natural selection could act on this relationship, favoring a tendency to respond positively to and encourage imitation by a child, and a corresponding tendency for infants to imitate adult caregivers spontaneously and respond to encouragement with further imitation. The positive emotions experienced by both caregiver and child following imitation, and the enhanced social relationship that results, may be nature's way of encouraging imitation to the benefit of both parties. Infant-directed speech may be a special case of this more general process. Natural selection may have favored the tendency for adults to produce infant-directed speech, and for children to respond to it, because it accelerates language learning.[78] Even everyday phenomena such as smiling and frowning may have been coopted to become part of our norm psychology; as cheap-to-produce but nonetheless highly effective signals of approval or disapproval, they can be easily used to teach others what they should or shouldn't do.

Gene-culture coevolution may also explain the existence of a curious relationship between imitation and cooperation. Like other animals, people (often inadvertently) copy each other's postures, mannerisms, and facial expressions. This form of social learning is known variously as "simple imitation,"[79] "response facilitation,"[80] "mimicry,"[81] or the "chameleon effect."[82] While experiments have established that such copying allows animals to acquire valuable life skills, in humans it also appears to enhance social interaction. Experimental investigations have established that simple imitation is causally related to the emergence of cooperative attitudes, and that the relationship between imitation and cooperation is bidirectional; being imitated makes individuals more cooperative, while being cooperative makes one more likely to imitate others.[83] The experiments show that if they are imitated, humans begin to like the imitator more,[84] find them more persuasive,[85] and describe their time together as more enjoyable than if they are not imitated.[86] Children as young as 18 months will rush to help adults (for instance, by picking up items that the adult has dropped) more readily when the adult has imitated the child, compared to when no such imitation has taken place.[87] Adults too are more willing to help others with simple tasks, and even donate more money to charity, after being imitated.[88]

This relationship is seemingly bidirectional. People imitate those they like more than those they don't like, and show a preference for imitating members of their own group over members of an out-group,[89] a finding that held for both religious and ethnic groups.[90]

Oxford psychologist Cecilia Heyes has argued that the bidirectional causal relationship between imitation and cooperation may function to maintain cooperation, collective action, and information sharing among members of a social group, via a "virtuous circle" of subconscious imitation and prosocial attitudes.[91] The fact that this virtuous circle functions to maintain group boundaries is consistent with these tendencies evolving through a cultural group selection mechanism. Like dialects, local physical mannerisms can spread through imitation to act as ethnic markers that subconsciously symbolize group membership.[92] A tendency to be well disposed toward individuals who behave like you, and even exhibit your mannerisms, may have been favored as an aspect of our evolved psychology, because it helps promote cooperation among the group's members. Conversely, mathematical models have found that in variable environments it pays to copy locals, because they are more likely to know the locally optimal behavior.[93] Therefore, a tendency to imitate cooperative individuals may have been favored because they are more likely to be locals. Another possibility is that these tendencies may be learned and socially transmitted.

Cultural group selection may also have favored social practices that foster the development of enhanced imitative capabilities.[94] Many societies possess traditions of dancing in synchrony, and train their military through extensive synchronous marching and fighting drills. Such groups may have been more successful than others in part because this synchronous activity trained individuals' neural circuitry for imitative proficiency, enabling them to connect the perception of others performing the action to their own performance,[95] thereby promoting within-group bonding.[96] Synchronous action that triggers endorphin release (e.g., a group of individuals that exercise together) may lead to individuals associating the simultaneous activity with positive reward, resulting in synchrony itself becoming rewarding.[97] Alternatively, if rewards are received at the end of synchronous action, such as hunting together, again a learned association may arise. If synchrony is rewarding, then social behavior that promotes synchronous action will be more likely to occur. The extensive use of rhythm (e.g., drumming) and music as a means of helping to coordinate the actions of large numbers of individuals and promote social bonding could be favored in this context.[98] Groups of soldiers that sing or chant when running can run further, faster, with less pain, and bond with each other in the process.

In fleshing out Allan Wilson's cultural drive hypothesis in chapter 6, I suggested that selection for more accurate and efficient forms of social learning might have in turn generated selection for enhanced imitative capabilities, as well as other aspects of cognition. Such imitative capabilities might, for instance, be underpinned by dedicated structures or networks in the human brain that allow us to solve the correspondence problem (the challenge of imitating when the perception of self and another individual performing the same action can be quite different), or at least confer the neural plasticity to be able to solve this problem with relevant experience. I believe that our potent capacity for imitation is itself an adaptation to cultural life.

This issue remains contentious, however. Cecilia Heyes, for instance, believes that the human capacity for imitation relies on an ancient associative learning capability, and that both the extent of our reliance on imitation, and our imitative proficiency, are socially constructed.[99] I agree that humans have constructed a world complete with mirrors and rich with synchronous activities that is unusually replete with opportunities to develop imitative competences through experience; Cambridge ethologist Patrick Bateson and I made this argument some years ago.[100] Experimental evidence demonstrates that imitative tendencies can be increased through positive reinforcement. For instance, when infants imitate parents, the latter typically respond with smiles and encouraging remarks.[101] However, I do not believe that other animals would become as competent at imitating as humans if given the same prior experience; indeed, the experiments of teaching apes to talk more or less demonstrate this. As the preceding chapter documented, there has been extensive natural selection acting on the human brain in the period since humans and chimpanzees shared a common ancestor, including evidence that our learning capabilities have been massively upregulated. Almost all human learning is socially guided and set in a

social context, and that is also likely to have been the case for our hominin ancestors for over at least the last two million years. Therefore, it
would seem more appropriate to regard the enhanced human capacity
for asocial learning as a side effect of selection for proficient copying,
rather than the other way around.[102]

Showing that a character is an adaptation has proved surprisingly
difficult for evolutionary biologists,[103] and demonstrating that certain
cognitive traits possessed by humans are adaptations to promote social learning is therefore no trivial task. Indeed, I consider the issue of
what precisely are human social learning adaptations to be one of the
great, unresolved questions in the field. Nonetheless, there are a sufficient number of strong candidates to leave me confident that humans
will eventually be shown to possess cognitive adaptations for cultural
learning.

The generalized capacity in humans for teaching, including the motivation to teach and be taught, and the ability to comprehend the state
of knowledge of the pupil, is one such candidate. If I am correct about
the original function of language to promote teaching, then language is
another adaptation for social learning. The tendency to produce, and to
attend to, infant-directed speech are further traits that may have evolved
to promote social transmission, as are other forms of pedagogical cueing.[104] The extraordinary social motivation to imitate, which anyone
with children will recognize as powerful and pervasive, and the social
bonding between individuals that this generates, constitutes another
strong case.[105] The tendency of young children to attend to the gaze of
others, and the motivation to share experience with others via joint attention is a further compelling candidate.[106] Even the whites of our eyes
may have evolved to make it easier for us to follow another individual's
gaze when in close-quarter communicative interactions.[107] The extraordinary human tendency to conform to the majority is another strong
candidate for a social learning adaptation.[108] Cultural evolution models,
but not associative learning theory, can explain why humans are more
likely to copy an action that is performed by three individuals one time
each, than an action performed by one individual three times.[109] Our
capacities for theory of mind and perspective-taking, and abilities to
read the intentions of others, are surely also part of this story.[110]

Finally, I have presented both neural and genetic evidence that humans possess elevated levels of plasticity that allows them to form cross-modal neural connections between sensory inputs and motor outputs; this plasticity thereby facilitates imitation and emulation.[111] As one would expect with the evolution of extended periods of juvenile dependency to promote the transfer of knowledge across generations, the period of synaptic plasticity characteristic of early brain development has been extended in humans.[112] Even our basic and ancient learning capabilities, including our ability to see connections among events, to discern the consequences of our actions, and to adjust our behavior flexibly have been substantially upgraded,[113] most probably through selection for enhanced social learning. This enhancement in general learning and plasticity is likely to play critical roles in human cognitive development. While I anticipate that multiple psychological adaptations for social learning may eventually be uncovered in our species, I expect the unlearned roots of such cognition to be subtle. Social learning adaptations are likely to be expressed as complex products of development that build upon evolved motivational, perceptual, or cognitive biases through very general learning processes that respond sensitively to a culturally constructed, symbolically encoded environment.[114]

In sum, there is strong evidence that the large-scale cooperation observed solely in human societies arises because of our uniquely potent capacities for social learning, imitation, and teaching, combined with the coevolutionary feedbacks that these capabilities have generated on the human mind. Culture took human populations down evolutionary pathways not available to noncultural species, either by creating conditions that promoted established cooperative mechanisms, such as indirect reciprocity and mutualism; or by generating novel cooperative mechanisms not seen in other taxa, such as cultural group selection. In the process, gene-culture coevolution seemingly generated an evolved psychology, comprising an enhanced ability and motivation to learn, teach, communicate through language, imitate, and emulate, as well as predispositions to docility, social tolerance, and the sharing of goals, intentions, and attention. This evolved psychology is entirely different from that observed in any other animal, or that could have evolved through genes alone.[115]

The world's people exhibit an extraordinary diversity of appearance, fashion, language, diet, method of subsistence, and custom, and while there are genetic differences between human populations, the genetic variation among humans is tiny compared with that found in other apes. What separates all human societies is not our genes, but the products of a few thousand years of cultural evolution. However, it would be a mistake to assume that our biology is irrelevant to understanding contemporary human adaptation and diversity. Through our culture we have built our world, but that is only possible because our minds are fashioned for culture.

CHAPTER 12

THE ARTS

We have all experienced technological innovation during our lifetimes and, depending on our age, can remember the first appearance of iPods in 2001, the World Wide Web in the 1990s, mobile phones in the 1970s, or color TV in the 1960s. Each of these influential innovations swept society as the cutting-edge advance of the day, only to be refined, elaborated, and improved upon by succeeding technology. The logic of cultural evolution is identical to that of biological evolution, even if the details differ.[1] New ideas, behaviors, or products are devised through diverse creative processes; these differ in their attractiveness, appeal, or utility, and as a result are differentially adopted, with newfangled variants superseding the obsolete. Technology advances and diversifies by refining existing technology, which in turn has bolted the innovations of earlier times onto their predecessor's standard. Through endless waves of innovation and copying, cultures change over time. The logic applies broadly, from the simplest of manufactured products like pins and paper clips, to the dazzling complexity of space stations and CRISPR gene editing, and even back through time to the stone knapping of our hominin ancestors and the creations of animal innovators from the distant past. Technological evolution is relentless for exactly the same reason that biological evolution is; where there is diversity, including diversity in functional utility and inheritance, then natural selection inevitably occurs.

Curiously, while the evolution of technology is apparent to many, the evolution of the arts is less widely accepted.[2] Yet the production of artistic works, and the manner in which art changes over time, owes a substantive debt to imitation that goes far beyond the copying of styles, techniques, and materials. The film and theatre industries illustrate what architecture, painting, and sculpture affirm. That is, in the absence of a mind fine-tuned by natural selection for optimal social learning, art simply could not be produced.

Having examined the little appreciated dues that the art world owes to our biological heritage, we will go on to consider the evolution of dance. The history of dance is particularly well documented, and provides a wonderful case study with which to illustrate how human culture evolves. We will see that cultural evolution is neither linear (constantly progressing from simple to more complex over time), as envisaged by nineteenth-century anthropologists,[3] nor treelike with independent lineages constantly branching, as Darwin portrayed biological evolution.[4] Cultural evolution is more of a melting pot, with innovation often the product of borrowing from other domains, such that cultural lineages come together as well as diverge. This can be seen in the richly cross-fertilizing coevolution of dance, music, fashion, art, and technology, whose histories are intimately entwined.

We will begin with the movies. In the critically acclaimed film *The Imitation Game*, Benedict Cumberbatch received plaudits for his brilliant portrayal of Alan Turing, the eccentric genius who cracked the cyphers of the Enigma machine, used by the Nazis to send secure wireless messages during the Second World War; in so doing, Turing devised the world's first computer. Turing's machine endeavored to imitate the human mind and perhaps a particular mind, the mind of his childhood friend and first love, Christopher Morcom, after whom he named his electromechanical code breaker. Turing was awarded the OBE by King George VI for his Bletchley Park services, which were estimated to have shortened the war by two to four years. But Turing's life ended in tragedy. Prosecuted for homosexual acts, still a criminal offense in Britain in 1952, he endured two years of aggressive hormone "treatment" before committing suicide by eating an apple laced with cyanide shortly before his 42nd birthday.[5] Only in 2013 did Queen Elizabeth II grant him a posthumous pardon and British Prime Minister Gordon Brown apologize for the appalling treatment that this brilliant scientist and war hero suffered at the hands of his nation.

Turing is widely regarded as the father of modern computing. According to artificial intelligence legend Marvin Minsky of MIT, Turing's landmark paper of 1937 "contains, in essence, the invention of the modern computer and some of the programming techniques that accompanied it."[6] The metaphor of the mind has inspired artificial intelligence research for half a century, fueling countless advances in computing technology. As far back as 1996, human conceit took a humiliating hit when a supercomputer called Deep Blue defeated Garry Kasparov, perhaps the greatest-ever chess grandmaster, to exert the superiority of the mechanical over the organic mind. The world's most powerful computer today, the Tianhe-2 supercomputer at China's National University of Defense Technology, is the latest in a long line of refined imitations of pre-existing technology that can be traced all of the way back to Hut 8 at Bletchley Park. One day soon, quantum computers are expected to supplant today's digital computers. Already the world's most accurate clock is the Quantum Logic Clock, produced by the National Institute of Standards and Technology (NIST), which uses the vibrations of a single aluminum atom to record time so accurately that it would neither gain nor lose as much as a second in a billion years. Yet, in homage to their humble beginnings, such technologies are known as "quantum Turing machines."

We readily recognize the role that imitation plays in technological evolution, just as we easily comprehend Turing's attempt to imitate the mind's computational power with a thinking machine. What is often overlooked is that, as every actor and actress earns a living by imitating the individuals portrayed, *all movies are in the imitation game*. The entire film industry relies on the ability of talented thespians to study their focal character's behavior, speech, and mannerisms in meticulous detail and to duplicate these with sufficient precision to render their portrayal a compelling likeness, and leave the storyline credible. Cumberbatch convinces us that he really is Alan Turing, just as Marlon Brando persuaded us that he was Vito Corleone in *The Godfather*, or Meryl Streep is the quintessential Margaret Thatcher in *The Iron Lady*. The magic of the movies would dissipate instantly if this pretense ever broke down. Academy awards and Golden Globes are the ultimate recognition handed out to honor the world's most gifted imitators. Tens of millions of years of selection for more and more accurate social learning has reached its pinnacle in the modern world's Brandos and Streeps. Yet such extraordinary acting talent was clearly not directly favored by natural selection. No amateur dramatic productions were performed in the Pleistocene, and being a proficient actor did not bring reproductive benefits to early hominins. Acting is not an adaptation, but rather an "exaptation,"—that is, a trait originally fashioned by natural selection for an entirely different role.[7] Acting proficiency is a byproduct of selection for imitation.

Among our distant ancestors, those individuals who were effective copiers did indeed enjoy fitness benefits, but their copying was expressed in learning challenging life skills, not the performing arts. We are all descended from a long line of inveterate imitators. By copying, our forebears learned how to make digging tools, spears, harpoons, and fish hooks; make drills, borers, throwing sticks, and needles; butcher carcasses and extract meat; build a fire and keep it going; pound, grind, and soak plant materials; hunt antelope, trap game, and catch fish; cook turtles, and make tools from their shells; mount a collective defense against ferocious carnivores; as well as learn what each sign, sound, and gesture observed in their society meant. These, and hundreds of others skills, were what shaped the polished, imitative capabilities of our lineage. Acquiring such proficiencies would have been a matter of life and death to the puny and defenseless members of our genus in their grim struggle to forge a living on the plains of Africa, the deserts of the Levant, or the Mediterranean coast.

Hundreds of thousands, perhaps millions, of years of selection for competent imitation has shaped the human brain, leaving it supremely adapted to translate visual information about the movements of others' bodies into matching action from their own muscles, tendons, and joints. Now, eons later, we effortlessly direct this aptitude to fulfill goals utterly inconceivable to our forebears, with little reflection on what an extraordinary adaptation the ability to imitate represents. Imitation is no trivial matter. Few other animals are capable of motor imitation, and even those that do exhibit this form of learning cannot imitate with anything like the accuracy and precision of our species.[8] For over a century psychologists have struggled to understand how imitation is possible.[9] Most learning occurs when individuals receive "rewards" or "punishments" for their actions,[10] like achieving a desired goal or else experiencing pain. This reinforcement encourages us to repeat actions that brought us pleasure and to avoid those activities that brought pain or stress, a process known as operant conditioning. The reward systems that elicit positive or negative sensations are ancient structures in the brain, fashioned by selection to train animals' behavior to meet adaptive goals.[11] However, when we learn to eat with chopsticks or to ride a bicycle by observing another individual, we have seemingly not received any direct reinforcement ourselves, so how do we do it? Even more challenging to understand, how do we connect the sight of someone else manipulating chopsticks, or peddling a bike, with the utterly different sensory experience that we encounter when we do these things? This correspondence problem has been the bugbear of imitation researchers for decades. Even today, there is little consensus as to how this is done.[12] One conclusion is clear, however. Solving the correspondence problem requires links, in the form of a network of neurons, between the sensory and motor regions of the brain. Years ago, when a postdoctoral fellow at Cambridge University, working with eminent ethologist Patrick Bateson, I explored the evolution and development of imitation using artificial neural network models. We found that we could simulate imitation and other forms of social learning, provided we pretrained the artificial neural network with relevant prior experience that allowed it to create such links between perceptual inputs and motor outputs.[13] Interestingly, our neural networks that simulated imitation possessed exactly the same properties as "mirror neurons."

Mirror neurons are cells in the brain that fire both when an individual performs an action, and when the individual sees the same action performed by others.[14] Mirror neurons are widely thought to facilitate imitation.[15] As the brain expanded during human evolution, those regions now known to be involved in imitation, such as the temporal and parietal lobe, grew disproportionately larger.[16] The parietal lobe is that precise region of the primate brain in which mirror neurons were first detected in monkeys, and brain-imaging studies show that the same regions of the human brain indeed possess these mirroring properties.[17] Plausibly, the mirror neuron system was the direct product of selection for enhanced imitation among our ancestors. These cognitive abilities continue to allow us to learn new skills today—for instance, how to drive, wield a hammer, or cook a meal. However, those same cognitive abilities are also what permits Jimmy Stewart to impress us every Christmas as George Bailey in *It's a Wonderful Life*.

Also overlooked, but far less obvious, is how reliant film, theatre, opera, and even computer games are on the audiences' abilities to imagine themselves part of the action, to experience the fear and the tension, and to share the main character's emotions vicariously. These capabilities were also likely fashioned in the sweaty heat of the African jungle, where the ability to take the perspective and understand the goals and intentions of those occupied in important tasks helped the observer to acquire the relevant technology. Here too, the ancestral sharing of emotions in social settings, such as responding with anxiety to the fear of another or drawing joy from the laughter of a child, helped to shape the empathy and emotional contagion that makes the movie a heartfelt experience. These sensitivities are also reliant on forms of social learning with adaptive functions, such as helping individuals learn the identity of predators or circumvent other dangers.[18] In the absence of these social learning abilities, we would all watch movies like sociopaths, utterly indifferent to the lead character's trauma, equally unmoved by the *Psycho* shower scene or Rhett Butler and Scarlett O'Hara's kiss. Global box office revenue was estimated at $40 billion in 2015. Without the human ability to imitate there would be no movie industry; and for that matter, no theatre or opera.

In fact, when we start to think about it, connections emerge between a surprising number of the arts and the imitative and innovative capabilities that drove the evolution of the human brain. Consider, for instance, sculpture. In order to complete his statue of *David* in 1504, Michelangelo had to solve a correspondence problem of his own. Rather than moving his own body to match David's posture, Michelangelo had to move his hands and arms, skillfully wielding hammer and chisel to transform a block of marble into an exact replica. To do this, Michelangelo had to translate the visual inputs corresponding to the sight of the male model into motor outputs that generated a matching form in the stone. That he exceled in this challenge, and produced one of the greatest masterpieces of the Renaissance, is testament not just to his talent but also to many years of practice in stonework. Michelangelo began his artistic training at the age of 13, and spent time as a quarryman in Carrara, where he learned to brandish a hammer to good effect. Those years of experience functioned to train the neural circuitry of his brain (just as we trained our artificial neural networks) to be sensitive to the correspondences between the movements associated with his masonry and the physical results in the stone. That training, however, could only be effective because Michelangelo possessed a brain uniquely designed to generate rich cross-modal mappings between the sensory and motor cortex when given the right experience; this was a legacy of ancestral selection for imitative abilities.

Admiring marvelous sculpture like *David* or the *Venus de Milo* can be a startlingly sensual experience, especially when one considers that we are confronted with, in essence, a block of stone. One is often left secretly wanting to reach out and touch the beautiful forms. Some cultures, such as the Inuit, even make small sculptures that are solely meant to be handled, rather than seen.[19] That we should experience such sensations again draws on those cross-modal neural networks. These connect physical representations of objects in our minds to the objects themselves, and from there to a pre-existing network of associations and, often intimate, memories.

Only a very large-brained species could ever have produced works of sculpture fashioned with such precision. Such works require meticulous and controlled hand movements, manual dexterity that evolved along with increased brain size. Mammalian brains changed in internal organization as they got larger, inevitably becoming more modular and asymmetrical with size,[20] as described in chapter 6. With increasing overall size, larger brain regions typically become better connected to other regions and start to exert control over the rest of the brain.[21] This occurs because neurons vie with each other to connect to target regions and this competition is generally won by those neurons that collectively fire the target cells, giving large brain regions an advantage. The net result is an increase in the ability of the larger brain regions to influence other regions. The dominant structure in the human brain is the neocortex, which accounts for approximately 80% of the brain by volume, more than in any other animal. In the primate lineage to humans, the neocortex (the thinking, learning, and planning part of the brain) has become larger over evolutionary time, and has exerted increasing control over the motor neurons of the spinal cord and brain stem; this has led to increased manual dexterity and more precise control of the limbs.[22] The cerebellum, the second largest region of the human brain, also plays an important role in motor control and has enlarged during recent human evolution as well.[23] This motor control is what makes humans exceptional at finely coordinated movements. If I am correct, and innovation and social learning have driven the neocortex and cerebellum to become larger over evolutionary time, then this natural selection may simultaneously have generated human greater dexterity, which could be expressed not just in painting and sculpture, but also acting, opera, and in particular, dance.

The motor control that allows humans to produce artistic works and performances spontaneously is a capability that no other animal shares. Granted, the internet is awash with reports and YouTube footage of artistic animals, but these have not stood up to close scrutiny from animal behavior experts. You may well be able to buy painting kits for your cat or dog, and your pet may well enjoy the experience, but little that is genuinely artistic is produced. Like most other animals that have been handed a paintbrush, dogs and cats lack both the inclination and motor control to produce representational art, and I strongly suspect that any abstract beauty observed in the colorful product is strictly in the eye of the pet owner.

Intriguingly, the Humane Society of the United States recently organized a Chimpanzee Art Contest, to which six chimpanzees submitted "masterpieces." The winner, Brent, a 37-year-old male from ChimpHaven in Louisiana, received a $10,000 prize from the stately hands of Jane Goodall. Brent, apparently, produced the work with his tongue, rather than bothering to use a paintbrush. The original works were then auctioned off on eBay with the many thousands of dollars raised going to support primate sanctuaries.[24] Yet, however much one admires this charming, clever, and well-motivated funding initiative, the claim that the chimpanzees concerned are artists, in any meaningful sense, is greeted with skepticism by animal behaviorists and art scholars alike. A generous reading of the artistic pretensions of these animals would at best acknowledge some pleasure in generating colorful compositions.

Elephants are considerably more interesting because to the astonishment of thousands of gullible tourists, they regularly produce realistic paintings of trees, flowers, or even other elephants in sensational public performances at several sanctuaries in Thailand (figure 11). The artwork, which the elephants sometimes even sign with their name, sells in droves. However, all is not as it seems. Each paintbrush is placed in the elephant's trunk by its trainer, who then surreptitiously guides the trunk movements by gently tugging at its ears. The elephant has been

FIGURE 11. Painting elephants are becoming a major tourist attraction in Thailand. The elephants regularly produce realistic paintings of trees, flowers, and other elephants in impressive public performances. However, all is not as it seems, and the tourists are being hoodwinked. By permission of Philippe Huguen/AFP/Getty Images.

trained to hold the brush to the paper and move it in the direction to which its ear is being pulled.[25] At the very least, one has to acknowledge an impressive piece of animal training,[26] and one cannot help but admire the precision and control that the painting elephants exhibit with their trunks. Yet, a trick has taken place, and the trainer gets away with it by cleverly positioning himself behind the elephant. The tourists nonetheless typically go home happy, even those who spot the ruse, since no one can say that their "priceless" artwork was not painted by an elephant![27]

Representational art is a uniquely human domain. That elephants can, with guidance, produce these pictures is nevertheless fascinating, precisely because it demonstrates that with training, they too are capable of building up cross-modal neural networks in their brains that translate tactile sensory inputs into matching motor outputs. The painting elephants have solved a correspondence problem of their own. It may be no coincidence that an Asian elephant from South Korea called Koshik was recently shown to be capable of vocal imitation, including mimicking human speech,[28] while Happy, another Asian elephant at the

Bronx Zoo in New York, was shown to be able to recognize herself in a mirror.[29] Almost certainly, these capabilities are related. Like sculpture, producing paintings (and mirror self-recognition) makes demands on the circuitry of the brain involved in imitation.[30]

Our big brains not only afford precise control of our hands, arms, legs, and feet, but also of our mouth, tongue, and vocal chords, which is what endowed our species with the vocal dexterity to speak and sing.[31] Without that cortical expansion, members of our species could neither have fashioned a work of art, nor vocally expressed their admiration for it. The evolution of language is surely central to the origins of art, since art is rife with symbolism. As described in chapter 8, symbolic and abstract thinking are widely regarded as foremost features of human cognition. The use of arbitrary symbols allows humans to represent and communicate a wide range of ideas and concepts through diverse mediums. We possess minds fashioned by natural selection to manipulate symbols and think abstractly through spoken language, but we also express this penchant for symbolism in numerous artistic endeavors.

Architecture is one such domain. Victor Hugo's 1831 masterpiece, *Notre Dame de Paris*, contains an extraordinary chapter entitled, "This Will Destroy That"; it echoes the enigmatic words of the evil Archdeacon Frollo, who rants against the invention of the printing press. Frollo expresses the terror of the church in the face of a rising new power—printing—that threatens to supplant it. The concern was not just that people might start to rely on books rather than priests to acquire their knowledge and advice, but also that the cathedral's magnificent gothic architecture, already in disrepair, would lose its power and symbolism:

> It was a premonition that human thought . . . was about to change its outward mode of expression; that the dominant idea of each generation would, in future, be embodied in a new material, a new fashion; that the book of stone, so solid and so enduring, was to give way to the book of paper.[32]

To the modern reader such fears appear irrational. Yet, in the preliterate world, powerful institutions literally wrote their authority in stone. From the Pyramids to St. Peter's Basilica in Rome or the Palace of Versailles, the magnificence, scale, wealth, and beauty percolated with the symbolism of God-given command and assuredness.

Human artwork has a long history, dating back some 100,000 years.[33] It exhibits all the hallmarks of cultural evolution.[34] While painting manifests multiple divergent styles, one ancient conceptual lineage sets out to represent the visual experience with accuracy. Consider, for instance, René Magritte's famous painting *The Treachery of Images*, which shows a pipe that looks as though it is a model for a tobacco advertisement. Much to the puzzlement of millions of admirers, Magritte painted below the pipe, "*Ceci n'est pas une pipe*," which is translated as "This is not a pipe." At first sight, this appears completely untrue. What we momentarily forget, of course, is that the painting is not a pipe, but an image of a pipe. When Magritte was once asked to explain this picture, he apparently replied that of course it was not a pipe; just try to fill it with tobacco! Magritte's point might appear trite to some, privileged as we are to live in an age where we can overdose on magnificent artworks that perfectly capture perspective and exhibit astonishing accuracy of portrayal. In the contemporary artistic movement of hyperrealism, the pictures of artists like Diego Fazio, Jason de Graaf, or Morgan Davidson use acrylics, pencil, or crayons with such astonishing accuracy that they are almost always mistaken for photographs. Their work can be placed in a long-standing tradition that sets out to produce precise, detailed, and accurate representation of the actual visual appearance of scenes and objects. This movement flourished at various periods, and has been known as "realism," "naturalism," or (with appropriate reference to imitation) "mimesis." Such hyperreal works allow the viewer to escape the correspondence problem by producing an image that exactly mimics what it represents. However, there can be no such escape for the artist, who must overcome this challenge in order to succeed.

Nowhere in the arts is the correspondence problem more clearly manifest than in dance, which again harnesses those same cognitive faculties that are necessary to integrate distinct sensory inputs and outputs. Following an excited conversation in a Cambridge pub in 2014, I recently began a collaboration with Nicky Clayton and Clive Wilkins to study the evolution of dance. Nicky is a professor of psychology at Cambridge University and expert of animal cognition; she is also a passionate dancer, and she merges this with her research as scientific director to the Rambert, a leading contemporary dance company.[35] Clive is equally impressive as a successful painter, writer, magician, and also a dance enthusiast. We rapidly converged on the hypothesis that dancing may only be possible because its performance exploits the neural circuitry employed in imitation.[36]

Dancing requires the performer to match their actions to music, or to time their movements to fit the rhythm, which can sometimes even be an internal rhythm, such as the heartbeat. This demands a correspondence between the auditory inputs the dancer hears and the motor outputs they produce. Likewise, competent couple or group dancing requires individuals to coordinate their actions, and in the process match, reverse, or complement each other. This too calls for a correspondence between visual inputs and motor outputs. That humanity is able to solve these challenges, albeit with varying degrees of ease and grace, is a testament to the neural apparatus that we uniquely possess as a legacy of selection for imitative proficiency. The same reasoning applies when individuals dance alone.

Contemporary theories suggest that while the potential for imitation is inborn in humans, competence is only realized with appropriate lifetime experience.[37] Early experiences, such as being rocked and sung to as a baby, help infants to form neural connections that link sound, movement, and rhythm, while numerous experiences later in life, such as playing a musical instrument, strengthen these networks. The suggestion that taking up the piano will make you a better dancer might seem curious, but that is a logical conclusion to draw from the neuropsychological data.

The relentless motivation to copy the actions of parents and older siblings that is apparent in young children may initially serve a social function, such as to strengthen social bonds. However, childhood imitation also trains the "mirroring" neural circuitry of the mind, leaving the child better placed later in life to integrate across sensory modalities.[38] Theoretical work suggests that the experience of synchronous action forges links between the perception of self and others performing the same movements.[39] Whether because past natural selection has tuned human brains specifically for imitation, or because humans construct developmental environments that promote imitative proficiency—or both—there can be no doubt that, compared to other animals, humans are exceptional imitators. A recent brain-scanning analysis of the neural basis of dance found that foot movement timed to music excited regions of the brain previously associated with imitation, and this may be no coincidence.[40] Dancing inherently seems to require a brain capable of solving the correspondence problem.

Comparative evidence is remarkably consistent with this hypothesis. A number of animals have also been characterized as dancers, including snakes, bees, birds, bears, elephants, and chimpanzees; the last of these perform a "rain dance" during thunderstorms, which has a rhythmic, swaying motion. However, whether animals can truly be said to dance remains a contentious issue,[41] which depends at least in part on how dance is defined. In contrast, the more specific question of whether animals can move their bodies in time to music or rhythm has been extensively investigated, with clear and positive conclusions. Strikingly, virtually all animals that pass this test are known to be highly proficient imitators, frequently in both vocal and motor domains.

This ability to move in rhythmic synchrony with a musical beat by nodding our head or tapping our feet, for instance, is a universal characteristic of humans,[42] but is rarely observed in other species.[43] The most prominent explanation for why this should be, known as the "vocal learning and rhythmic synchronization" hypothesis,[44] is broadly in accord with the arguments presented here.[45] This hypothesis suggests that moving in time to the rhythm (known as "entrainment") relies on the neural circuitry for complex vocal learning; it is an ability that requires a tight link between auditory and motor circuits in the brain.[46] The hypothesis predicts that only species of animal capable of vocal imitation, such as humans, parrots and songbirds, cetaceans, and pinnipeds, but not nonhuman primates and not those birds that do not learn their songs, will be capable of synchronizing movements to music.

The many videos of birds, mostly parrots, moving to music on the internet are consistent with the hypothesis, but compelling footage of other animals doing the same is comparatively rare. Some of these "dancing" birds have acquired celebrity status; the best known is Snowball, a sulphur-crested cockatoo,[47] whose performances on YouTube have "gone viral." Snowball can be seen to move with astonishing rhythmicity, head banging and kicking his feet in perfect time to Queen's "Another One Bites The Dust" or the Backstreet Boys (figure 12).[48] Home videos can be faked, and parrots also have the ability to mimic human movements,[49] so the footage alone cannot show that Snowball is keeping time

FIGURE 12. Snowball, a sulphur-crested cockatoo, performs dances on YouTube that have thrilled millions. Careful experiments have demonstrated that Snowball adjusts his movements to keep time to the music. By permission of Irena Schulz.

to music directly. For this reason, a team of researchers led by Aniruddh Patel at The Neurosciences Institute in San Diego brought Snowball into the laboratory to carry out careful experiments.[50] Manipulating the tempo of a musical excerpt across a wide range, the researchers conclusively demonstrated that Snowball spontaneously adjusts the tempo of his movements to stay synchronized with the beat.

Thus far, evidence for spontaneous motor entrainment to music has been reported in at least nine species of birds including several types of parrot, and the Asian elephant, all of whom are vocal imitators,[51] and several of which show motor imitation.[52] Entrainment has also been shown in a chimpanzee,[53] a renowned motor imitator.[54] The sole exception to this association is the California sea lion,[55] which is not known to exhibit vocal learning. However, the fact that related species show vocal learning, including several seals and the walrus,[56] raises the possibility that this capability or a relevant precursor may yet be demonstrated. Lyrebirds have not been subject to entrainment experiments, but males are famous for their ability to imitate virtually any sounds, including dog barks, chainsaws, and car alarms. They can match subsets of songs from their extensive vocal repertoire with tail, wing, and leg movements to devise their own "dance" choreography.[57] Clearly, there is more to dance, at least social or collective dance, than entrainment to music. There must also be coordination with others' movements, which would seemingly draw on the neural circuitry that underlies motor, rather than vocal, imitation.[58] However, a recent analysis of the avian brain suggests that vocal learning evolved through exploitation of pre-existing motor pathways,[59] implying that vocal and motor imitation are reliant on similar circuitry. The animal data provide compelling support for a causal link between the capabilities for imitation and dance. Whether this is because imitation is necessary for entrainment, or merely facilitates it through reinforcing relevant neural circuitry, remains to be established.

Dance often tells a story, and this representational quality provides another link with imitation. For instance, in the "astronomical dances" of ancient Egypt, priests and priestesses accompanied by harps and pipes mimed significant events in the story of a god, or imitated cosmic patterns such as the rhythm of night and day.[60] Through dance, Australian Aborigines depict the spirits and ideas associated with every aspect of the natural and unseen world.[61] There are animal dances for women,

FIGURE 13. Dancers from the Rambert Dance Company in *The Comedy of Change*. Several lines of evidence connect the ability to dance with imitation. By permission of Hugo Glendinning.

which are thought to function like love potions or fertility treatments to make a lover return, or to induce pregnancy, while male dances are more often about fishing, hunting, and fighting. Africa, Asia, Australasia, and Europe all possess long-standing traditions for mask-culture dances, in which performers assume the role of the character associated with the mask and, often garbed in extravagant costumes, enact religious stories.[62] Native Americans are famed for their war dances, which were thought so powerful and evocative they were banned by the United States government—the law was not repealed until 1934.[63] A variety of animal dances are also performed by Native Americans, and include the *buffalo dance*, which was thought to lure buffalo herds close to the village, and the *eagle dance*, which is a tribute to these venerated birds.[64] This tradition continues right through to the present. For instance, in 2009, the Rambert Dance Company marked the bicentenary of Darwin's birth by collaborating with Nicky Clayton to produce *The Comedy of Change*, which evoked animal behavior on stage with spellbinding accuracy (figure 13). In all such instances, the creation and performance of the dance requires an ability on the part of the dancer to imitate the movements and sounds of particular people, animals, or events. This reproduction contributes importantly to the meaning of the dance in the community, and imparts a bonding or shared experience. Such dances reintroduce the correspondence problem, since the dancer, choreographer, and audience must be able to connect the dancers' movements to the represented target phenomenon.

The most transparent connection between dance and imitation, however, will be readily apparent to just about anyone who has ever taken or observed a dance lesson; that is, dance sequences are typically learned through imitation. From beginner ballet classes for infants to professional dance companies, the learning of a dance routine invariably begins with a demonstration of the steps from an instructor or choreographer, which the dancers then set out to imitate. It is no coincidence that dance rehearsal studios around the world almost always have large mirrors along one wall. These allow the learner to flit rapidly between observing the movements of the instructor or choreographer and observing their own performance. This not only allows them to see the correspondence, or lack of correspondence, between the target behavior and what they are doing, but also allows the dancers to connect feedback from their muscles and joints to visual feedback on their performance, allowing error correction and accelerating the learning process.[65]

Prospective new members of professional dance companies are given challenging auditions in which they are evaluated partly on their ability to pick up new dance routines with alacrity, an essential skill for a dancer. Dancing is not just about body control, grace, and power, but it also demands its own kind of intelligence.[66] A key element in whether or not a dancer makes the grade essentially comes down to how good they are at imitating. A professional dancer at the Rambert once told Nicky and me that she had recently taken up sailing, and her instructor was flabbergasted at how quickly she had picked up the techniques involved. What the instructor had failed to appreciate was that dancers earn their living by imitation.

Imitation is not the only cognitive faculty that is necessary for learning dance. Also important is sequence learning, particularly in choreographed dances, which require the learning of a long, and often complex, sequence of actions. Even improvised dances such as the Argentine *tango* require the leader to plan a sequence of movements that provide the basis for the exquisite conversation between leader and follower. As we have learned, long strings of actions are very difficult to learn asocially, but social learning substantially increases the chances that individuals will acquire the appropriate sequence.[67] Our ancestors were predisposed to be highly competent sequence learners because many of their tool-manufacturing and tool-using skills, as well as food-processing techniques, required them to carry out precise sequences of actions, with each step in the right order. The fact that these sequence-learning capabilities are clearly exploited in dance provides further evidence of the extent of the surprising connection between imitation and dance.

Let us now turn to the history of dance, and consider whether this particular art form can be said to evolve in any rigorous sense. In principle, any system that possesses variation, differential fitness, and inheritance is expected to change over time through a selective process.[68] These necessary conditions are well recognized in biological evolution, where species evolve by natural selection, but they occur in many other spheres of life too. The vertebrate immune system, for instance, responds to disease by first generating diverse antibodies, then determining which are most effective at countering the pathogen, and finally reproducing in large numbers the most effective disease-combating molecules; variation, differential fitness, and inheritance thus lead to a within-lifetime "adaptation" in immunity.[69] The central nervous system, vascular system, and muscular system all operate in a similar manner, throwing out axons, blood vessels, or muscle cells, respectively, and then strengthening those connections that prove useful, while letting the less helpful ones die away.[70] Charles Darwin himself pointed out that one aspect of culture—namely language—evolves, and this has been confirmed by extensive research.[71] Is the same true of dance?

It is said that among the Bantu peoples of Central Africa, when an individual from one tribe meets someone from a different group, they ask, "What do you dance?"[72] Throughout time, communities have forged their identities through dance rituals that mark important events in the life of individuals, including birth, coming of age, marriage, and death—as well religious festivals, and critical points in the seasons, such as the cycle of crops.[73] The social structure of many communities, from Africa tribes to Spanish gypsies, and to Scottish clans, gain much of their cohesion from the group activity of dancing. Historically, dance has been a strong, binding influence on community life, a means of expressing the social identity of the group, and participation allows individuals to demonstrate a belonging. As a consequence, in many regions of the world there are as many types of dances as there are communities with distinct identities.

Dances certainly exhibit variation, in no small part because they are often an important part of the cultural identity of a people. For instance, documenting the vast corpus of folk dances around Europe is a monumental task that has occupied scholars and amateur enthusiasts for decades.[74] The many resulting volumes detail literally thousands of different dances, or dance variants. Many are now forgotten, but others remain popular. One fragment of this immense body are the diverse European sword dances, forms of which are listed in fifteen separate countries.[75] In current performances, the weapons are entirely symbolic. For instance, in Scotland, where I live, two swords are placed crossed on the ground, marking out four quadrants between which the kilted dancer must hop and skip lightly. However, the origin of such dances can be traced all of the way back to ancient Greece; they were brought to Britain by the invading Danes and Vikings, who performed them prior to battle with vigorous swordplay.[76] Similar diversity exists among the hundreds of different masked dances performed throughout Asia, from the Kathakali dancers of India to the Kabuki dramas of Japan.[77] Another example concerns the regional Spanish folk dances, the best known of which is the gypsy's *flamenco*, but there are also *jotas* from the north of Spain, *seguidillas* from the central regions, *sardanes* from the east, the *fandangos* of Andalucia, and the *zapateados* of the south. Dance diversity is also manifest at different scales, with closely related communities often exhibiting similar dances, but with larger differences found between more distinct regions of the world. For instance, in much Western dancing, exemplified by ballet, the dancer seeks to exhibit lightness of movement, and tries almost to fly, to disavow gravity through athletic leaps, elongation of the body, and the pointe work of ballerinas. In contrast, dances of Eastern origin often assert the importance of gravity, in which feet may stamp and thump the ground, but contact with the earth is constant.[78]

Innovation in dance often results from the recombination of distinct elements of culture from different communities.[79] When the Moors invaded southern Spain, they brought a sinuosity of torso and use of hands and fingers to local dance. Likewise, when the Russians dominated ballet, they introduced high leaps and other extreme feats of male athleticism, in keeping with the folk dancing traditions of the Caucasus that contain many leaping steps through which young men demonstrate their virility. Often the extraction of dance elements from other communities is deliberate, as in the reproduction of Roman dance spectacles in the Renaissance, or Martha Graham's famous *Primitive Mysteries* that drew on Native American themes.[80] Sometimes change has been deliberately institutionalized, as in the formalization of the five positions in ballet, or the codification of ballroom dances by the Imperial Society of Teachers of Dancing in the early twentieth century. Moreover, in the same ways that the Impressionists were moved to produce blurry paintings as a rebellious reaction against the realism that dominated European art in the nineteenth century, and that Debussy and Stravinsky shifted away from the dominant tonal tradition in classical music around the same time, modern dance pioneers like Isadora Duncan and Martha Graham devised compositions that contrasted to the stylized dance strictures exemplified in ballet. Vital though imitation is to dance, the inspiration for much dance innovation has been precisely a reaction against "mere imitation."

The rapid spread of such innovations illustrates the second condition necessary for adaptive cultural evolution, namely "differential fitness," which in this instance means different rates of uptake. Few dance innovations had more impact than the waltz. At the end of the eighteenth century, the *waltz* took the many hundreds of dance halls in Paris by storm. The word "*waltz*" means to turn, and the dance originated through refinements of pre-existing turning dances, such as the *länder*, an extremely popular dance among poorer people of Germany. Refined and modified in ballrooms, both the *Viennese waltz* and the slower German variant rapidly swept Europe, and easily charmed the dance-crazed middle classes bored with stuffy, aristocratic minuets. The dance's intoxicating swirling, and the dangerously intimate contact between male and female were a major draw. Nowadays, to think of the *waltz* as debauched is difficult, but at the time it was a source of great controversy, and many people of prominence spoke out against this "vulgar new craze." Nevertheless, the attraction of its music and its emotional appeal to the courting young left the *waltz* unstoppable. The dance's pinnacle was in the second half of the nineteenth century, when Johann Strauss II was composing classics such as *The Blue Danube*. But the allure of the *waltz* has never completely faded, and it remains a staple of the ballroom.

Relative to other dances in the late eighteenth century, the *waltz* could be said to possess high "cultural fitness," which really means little more than it was unusually appealing and as a result increased readily in frequency. Of course, the *waltz* is far from the only dance to sweep the populace. Shortly afterward, the Polish *mazurka* conquered the ballrooms, and subsequently *polka*-mania inflamed the masses, spreading like wildfire from the Czech Republic where it originated to Paris and London in the 1840s. In the nineteenth century, an influx of immigrants brought European social dances to America, where the *waltz*, *polka*, and many others were rapidly adopted. A few decades later, the Americans returned the gift when barn dances and *"two steps"* swept Europe, and the export of dance from the United States has continued ever since. In the twentieth century, the Charleston, the jitterbug, *rock 'n roll*, *disco*, and breakdancing all crossed the Atlantic to overwhelm their predecessors.

Dances are also clearly inherited, being transmitted both among individuals and across generations, often with remarkable stability. Circle dances, for instance, have a history that can be traced back, quite literally, thousands of years.[81] The book of Exodus recounts how David and the people of Israel danced around the golden calf in a ring. Scholars have mapped this particular circle dance back to the ancient Egyptian cult of Apis, and traced it forward to the religious ring dances performed around Europe in the Middle Ages. The circling of a central object representative of a divine force (e.g., a sacrificial stone, altar, tree, or fire) was a recurrent theme in countless folk dances across the whole of Europe, and from the twelfth century to the present. Such circles often represented the cycle of life, the seasons, or daily rhythms. These round dances, in turn, gave birth to the *"carole,"* which is still performed today through Europe, and to the numerous *maypole dances* that stretch across Europe from north to south, and even into Mexico as a result of the Spanish conquest. Here the totemistic nature of the pole combines the aforementioned symbols of divinity with those of a fruitful and protective tree, where the ribbons represent the tree's branches and connect the dancers to the central source of fertility.

The ancient Greeks believed that the oldest dance was connected to the birth of Zeus; the dancers impersonated the mythical rescuers of the infant god who had saved the child from being eaten by his father, Kronos.[82] In fact, the origins of this dance have been found to relate to an even more ancient fertility ritual, in which leaping dancers, shouting and beating weapons, invoked prosperity for their crops or banished evil spirits. These same traditions, which can be still be seen in certain African tribal dances today, repeatedly recurred throughout the Middle Ages and in the many Christian religious dances observed in Europe in the sixteenth to the eighteenth centuries.[83]

The rather curious *English morris dance*, with its ringing leg bells, high jumps, and clashing sticks is also thought to have originated as an ancient fertility ritual, but one that acquired an entirely new identity following the Crusades.[84] The stamping and leaping refer back to ageless religious rites invoking divine intervention to encourage rich harvests; the crops were thought to grow to the height that the dancers jump, and the jingling bells served to ward off evil spirits. However, the name "morris" is a corruption of "Moorish," and the clashing of sticks represents the Christian armies' attempts to beat down the Muslim enemy and free the Holy Land from their rule. *Morris dances* are still performed throughout England today, although audiences rarely appreciate their symbolism.

There are even "rave"-like traditions that recur with surprising frequency throughout history.[85] These began with the crazed and intoxicated Dionysiac dances of ancient Greece, where after harvesting the grapes, villagers celebrated with a drunken orgy in honor of Dionysus, god of wine. Their feverish stomping is a favorite scene on Greek vases, and these drunken dancing girls were immortalized in a tragedy by Euripides, the *Bacchae*, in which their frenzy led them to commit murder. Countless societies since have witnessed similar traditions for frantic dancing in a mood of exhilaration heightened by pounding rhythm and flowing alcohol. These occur in dances as distant in space and time as those in ancient Rome, among Native Americans, in the West Indies (the *Voodoo* dances), in Turkey (whirling dervishes), right up to those in the techno and acid house scene that hit western Europe and North America in the 1990s.

Once we recognize that dance possesses the properties of ample variation, differential fitness, and inheritance, we can start to comprehend how even the richest and most challenging dance forms could evolve. The earliest agricultural societies that emerged in the Near and Middle East—in Sumeria, Assyria, Babylon, Egypt, and the cultures of the Mediterranean—had dance at their very heart. In each instance, the society's beliefs were enacted in rituals of dance and drama, partly as acts of celebration, partly as worship, and partly to ensure good harvests. In both ancient Greece and Rome, dance was so fundamental to worship and social life that it was considered a necessary part of a young man's education. In *The Odyssey*, Homer describes how, after eating, the suitors to Penelope "turned their thoughts to . . . the music and dancing without which no banquet is complete."[86] Sacred occasions, such as the games at Olympia from the eighth century BCE, were inaugurated with dancing by the temple virgins. In the East, the earliest Hindu book on dance, the *Natya Shastra of Bharati*, dated to between the second century BCE and the third century AD is considered a sacred text. As far back as records go, numerous communities from African tribes to Native American peoples, to Australian Aborigines, mark key events in the lives of individuals or the annual cycle, with dancing.

Dance originated as a cohesive symbol of tribal identity, choreographed in religious ritual. However, in many larger stratified societies, at least from ancient Egypt onward, dance evolved more specialized manifestations in which priests or professional dancers became the representative of the people who would communicate directly with the gods. From this juncture, dance came to be concerned with creating effects upon the audience, from inspiring religious devotion to ensuring that legends and histories were preserved. The significance attributed to dance also created another professional position at this early stage, that of the dance instructor, whose job was to ensure their pupils received an appropriate education. Once dancers became professionals, dance simultaneously moved into a parallel realm as display; here, dance sought to communicate with the audience, but did not involve its members as direct participants. Gradually, in this domain, the desire for impressive performance led to arduous training and the development of exceptional physical skills in the dancers, through which the twin ideals of beauty and athleticism could be expressed.

Dance may have been steeped in religious symbolism from the outset, but it has nonetheless endured a long-standing rocky relationship with the authorities. Throughout time, religious leaders have sought to control and regulate dance, attempting to constrain it to be a pious expression of doctrine or faith, or a symbolic casting out of demons. Yet participation dances have seemingly always been capable of evoking positive moods, particularly when heightened by pounding rhythms and intoxicating drugs, leading some practitioners to get carried away. As early as AD 554, Childebert, King of the Franks, was forced to ban dancing because of its tendency to descend into depravity, and throughout the following 1,500 years the Christian church repeatedly sought to control dance and rid it of any licentious leanings. These efforts shaped the evolving medium, thereby imposing a top-down structure and formalism on dances in the form of acceptable steps or appropriate forms of contact between the sexes. Change through regulation by authoritative institutions has been a feature of dance through its history, from the strictures of the church, to the codification of ballet positions in the seventeenth century, to the production of dance manuals in recent times.

Dance often generates feelings of release, arousal, pleasure, and excitement. Let us, as an aside, reflect briefly on why dance should be so enjoyable. Part of the explanation for the positive mood elicited by dance may be the release of endorphins that accompanies exercise, and of hormones such as oxytocin that circulate with increased arousal and social behavior.[87] Another factor is the thrill of courtship in dancing with someone attractive, or for the observer, the voyeurism associated with observing lithe, athletic, and appealing young bodies move with grace and beauty. Yet, in many cultures, individuals of the same sex dance together in groups and happily share the experience with individuals to whom they are not sexually attracted. Moreover, people still enjoy dance when the physical demands are too modest to lead to an endorphin rush. Of particular interest here is social dance (dancing with a partner or in a group), especially when the dancing is coordinated and synchronized, as in *ceilidh* or "*river dance.*" Such dance often appears to lead to a sense of bonding, or at least shared pleasure, and can induce positive emotions in an audience.[88] While some properties of dance that make people feel socially close are very general, such as sharing attention and goals with others,[89] others may be dance-specific, such as the externalization through music making of predictable rhythms, which helps people to synchronize their movements.[90] Indeed, synchronized activities have long been associated with social bonding.[91]

Here the intriguing relationship between imitation and cooperation, described in the preceding chapter, may be relevant. We saw how imitation enhances social interaction and induces positive moods, even when the imitated individual is unaware of being copied and the imitator does so unintentionally. Recall that the relationship between imitation and cooperation is bidirectional; being imitated makes individuals more cooperative, while being in a cooperative frame of mind makes one more likely to imitate others.[92] These bidirectional, causal relationships may function to maintain cooperation, collective action, and information sharing among members of a social group.[93] If positive rewards to synchronous behavior have been favored by selection to facilitate cooperation, then dancing in a synchronous manner would be expected to induce warm feelings.[94] The same imitative neural networks in our brains that link sound and rhythm, and thereby allow us to dance to music, are also almost certainly what explains our tendency to tap or clap to music, and the pleasure that experience affords.

Just as the fossil record allows biologists and paleontologists to reconstruct the histories of biological species,[95] and to comprehend the patterns of diversity observed as a combination of historical legacy and local adaptation, so historical records have allowed dance scholars to document the journeys that individual dances have undertaken, to make sense of the variation in form observed, and to understand how the structure and complexity of some complicated dance forms have arisen over time. For instance, the origins of most ballroom dances have been traced back to earlier folk dances, which were adapted and codified by dancing masters and subsequently exported to other countries as dance hall fashions. Latin dances have a similar, if more exotic, history. The origins of the *tango* can be traced to a slave dance in Cuba, from where it spread to the River Plate and was transformed, around the turn of the twentieth century, into the Argentinian classic. *Samba* is known as the dance of Brazil, but was originally taken there by African slaves, where it was transformed into the national dance that explodes onto the streets of Rio during carnival season. Subsequently, samba was exported around the world, via film and television, and codified by dance masters as a ballroom complement. The *paso doble*, which contains steps that imitate the bullfight, originates from Spain, but was refined for the ballroom in southern France. *Salsa* dancing, so popular today, is a descendant of the Cuban *mambo,* a pedigree that it shares with the *cha cha cha*.[96]

The history of ballet is also well researched, with a gradual evolution that began in fifteenth-century Italy, was formalized and professionalized in seventeenth-century France, and polished in nineteenth- and twentieth-century Russia. Ballet originated in the Italian Renaissance courts, where noblemen and women were treated to lavish events; especially at wedding celebrations, dancing and music created an elaborate spectacle.[97] Several important innovations critical to modern ballet were devised in this period, including the creation of dance rhythms, codification of movements, emergence of professional dancing masters and choreographers, and the appearance of dancers in allegorical or mythical narratives with costumes and settings.[98] Dancing masters taught the steps to the nobility, and the court participated in the performances.

In the sixteenth century, Catherine de Medici, an Italian noblewoman who was the wife of King Henry II of France and a great patron of the arts, began to promote ballet in France. She devised elaborate festivals that combined dance, costume, song, and music. These were known as *ballet de cour* because they were performed in the French courts. Between 1583 and 1610 there are records of more than 800 ballets being performed in France.[99] Later, King Louis XIV, himself an enthusiastic dancer, helped to popularize and standardize ballet, and took the first steps toward professional ballet by forming the Académie Royale de Danse, as well as the Académie Royale de Musique. In the 1680s, French opera began to incorporate ballet elements into their performances, creating an opera-ballet tradition. Women participated as theatrical dancers, but played a secondary role to men, partly because their costumes prohibited any great agility.[100]

By the mid-1700s, ballet began to stand on its own as an art form, incorporating expressive, dramatic movements capable of telling a story. Early classical ballets such as *Giselle* and *La Sylphide* were created during the Romantic Movement in the first half of the nineteenth century, which emphasized the supernatural world of spirits and fairy tales. The dancer Marie Taglioni is credited with transforming the image of the ballerina to one exemplifying grace, demure charm, and delicacy, which fit with the Romantic period's portrayal of women as passive and fragile. Taglioni introduced dancing on the tips of the toes, known as pointe work, which subsequently became the norm for the ballerina. The tutu, at that stage a calf-length skirt, was introduced to show off the ballerinas' intricate footwork. Gradually the ballerina became the central figure on stage, and it was the male dancer's turn to be sidelined.[101]

In the nineteenth century, ballet became a declining art in France and the baton was passed to Russia, where its popularity, in contrast, soared. Russian choreographers and composers took ballet to new heights, producing some of the best-loved classical ballets, including *The Nutcracker*, *Sleeping Beauty*, and *Swan Lake*. The Russians introduced their own acrobatic style, incorporating complicated sequences with demanding steps, leaps, and turns. This set the scene for the return of gifted and athletic male dancers, such as Vaslav Nijinsky and Rudolf Nureyev. A much shorter and stiffer classical tutu was introduced at this time to allow the ballerina greater range of movement. The plots of the stories were also altered to make them ideologically suited to Soviet Russia. Subsequently, ballet has thrived in many countries around the world, intertwining classical forms, traditional stories, and choreographic innovations.

The history of contemporary dance has also been carefully studied. The brilliant but heartrending Isadora Duncan is widely considered the mother of modern dance. Duncan was born in San Francisco in 1878, but established herself as a dance pioneer in Paris and later moved to Moscow, inspired by the idea of the Russian Revolution. Duncan rejected conventional dance in favor of flowing movement in modern dress, which better expressed her unbridled passions. She was moved by the ideals of freedom and scholarship that she drew from ancient Greece, and would dance to classical music, including Beethoven, Wagner, and Brahms. Initially her dances were light in texture, but from 1913, when her two children tragically drowned, her art took on a darker form. Duncan's short life was marred by drunkenness and marital failure, and ended prematurely, in calamitous fashion. When she was just 49, Duncan was strangled by her own scarf when it was caught in the wheels of her car. However, Duncan left behind a legacy that truly revolutionized dance. She possessed enormous power to excite and inspire others, and created a vast following.

Another pioneering American dancer of this period was Ruth St. Denis, who equally felt the need to break away from the constraints of ballet, but in her case, she drew inspiration from Asian dances. In 1914, she married fellow dancer Ted Shawn, and the two of them created the influential Denishawn School, which integrated ballet, European, and Oriental dance forms. The most famous Denishawn student was Martha Graham, who in turn reacted against its Orientalism to usher in a new expressionist style of contemporary dance relevant to twentieth-century America. Graham's principle innovation was a particular technique, in which the lower back and pelvis are the source of all movement, and where traveling into space is the result of a subtle off-balancing of the dancer's weight.[102] Graham effectively devised and codified the language of modern dance, utilized by the generations that followed her, and continuing up to the present. Her influence was absolutely central to the expansion of American modern dance. Indeed, a survey of a hundred prominent contemporary dance companies identified "parental" influences that traced back to seminal troupes; the most prominent of these was Martha Graham's company.[103]

Thus far, I have focused on dance as if it were an isolated and blinkered province of culture, but any such impression would of course be entirely misleading. From the outset, dance traditions have been enriched by tremendously creative interactions with music, fashion, art, technology, and many other aspects of culture. Dance began in alliance with music, frequently through instruments of religious ritual. In traditional African, Asian, and Native American dances, the drum provides an essential rhythm, which is the heartbeat of the ceremony. European folk dancers are variously accompanied by musicians and singers. Flamenco dance, for instance, requires an auxiliary troupe of guitarists, singers, clappers, and stampers. In fact, music and dance have always gone hand in hand, with innovation in each art form inspiring change in the other. From at least the sixteenth century onward, the demand for public dancing has created an industry of dance music, as witnessed by the publication of sheet music and, centuries later, of recordings. The popularity of the *waltz* stirred Johan Strauss, Frederic Chopin, and Franz Schubert, among others, to compose some of the most evocative and romantic classical music, and that music, in turn, fueled the dance craze. The intimate relationship between music and dance is seen in the coevolution of the *one step* and ragtime music, or the *jitterbug* and swing music, but is most overt in dance forms that share their name with the music, such as *rock 'n roll, disco, hip hop*, or *salsa*. Indeed, the connection is so strong that some aspects of musical rhythm, such as groove and syncopation, can only be fully understood in the context of dance.[104] Classical music and dance are also strongly connected. Tchaikovsky, Stravinsky and Prokofiev composed the scores for the classical ballets *Swan Lake, Petrushka*, and *Romeo and Juliet*, respectively. Likewise, for nearly a century, pioneering modern dancers have commissioned original music for their performances.

Equally, dress and other forms of adornment have served to heighten the visual, symbolic, and dramatic power of dance.[105] From the earliest times, performers dressed to dance by painting, decorating, marking, and even mutilating themselves for their performances, as well as frequently donning ornate costumes and masks.[106] How one dresses has been a critical aspect of social dancing for centuries. Likewise, clothes and fashion designers have long been inspired by the way that dancers look.[107] When Marie Taglioni and her contemporaries learned to dance on their toes they created a market for the blocked pointe shoe. The subsequent development of pointe technique depended intimately on the interaction between the materiality of the shoes and the training methods of the dancers, and these elements coevolved.[108] Numerous other forms of footwear, from ballet slippers to tap shoes, are of course now widely manufactured, while under the influence of modern dance choreographers such as Merce Cunningham, contemporary dance has gravitated toward a simple uniform of leotards and tights.

Clothing fashions have also shaped dance. For instance, in the late eighteenth century, a neoclassical style swept the world of fashion, leading to lighter dresses and flatter shoes. These indirectly revolutionized choreography by permitting greater speed and range of movement.[109] The relationship was reciprocal, and many famous dancers, including Isadora Duncan and Margot Fonteyn, became fashion icons who greatly influenced the clothing industries.[110] The influence of dancers on fashion continues to the present.[111] Christian Dior, for instance, was explicitly inspired by the tutus and pointe shoes of ballerinas, while Yves Saint Laurent was influenced by the Orientalism of the Ballet Russes.[112] Just as professional dance companies commissioned music from leading composers, so the companies forged alliances with fashion designers. As early as 1924, Coco Chanel designed costumes for the Russian ballet *Le Train Blue*, while in 1965 Yves Saint Laurent designed for the ballet *Notre Dame de Paris*. In recent years, it would appear that virtually everyone in the fashion world, from Valentino to Vivienne Westwood, is designing dance costumes.[113] The tension between the demands of performance and societal decorum have led to some famous fashion battles. In 1885, the great nineteenth-century ballerina Virginia Zucchi caused a scandal by insisting on performing in a shortened tutu, while in 1910, Vaslav Nijinsky was sacked after refusing to cover up his tights![114]

A long-standing relationship between art and dancing exists. Edgar Degas devoted much of his career to painting and sculpting dancers, and Toulouse Lautrec was commissioned to paint the posters for the Moulin Rouge. Pablo Picasso and Henri Matisse each designed sets for various ballets and dance productions, while Martha Graham collaborated with the Japanese-American sculptor Isamu Noguchi, whose settings were thought to enhance a dance's visual power.[115] Dance also shares entwined histories with theatre, opera, and comedy. Cinema too has played important roles in popularizing dances, from the ballroom and tap era of Fred Astaire, Ginger Rogers, and Gene Kelly with classics like *Top Hat* and *Singin' in the Rain*. Polished film adaptations of dance musicals have proliferated, such as *Oklahoma*, *Guys and Dolls*, and *West Side Story*, through to the *rock 'n roll* and *disco* eras, with *Rock Around the Clock*, *Saturday Night Fever*, and *Grease*.

The evolution of technology has shaped the medium too, most obviously through the manufacture of music recording and playing devices, beginning with the gramophone and records in the early part of the twentieth century, right up to today's iPods. These advances allowed the owner to produce music and dance anywhere. The invention of radio, cinema, and television propagated dance and music, and transformed regional and national dances into global exports. Printing and lithography allowed sheet music and dance steps to be formalized and widely distributed, and also led to the widespread circulation of prints of dancing icons like Taglioni and Duncan, as well as more disreputable pictures of scantily dressed dancers. Then there are various dance industries. These include not just professional dance companies but also the pursuits of dance instruction, choreography, commercial dance production, set design, and costume manufacture; the dance hall, ballroom, and night-club industries; composition and sales of dancing books, manuals, and dance music; as well as businesses in dance exercise and dance therapy. Dance has certainly not evolved in a vacuum, but has been subject to constant stimulation and cross-fertilization from many other domains.

This brief *précis* of how dance has changed over millennia, while inadequate as a historical analysis, serves I hope to illustrate two important points. First, in spite of the fact that dance exhibits bewildering variation in form, the manifest diversity can nonetheless be understood by tracing back the interwoven histories of each lineage, recognizing the impact of miscellaneous internal and external influences that fueled dance innovations, as well as the social contexts to which the dances were adapted. The chronicle is beguiling, and there are certainly details of history that have been lost or remain unclear, but the processes by which such a vast corpus of dance could arise are, at least in general terms, not at all mysterious. In spite of awesome diversity, we have no problem comprehending how so much dance could have arisen. This is significant, since if we can, at least in principle, understand the processes that explain the diversity of dance, then we should not be flummoxed by the diversity of foods, medical treatments, motorized vehicles, or languages. Each cultural sphere may well change through unique details, but if we think in broad-brush terms about the three prerequisites of variation, differential fitness, and inheritance, then sheer diversity is no impediment to comprehending any sphere's existence.

Second, we can also start to see how, gradually over time, complexes of complementary ideas have become cobbled together to generate highly intricate, coherent, and sophisticated dance forms; these are often deeply embedded in other aspects of culture, particularly music, technology, costume, and performance. The evolution of ballet, for instance, is not easily described as the spread of a single ballet "meme," but required the painstaking accumulation of countless innovations over several centuries, each triggering a cascade of reverberations in the other cultural elements central to the developing structure. In many respects, the emergence of complex, integrated cultural forms resembles the evolution of composite biological adaptations—such as the eye—which typically require the fixation of numerous individual mutations to exhibit any semblance of adaptedness and design. If, notwithstanding the awesome talent on show, we can understand how something as magical as *Swan Lake* could have come into existence, then the complexity of satellites, financial markets, or Catholicism are also rendered potentially comprehensible.

Culture evolves in two senses: the observation that cultural phenomena change over time, and the evolution of the capacity for culture. Evolutionary biology can shed light on these issues by helping to explain how the psychological, neurological, and physiological attributes necessary for culture came into existence. In the case of dance, evolutionary insights explain how humans are capable of moving in time to music; how we are able to synchronize our actions with others or move in a complementary way; how we can learn long, complex sequences of movements; why it is that we have such precise control of our limbs; why we want to dance what others are dancing; and why both participating in dancing and watching dance is fun. Armed with this knowledge, we can make better sense of why dance possesses some of the properties that it does, and why dances changed in the manner they did. As it is for dance, so it is for sculpture, acting, music, computer games, or just about any aspect of culture. Biology provides no substitute for a comprehensive historical analysis. However, our understanding of the underlying biology feeds back to make the historical analysis so much richer and intelligible.

AWE WITHOUT WONDER

This book began with an innocent glance out of the window to contemplate the entangled bank of human culture. Like many before me, I was inspired to ask whether evolutionary theory, such a powerful source of explanation for the natural world, could also account for the existence of cars and houses, hospitals and factories, road networks and electricity grids, or theatrical productions and orchestral symphonies. Was there a scientific explanation for the origins of technology, engineering, the arts, and science itself, with roots that could be traced back to the realm of animal behavior?

Nearly thirty years ago, as a graduate student at University College London, I first asked these questions, and found myself surprised to be stymied for an answer. To simply state that building contractors erected the structures seen (the immediate mechanical explanation) was inadequate. I wanted to know how we evolved the underlying capability to build, plan, coordinate, and work together in teams that such extraordinary edifices demanded. Nor was merely attributing human successes to our culture, language, intelligence, or cooperation satisfactory, because these attributes were themselves mysterious, and utterly unprecedented in the natural world.

The more that I contemplated the problem, the more the multifaceted richness, complexity, and diversity of human culture appeared to elude scientific analysis. Rather than leaving me satisfied that evolutionary principles could explain these most human aspects of our existence, more and more challenging questions began to emerge, like layers of an onion. Far from the confident satisfaction that I had expected the biological sciences to deliver, I was left with an overwhelming sense of awe and wonder. How on earth were we to account for the majesty of human culture? The very characteristic that marked our species out as exceptional, that accounted for our extraordinary ecological and demographic success as a species, also seemed to separate us from the rest of nature. Not only were the origins of culture hard to explain, but there was a huge gulf between the cognitive abilities of humans and other animals, which far from shrinking as new research emerged over the decades, was being cemented. How could science bridge that gap?

Now, having dedicated my career to addressing this challenge, and abetted by the many members of my research group and numerous other professional colleagues, I finally feel like I have some semblance of an answer. Certainly it is not the whole story. Probably there are elements that I have got wrong. Undoubtedly I have not done justice to the contributions of others. What is important to me, however, is that the scientific method should have delivered enough insight for the utter mystery of culture's origins to have dissipated. Researchers may never know exactly how the psychological capabilities that underlie our capacity for culture evolved. A perfect reconstruction of how that evolved psychology combined with a rich social environment to spawn myriad ideas, behaviors, and artifacts may also remain evasive. Nonetheless, a great deal of satisfaction can be derived from the fact that science now has a credible account of the genesis of central aspects of the human mind, intelligence, and culture, and in a world in which a large proportion of the population disputes the fact that humans have evolved, that is very much worth having. Here is my explanation.

We have learned through extensive experimental work by behavioral scientists that both copying and innovation are widespread among animals, but that animals can be highly strategic about the manner in which they exploit learned information. The social learning strategies tournament explained much about this learning by demonstrating a selective advantage to copying when implemented with accuracy and efficiency. Strategic, high-fidelity copying confers fitness benefits. This theoretical insight leads to the expectation that natural selection would have favored more efficient and higher-fidelity forms of social learning, as well as those neural structures and functional capabilities in the primate brain capable of bringing this about. In the process, natural selection will have shaped the evolution of the primate brain and intelligence.

Comparative data across primates support this suggestion and reveal strong associations between social learning, innovativeness, and brain size in primates; social learning also covaries with a number of measures of intelligence, including naturalistic measures like tool use and performance in laboratory tests of learning and cognition. The findings imply that a "cultural drive" process may have been operational across several distinct primate lineages, whereby natural selection favored efficient copying. Selection for high intelligence in primates almost certainly derives from multiple sources; however, comparative analyses suggest that widespread selection for social intelligence in monkeys and apes was followed by more restricted selection for cultural intelligence in the great apes, capuchins, and macaques, mediated by conferred increases in longevity and diet quality. This selection is thought to have enhanced several aspects of cognition, including learning, perspective-taking, computation, tool use, and in particular, collaborative social interaction.

The comparative analyses, in turn, raise the question of why humans alone should exhibit a culture that ratchets up in complexity. The answer, derived from theoretical work, is that complex culture requires high-fidelity information transmission. Analyses show that small increases in the accuracy of social transmission can lead to big increases in the amount and longevity of culture, and that high-fidelity knowledge transfer is necessary for cumulative culture. In addition, the tournament taught us that high levels of reliance on social learning automatically generated extreme longevity of cultural knowledge. Populations appeared to pass a threshold level of reliance on social learning, above which cultural knowledge became extremely stable and persisted almost indefinitely. With increasing social, as opposed to asocial, learning, our ancestors' behavior also became more conformist, and started to exhibit fads and fashions as we commonly see in human populations today. Put together, these theoretical results suggested that once our ancestors had evolved sufficiently strategic and accurate forms of copying, many aspects of the cultural capability witnessed in modern humans would arise.

How did our ancestors achieve high-fidelity information transmission? The obvious answer is through teaching, which is rare in nature but universal in human societies, once the many subtle forms it takes are recognized. Mathematical analyses reveal tough conditions that must be met for teaching to evolve, but show that cumulative culture relaxes these conditions. This implies that teaching and cumulative culture coevolved in our ancestors, creating for the first time in the history of life on earth a species that taught their relatives across a broad range of contexts. Humans are unique in the extensiveness of their teaching mainly because cumulative culture makes knowledge that is otherwise difficult to acquire available in the population to be taught.

Teaching is expected to evolve when (1) its costs are low or can be offset against the costs of provisioning; (2) instruction is highly accurate and effective in transmission; and (3) there is a strong degree of relatedness between tutor and pupil. Any adaptation that reduces the costs of teaching ought to be favored by selection provided that it does not seriously diminish teaching efficacy. It was here, in the unprecedented context of the widespread teaching exhibited by our hominin ancestors, that language first evolved as an adjunct to teaching. Language is an adaptation, fashioned by natural selection to reduce the costs, increase the accuracy, and expand the domains of teaching. This explanation has the advantage that it accounts for the honesty, cooperativeness, and uniqueness of language, as well as its power of generalization, how it was grounded, and why it was learned. Human language is unique, at least among extant species, because only humans constructed a sufficiently diverse, generative, and changeable cultural world that demanded talking about. Once our ancestors evolved a socially transmitted system of symbolic communication, other features of language, such as compositionality, came along for free.

Experimental studies support the hypothesis that a gene-culture coevolutionary dynamic arose between socially transmitted skills, including tool use, and aspects of human anatomy and cognition. This interaction was ongoing in human evolution from at least 2.5 million years ago, and has continued to the present. Theoretical, anthropological, and genetic studies all attest to the importance of gene-culture coevolutionary feedback in recent human evolution, which has shaped both our anatomy and our cognition, and speeded up rates of change. As expected, the brain regions associated with imitation, innovation, and tool use are among those that expanded during recent hominin evolution. Just as biological evolution gave way to gene-culture coevolution, cultural evolution then took over the reins of human adaptation, and the pace of change experienced by the members of our evolutionary lineage accelerated further. Culture provided our ancestors with food-procurement and survival tricks, and as each new invention arose, the population was able to exploit its environment more efficiently. This not only fueled brain expansion but population growth as well.

Human numbers and societal complexity both increased dramatically with the domestication of plants and animals and the advent of agriculture. These freed societies from the constraints imposed on hunter-gatherers by the requirement to be constantly on the move. Agricultural societies flourished both because they outgrew hunter-gatherer communities, through generating an increase in their environments' carrying capacity, and because agriculture triggered a raft of follow-on innovations that dramatically changed human society. In the larger populations that were supported by agricultural yields, beneficial innovations were more likely to spread and be retained. Agriculture precipitated a revolution through not only triggering the invention of related technologies, but by spawning entirely unanticipated initiatives, such as the wheel, the development of city-states, and religions. Through oral traditions, dance, and ritual, historical accounts were preserved and supplemented by externalized cultural memory stores, from written records and books to today's computer banks, which left cultural knowledge increasingly difficult to lose.

The scale and complexity of human cooperation is unprecedented. Theoretical and experimental data suggest that large-scale cooperation arose in human societies because of our uniquely potent capacities for social learning, imitation, and teaching. Culture took human populations down novel evolutionary pathways, both by creating conditions that promoted established cooperative mechanisms, such as indirect reciprocity and mutualism, and by generating novel cooperative mechanisms not seen in other animals, such as cultural group selection. In the process, gene-culture coevolution generated an evolved psychology entirely different from that observed in any other animal. This evolved psychology comprises a motivation to teach, speak, imitate, emulate, and share the goals and intentions of others, as well as a massively upgraded capacity for learning and computation. Theoretical and experimental studies suggest that both human cognition and culture differ from those found in other apes because our species uniquely possesses a package of sociocognitive capabilities that underlie human culture, including teaching, language, superior imitation, and enriched prosociality. These capabilities have coevolved with cumulative culture because they enhance the fidelity of information transmission.

Evolutionary biology can shed light on both the manner in which contemporary cultural phenomena change over time, and the origins of the psychological, neurological and physiological attributes necessary for culture to come into existence. This was illustrated with the example of the evolution of dance, which revealed, for instance, why humans are capable of moving in time to music, how we are able to synchronize our actions with others, and how we can learn long sequences of movements. In spite of the fact that dance exhibits bewildering variation in form, this diversity can be understood by tracing back lineages, recognizing diverse influences underlying dance innovations, and the social contexts to which dances adapted. We can also see how, gradually over time, complexes of complementary ideas have been drawn together to generate highly intricate dance forms that resemble the evolution of composite biological adaptations. As it is for dance, so it is for other aspects of culture, whether associated with the arts, the sciences, or technology; in all elements of culture, new forms arise as refinements or combinations of existing forms. Extremes of diversity and complexity are no impediment to scientific investigation. Far from destroying culture, our understanding of the underlying science feeds back to make the historical analysis richer and less mysterious. Human culture is indeed amenable to evolutionary analysis.

With the benefit of hindsight, we can understand now why illuminating the origins of human cognition and intelligence has proven such a tough task, one that thwarted the efforts of some of history's greatest minds, including Darwin's. Three factors in particular made this challenge acutely mountainous. First, the origins of none of the critical elements of human cognition (our cultural learning, intelligence, language, cooperation, or powers of computation) can be completely understood in isolation, because each shaped the others in a nexus of complex coevolutionary feedbacks. Second, the human mind did not evolve in a straightforward, linear manner, with changes in the external environment generating natural selection favoring cognitive adaptations. Rather, our mental abilities evolved through a convoluted, reciprocally caused process, whereby our ancestors constantly constructed aspects of their physical and social environments that fed back to impose selection on their bodies and minds, in endless cycles. Third, to understand the intricate dynamic process through which the human mind evolved required the tools of modern genomics, population genetics, gene-culture coevolution, anatomy, archaeology, anthropology, and psychology in a concerted multidisciplinary effort. The elements of this tool box were not available to Darwin or to any of his predecessors until recently. Our abilities to think, learn, understand, and communicate leave humanity genuinely different from other animals. Scientists can now comprehend that divergence as reflecting the operation of a broad array of feedback mechanisms in the hominin lineage, through which key elements of human cognition and culture accelerated together in a runaway, autocatalytic process.

As I look back now, I can see that my research group's struggles to comprehend the evolution of culture proved difficult for the very same reasons as understanding the origins of the human mind has proved challenging to the broader scientific community. The human cultural capability did not evolve in isolation but in intricate coevolution with central aspects of cognition and behavior, including our language, teaching, intelligence, perspective-taking, powers of computation, cooperative tendencies, tool use, memory, and control of the environment. An unanticipated bonus for our efforts is that in struggling to understand the origins of culture, we may in the process have shed some light on the origins of the human mind, language, and intelligence.

I would like to believe that my efforts, and those of the many talented scientists that shared this journey with me, have gradually chipped away at some of the bewilderment that the dazzling complexity of human culture originally inspired. Yet the creative and analytical power of the human mind, and the insatiable potency with which cultural evolution generates ever more richness to our cultural lives, remain just as impressive. The magic of a Mozart, a Shakespeare, or a da Vinci, lingers every bit as poignantly in light of the deeper understanding that an evolutionary analysis of cultural origins affords. The genius of Alan Turing, Marie Curie, and Isadora Duncan continues to be just as inspirational. The awe remains, even if a small part of the wonderment subsides.

NOTES

CHAPTER 1: DARWIN'S UNFINISHED SYMPHONY

1. Darwin 1859, p. 459.
2. Cultural evolution is the idea that changes in human cultural beliefs, knowledge, customs, skills, and languages can be understood as an evolutionary process. Mesoudi 2011, and Richerson and Boyd 2005 provide recent overviews, while Cavalli-Sforza and Feldman 1981, and Boyd and Richerson 1985 give pioneering and important formal treatments.
3. In *The Selfish Gene* (1976), Richard Dawkins introduces the notion of the "meme," a cultural replicator with gene-like properties. However, the modern science of cultural evolution derives very little from memetics. For an introduction to the now extensive experimental and theoretical work that underpins this field, see Mesoudi 2011, Richerson and Boyd 2005, or Henrich 2015. For a critical evaluation of the field, see Lewens 2015.
4. The uniqueness of human cooperation is discussed in Boyd and Richerson 1985, Richerson and Boyd 2005, Henrich and Henrich 2007, and Henrich 2015.
5. Caro and Hauser 1992, Hoppitt et al. 2008, and Thornton and Raihani 2008 provide reviews of animal teaching. Thornton and McAuliffe 2006 describe teaching in meerkats, one of the most compelling examples of animal teaching.
6. Cooperation in animals is not restricted to kin selection, and can occur through a variety of mechanisms. See Nowak and Highfield 2011 for a recent overview.
7. Darwin 1871, p. 160.
8. Currie and Fritz 1993.
9. Winterhalder and Smith 2000, Brown et al. 2011.
10. Gagneux et al. 1999.
11. Klein 1999, Boyd and Silk 2015.
12. Sterelny 2012a makes this case in more detail.
13. Boyd and Richerson 1985, Tomasello 1994, Richerson and Boyd 2005, Boyd et al. 2011, Henrich 2015.
14. Pinker 2010.
15. Boyd and Richerson 1985, Tomasello 1994, Richerson and Boyd 2005; Boyd et al. 2011, Henrich 2015.
16. Basalla 1988, Petroski 1992.
17. Petroski 1992.
18. Ibid.
19. See Petroski 1992 for a detailed treatment.

20. Cumulative culture is also described as "ratcheting"; see Tomasello 1994.

21. Boyd and Richerson 1985, 2005; Tomasello 1999; Whiten and Van Schaik 2007; Pagel 2012.

22. Zentall and Galef Jr. 1988, Avital and Jablonka 2000, Leadbeater and Chittka 2007, Hoppitt and Laland 2013.

23. My use of the term "copying" throughout this book refers to any form of "social learning,"—that is, any means by which one animal comes to learn as a consequence of observing or interacting with another animal or its products.

24. Warner 1988, Whiten et al. 1999, Van Schaik et al. 2003, Perry et al. 2003, Rendell and Whitehead 2001, Fragaszy and Perry 2003.

25. Fragaszy and Perry 2003, Hoppitt and Laland 2013.

26. On whales, see Rendell and Whitehead 2001, and Whitehead and Rendell 2015. On birds, see Mundinger 1980, Avital and Jablonka 2000, Emery and Clayton 2004, and Emery 2004.

27. For chimpanzees, see Whiten et al. 1999, 2009; for orangutans, see Van Schaik et al. 2003; for capuchin monkeys, see Perry et al. 2003.

28. Whiten 1998, Whiten et al. 2009.

29. Tomasello and Call 1997.

30. Claims of cumulative culture in animals remain rare and controversial (Tennie et al. 2009, Dean et al. 2012, 2014). One possibility is provided by the complex fishing tools of chimpanzees (Sanz et al. 2009), while a second are the stepped pandanus foraging tools of New Caledonian crows (Hunt and Gray 2003).

31. Boesch 2003.

32. Tennie et al. 2009.

33. Ectoparasite manipulation, well digging, and the use of sets of tools in sequence are all candidates for cumulative culture. See Boesch 2003, and Sanz et al. 2009.

34. Hunt and Gray 2003.

35. A recent study failed to find evidence for the observational learning of problem solving in these birds (Logan et al. 2015), but it is difficult to rule out the possibility that social learning might play a role in the acquisition of foraging skills and dietary preferences in more naturalistic settings, particularly as these birds are known to acquire a wide variety of rich foods through tool use (Rutz et al. 2010).

36. Basalla 1988, Ziman 2000.

37. McBrearty and Brooks 2000, D'Errico and Stringer 2011.

38. McPherron et al. 2010.

39. Stringer and Andrews 2005.

40. Ibid., Klein 2000.

41. Stringer and Andrews 2005.

42. Thieme 1997.

43. James 1989.

44. Movius 1950, Stringer and Andrews 2005.

45. Mellars 1996.

46. McBrearty and Brooks 2000.

47. For hafting implements, see Boeda et al. 1996. For awls, see Hayden 1993.
48. These dates are usually associated with Europe, but there is increasing evidence for this "Upper Paleolithic" technology in other regions of the world, and at earlier times. See McBrearty and Brooks 2000.
49. Stringer and Andrews 2005.
50. Ibid.
51. Bronowski 1973, Diamond 1997.
52. Ibid.
53. Laland and Galef Jr. 2009, Whiten et al. 2011.
54. Boyd and Richerson 1985, 1996; Galef Jr. 1992; Heyes 1993; Boesch and Tomasello 1998.
55. For a readable introduction to the history of the field of human evolution research, see Lewin 1987; for a more up-to-date summary, see Boyd and Silk 2015.
56. Darwin 1859, p. 458.
57. Darwin 1871, p. 327.
58. See the entry for René Descartes in the *Stanford Encyclopedia of Philosophy* (Hatfield 2016).
59. Darwin 1872.
60. Wallace 1869; see also "The Limits of Natural Selection as Applied to Man (S165: 1869/1870)" on the Alfred Russel Wallace Page (http://people.wku.edu/charles .smith/wallace/S165.htm).
61. However, a variety of interpretations can be found; see http://wallacefund.info /wallace-biographies.
62. Darwin 1871, p. 158.
63. See Miller's *The Mating Mind* (2001) for an account of the evolution of intelligence that is heavily reliant on Darwinian sexual selection theory.
64. Tomasello and Call 1997.
65. See Shettleworth 2010 for an excellent introduction.
66. Witness, for instance, the chapters in Kappeler and Silk 2009, in which the existence of a gap between the intellectual abilities of humans and other animals is acknowledged. See also Suddendorf 2013.
67. Linden 1975; Wallman 1992; Radick 2007; Byrne and Whiten 1988; Whiten and Byrne 1997; de Waal 1990, 1996, 2007, 2010.
68. See, for instance, the writings of philosopher Peter Singer or primatologists Frans de Waal and Jane Goodall.
69. Morris 1967.
70. Lorenz 1966, Ardrey 1966.
71. Diamond 1991.
72. Lewin and Foley 2004, Stringer and Andrews 2005.
73. Glazko et al. (2005) report that 80% of proteins are different between humans and chimpanzees.
74. Frazer et al. 2002.
75. Haygood et al. 2007.

76. Fortna et al. 2004.
77. Hahn et al. 2007.
78. Calarco et al. 2007.
79. King and Wilson 1975.
80. Carroll 2005, Müller 2007.
81. Birney 2012.
82. Other minor differences exist too, such as the eszett, which is a letter that exists only in German.
83. Carroll 2005.
84. Voight et al. 2006, Wang et al. 2006. See Laland et al. 2010 for an overview.
85. Caceres et al. 2003.
86. Enard, Khaitovich, et al. 2002.
87. Taylor 2009.
88. Striedter 2005.
89. Other recently discovered members of our genus include *Homo naledi* and *H. floresiensis.*
90. Henrich et al. 2001.
91. For an overview of human cooperation see Henrich and Henrich 2007, or Henrich 2015.
92. Jensen et al. 2007.
93. Fehr and Fischbacher 2003, Richerson and Boyd 2005, Henrich and Henrich 2007.
94. Tomasello and Call 1997.
95. Povinelli et al. 1992, Tomasello and Call 1997, Tomasello 2009.
96. Premack and Woodruff 1978.
97. Call and Tomasello 2008.
98. Call et al. 2004, Call and Tomasello 1998.
99. Heyes 1998, Seyfarth and Cheney 2000.
100. Call and Tomasello 2008.
101. Onishi and Baillargeon 2005.
102. Dennett 1983.
103. See, for instance, Call and Tomasello 2008, Herrmann et al. 2007, and Whiten and Custance 1996.
104. Herrmann et al. 2007.
105. Whiten 1998, Whiten et al. 2009, Tomasello and Call 1997.
106. Horner and Whiten 2005, Dean et al. 2012.
107. Byrne and Whiten 1988, Whiten and Byrne 1997, Dunbar 1995, Tomasello 1999.
108. Hauser 1996.
109. Seyfarth et al. 1980.
110. Caesar et al. 2013.
111. Janik and Slater 1997; Wheeler and Fischer 2012, 2015.
112. Bickerton 2009.
113. Ibid.

114. For more on attempts to teach apes to talk, see Gardner and Gardner 1969, Terrace 1979, and Radick 2008.

115. Radick 2008, Bickerton 2009, Fitch 2010.

116. Herbert Terrace famously "changed his mind" about apes possessing language. After many years of working with chimpanzee Nim Chimpsky, he was forced to conclude that Nim's behavior could be understood as the product of established learning processes and did not imply any linguistic ability. See Terrace 1979.

117. de Waal 1990, 1996, 2007, 2010.

118. Dawkins 2012.

119. Silk 2002.

120. This is also known as vicarious instigation (Galef Jr. 1988). Emotional contagion is discussed in Berger 1962, Curio et al. 1978, Kavaliers et al. 2003, Olsson and Phelps 2007, and Hoppitt and Laland 2013.

121. de Waal 1990, 1996, 1999, 2007, 2010.

122. See Bshary et al. 2002, and Bshary 2011; see also Abbott 2015.

123. de Waal 1990, 1996, 2007, 2010.

124. Lewin and Foley 2004, Stringer and Andrews 2005.

125. Ibid.

126. Caceres et al. 2003.

127. Enquist et al. 2008, 2011.

128. Barrett et al. 2001.

129. The World Intellectual Property Organization's 2013 report described 2.35 million patent applications filed worldwide in 2012 alone. They also reported that 24 million trademark registrations were active worldwide. See http://www.wipo .int/edocs/pubdocs/en/intproperty/941/wipo_pub_941_2013.pdf.

CHAPTER 2: UBIQUITOUS COPYING

1. Galef Jr. 2003, p. 165.

2. Galef Jr. 2003; Barnett 1975.

3. Darwin 1871.

4. Hoppitt and Laland 2013.

5. Darwin 1871, p. 50.

6. Twigg 1975; Barnett 1975.

7. Steiniger 1950.

8. Hepper 1988.

9. Galef Jr. and Henderson 1972.

10. Galef Jr. and Clark 1971b.

11. Ibid.

12. Galef Jr. and Clark 1971a.

13. Galef Jr. and Buckley 1996.

14. Galef Jr. and Heiber 1976, Laland and Plotkin 1991, Galef Jr. and Beck 1985.

15. Laland 1990, Laland and Plotkin 1991.
16. Laland and Plotkin 1993.
17. Galef Jr. and Wigmore 1983, Posadas-Andrews and Roper 1983, Galef Jr. et al. 1988.
18. Galef Jr. et al. 1984.
19. Laland and Plotkin 1993.
20. Galef Jr. and Allen 1995.
21. Valsecchi and Galef Jr. 1989, Galef Jr. et al. 1998, Lupfer et al. 2003, McFayden-Ketchum and Porter 1989, Lupfer-Johnson and Ross 2007, Ratcliffe and Ter Hofstede 2005.
22. Atton 2013.
23. Dornhaus and Chittka 1999, Leadbeater and Chittka 2007.
24. Jablonka and Lamb 2005, Hoppitt and Laland 2013.
25. Fisher and Hinde 1949, Hinde and Fisher 1951.
26. Marler 1952.
27. Marler and Tamura 1964; Catchpole and Slater 1995, 2008.
28. Von Frisch 1967.
29. Kawai 1965.
30. Kawai 1965.
31. Wilson 1975, Bonner 1980.
32. Kummer and Goodall 1985.
33. Huffman and Hirata 2003, Reader et al. 2011.
34. McGrew and Tutin 1978, McGrew 1992.
35. de Waal 2001, Galef and Laland 2009.
36. Whiten et al. 1999, 2001.
37. Ibid.
38. Ibid.
39. Hoppitt and Laland 2013, chapter 7.
40. Lonsdorf 2006.
41. Lonsdorf et al. 2004.
42. *Pongo* spp.
43. Van Schaik et al. 2003, Van Schaik 2009.
44. Ibid.
45. Ibid.
46. *Cebus capuchinus*.
47. Perry et al. 2003, Perry 2011.
48. Huffman 1996, Huffman and Hirata 2003, Leca et al. 2007.
49. Darwin 1841.
50. Leadbeater and Chittka 2007 point out that there are several routes by which honeybees could have acquired this behavior, only some of which involve social learning.
51. Blue tits (*Cyanistes caeruleus*). Great tits (*Parus major*).

52. Slagsvold and Weibe 2007, 2011.
53. Ibid.
54. Slagsvold et al. 2013.
55. Slagsvold et al. 2002, Hansen et al. 2008.
56. Johannessen et al. 2006.
57. Slagsvold and Hansen 2001.
58. Sargeant and Mann 2009.
59. Rendell and Whitehead 2001, Baird 2000.
60. Schuster et al. 2006.
61. Thornton et al. 2010, Kirschner 1987.
62. Cloutier et al. 2002.
63. Elgar and Crespi 1992.
64. Mery et al. 2009.
65. Dugatkin 1992, Witte and Massmann 2003, Godin et al. 2005.
66. White and Galef Jr. 2000, White 2004, Swaddle et al. 2005.
67. Galef 2009.
68. Little et al. 2008, and Jones et al. 2007.
69. Fruit flies (*Drosophila melanogaster*). Mery et al. 2009.
70. See, for instance, Goldsmidt et al. 1993, Kraak and Weissing 1996, and Forsgren et al. 1996.
71. Largiader et al. 2001.
72. *Poecilia reticulate.*
73. Dugatkin 1992.
74. Dugatkin and Godin 1992.
75. *Poecilia mexicana.*
76. Witte and Ryan 2002.
77. Plath et al. 2008.
78. Catchpole and Slater 1995, 2008.
79. Caldwell and Caldwell 1972, Janik and Slater 1997.
80. *Tursiops* spp.
81. *Orca orca.*
82. *Megaptera novaeangliae.* See Janik and Slater 1997 for much of the research on these species.
83. Payne and Payne 1985.
84. Noad et al. 2000.
85. Garland et al. 2011.
86. Throughout this book I adopt the common scientific convention of using the term "fishes" to refer to multiple species of fish, with the term "fish" being the plural applied to multiple individuals of the same species.
87. Brown and Laland 2003, 2006.
88. *Thalassoma bifascatum.*
89. Warner 1988.

90. Warner 1990.
91. Boyd and Richerson 1985.
92. Diamond 2006.
93. Laland and Williams 1997, 1998.
94. Couzin et al. (2005) show that as the number of informed individuals in an animal group (such as a shoal of fish) increases there is a corresponding increase in the ability of these knowledgeable individuals to lead a group to a desired location, such as a food source. Nonetheless, these authors found that the proportion of knowledgeable individuals can be strikingly small and yet are still able to lead the group effectively. At the extreme, less than 5% of the individuals can have the information and, provided they deploy some simple rules, such as leading the others to aggregate and move together, all of the group will reach the destination. A recent analysis of baboons confirmed these earlier findings, with the movements of individuals shaped more by a small number of individuals moving in a particular direction together than by high-status individuals imposing their decisions on others when conflicts over movement patterns arise (Strandberg-Peshkin et al. 2015).
95. Laland & Williams 1997, 1998.
96. Stanley et al. 2008.
97. Warner 1988, 1990; Helfman and Schultz 1984.
98. Lachlan et al. 1998, Laland and Williams 1998.
99. Lachlan et al. 1998, Day et al. 2001, Pike and Laland 2010.
100. Laland and Williams 1998.
101. Warner 1988.
102. Mueller et al. 2013.
103. Biesmeijer and Seeley 2005.
104. *Macaca mulatta.*
105. Mineka and Cook 1988, Mineka et al. 1984.
106. Mineka and Cook 1988.
107. Ibid., Mineka et al. 1984.
108. Mineka and Cook 1988.
109. *Turdus merula.*
110. Curio 1988.
111. Hoppitt and Laland 2013, Olsson and Phelps 2007, Leadbeater and Chittka 2007.
112. Brown and Laland 2001, Hirvonen et al. 2003, Brown and Day 2002.
113. Brown, Markula, et al. 2003; Brown, Davidson, et al. 2003; Brown and Laland 2002.
114. I place the terms "instinct" and "innate" in quotes to highlight the fact that these are problematic terms. Not only are they ambiguous, with multiple meanings (unlearned, species-specific, evolved, unchanging, etc.) attached to each, but these meanings have been found frequently to not co-occur. See Bateson and Martin 2000 for discussion.
115. Gerull and Rapee 2002.

116. Bandura and Menlove 1968, Mineka and Zinbarg 2006, Olsson and Phelps 2007.
117. Rogers 1988.

CHAPTER 3: WHY COPY?

1. Pennisi 2010 (the commentary), p. 167; Rendell et al. 2010.
2. Hoppitt and Laland 2013.
3. Bandura et al. 1961.
4. Carpenter 2006.
5. Boyd and Richerson 1985, Tomasello 1999.
6. Anderson and Bushman 2001.
7. Halgin and Whitbourne 2006.
8. Horner and Whiten 2005.
9. Whiten et al. 2009.
10. Lyons et al. 2007.
11. Over and Carpenter 2012.
12. Evans 2016.
13. Flynn 2008.
14. Boyd and Richerson 1985, Rogers 1988, Feldman et al. 1996, Henrich and McElreath 2003, Enquist et al. 2007.
15. Barnard and Sibly 1981, Giraldeau and Caraco 2000.
16. Or at least, this is what has generally been assumed. In a study carried out by one of my students, Alice Cowie, on social foraging in budgerigars (small parrots), we found that producer birds in fact consistently garnered more food than scroungers (Cowie 2014). This may reflect a tendency for budgerigars to satisfice—that is, settle for a reasonable return. It may imply that the birds do not have good comprehension of the returns available for other strategies, or perhaps there are constraints, such as ability or personality differences, that render switching strategies difficult for some individuals.
17. Kameda and Nakanishi 2002.
18. In fact, more recent theoretical work suggests that this conclusion may not always be robust, and there are a variety of circumstances under which social and asocial learning will not have equal fitness at equilibrium (Boyd and Richerson 1985; van der Post and Hogeweg 2009; Rendell et al. 2010; Rendell, Boyd, et al. 2011).
19. Rogers 1988.
20. A strategy is an ESS if, were it adopted by the whole population, it cannot be invaded by any other strategy that is initially rare.
21. Giraldeau et al. 2002, Henrich and McElreath 2003.
22. Boyd and Richerson 1985.
23. Tellier 2009.
24. Caselli et al. 2005.
25. This conclusion is reinforced by theoretical models that allow for more strategic copying behavior—for instance, only copying where asocial learning fails, or

copying in proportion to the payoff to the copied individual—which reveals fitness benefits to copying over asocial learning (see Boyd and Richerson 1995, Laland 2004, and Enquist et al. 2007).

26. Boyd and Richerson 1985, Rogers 1988, Feldman et al. 1996, Giraldeau et al. 2002.

27. Boyd and Richerson 1985, Henrich and McElreath 2003, Laland 2004.

28. Laland 2004.

29. Ibid.

30. Ibid.

31. Henrich and McElreath 2003; Laland 2004; Kendal et al. 2005; Kendal et al. 2009; Rendell, Fogarty, Hoppitt, et al. 2011; Hoppitt and Laland 2013.

32. There is circumstantial evidence consistent with all of these ideas, and more, although that data (a combination of experimental and observational findings) is rarely sufficiently detailed to be able to specify a unique strategy that is supported. More frequently, the findings are consistent with multiple alternative strategies (Rendell, Fogarty, Hoppitt et al. 2011; and Hoppitt and Laland 2013).

33. See, for instance, Kendal et al. 2009.

34. The term "optimal" is given in quotes here to signify the fact that there is a subtle difference between those traits that are expected to evolve in a given circumstance and those that are optimal. In reality, most theoretical analyses of this nature—for instance, those using evolutionary game theory—are seeking ESS's, which are better characterized as "uninvadable" rather than globally optimal (Maynard-Smith 1982).

35. Axelrod 1984.

36. European Commission contract FP6–2004-NESTPATH-043434.

37. The payoffs were drawn from an exponential distribution.

38. When the payoffs of a multi-armed bandit change over time, the bandit is termed "restless." Restless multi-armed bandits are widely recognized as hard problems, and satisfactory analytical solutions have not been found (Papadimitriou and Tsitsiklis 1999).

39. New behaviors were selected at random.

40. This is known as an exploration-exploitation trade-off.

41. The set of rules was specified either verbally, in "pseudocode," or in a computer language (Matlab).

42. To do this, we established pair-wise contests in which initially all agents in the population utilized a single strategy, and then a small number of agents using the alternative strategy would be introduced. We then looked to see whether the introduced strategy could invade and take over the population by outcompeting the resident. The strategy that was more effective at performing high-payoff behavior would, on average, be able to reproduce more frequently than the alternative, and would come to dominate. Each pair-wise match comprised repeated 10,000-round contests between a pair of strategies, which took turns in being the resident and invader. We recorded the average frequency of each strategy in the population

over the last 2,500 rounds of each simulation and gave each strategy a score that was the mean of these values over the simulations in which it participated.

43. The euro to USD exchange rate has changed since the time of the tournament, but the value given represents the amount around the time of the contest.

44. The full list of fields is anthropology, biology, computer science, engineering, environmental science, ethology, interdisciplinary centers, management, mathematics, philosophy, physics, primatology, psychology, sociology, and statistics.

45. As in Axelrod's tournament, the best-performing strategy had also come from Canada.

46. In the melee, *DISCOUNTMACHINE* won 35% of its contests, substantially more than the second-placed strategy *INTERGENERATION*, which won 24% of contests.

47. Indeed, many of the top strategies relied more heavily on recently acquired information.

48. Here, uniquely among the finalists, *DISCOUNTMACHINE* used a proxy of geometric discounting.

49. Hoppitt and Laland 2013.

50. Asch 1955, Latane 1981, Boyd and Richerson 1985, Henrich and Boyd 1998, Morgan et al. 2012, Morgan and Laland 2012.

51. Rogers 1988.

52. Rogers 1988, Feldman et al. 1996, Wakano et al. 2004.

53. Rendell et al. 2010.

54. Furthermore, the mean lifetime payoff in the population when all strategies competed together under the same conditions was lower than the levels achieved by lower-ranking strategies when playing alone.

55. Tilman 1982.

56. Kendal et al. 2009.

57. The average lifetime of agents in our tournament was 50 rounds. Life expectancy at birth for humans is currently around 67 years globally, which has increased from 26 years since the Bronze and Iron ages (*Encyclopaedia Britannica*, 1961). This affords a crude comparison between tournament rounds and human years.

58. Rendell, Boyd, et al. 2011; Laland and Rendell 2013.

59. In my judgment, the success of the tournament was in part attributable to the use of a multi-arm bandit. Multi-armed bandits have been widely deployed to study learning across a host of academic disciplines, including biology, economics, artificial intelligence, and computer science, because they mimic a common problem faced by individuals who must make decisions about how to allocate their time in order to maximize their payoffs (Schlag 1998, Koulouriotis and Xanthopoulos 2008, Gross et al. 2008, Bergemann and Valimaki 1996, Niño-Mora 2007, and Auer et al. 2002). They capture the essence of many difficult problems in the real world—for instance, where there are many possible actions, only a few of which yield a high payoff; where it is possible to learn asocially or through observation of others; where copying error occurs; and where the environment changes. Of

course, the simulation framework remains a simplification of the real world where, for instance, individuals may choose demonstrators with particular characteristics and where direct interactions between individuals operate (Apesteguia et al. 2007, Boyd and Richerson 1985, and Laland 2004). It remains to be established to what extent our results will hold if these are introduced in future tournaments, where the specific strategies that prospered in ours may not do so well. (At the time of writing, my collaborators and I are currently analyzing the findings of the second social learning strategies tournament, which has three extensions of the original framework, allowing for model-based biased, spatially variable environments, and cumulative cultural learning.) Nonetheless, our tournament offers greater realism than past analytical theory, which is why I place particular weight on its findings. The basic generality of the multi-armed bandit problem we posed lends confidence that the insights derived from the tournament will prove to be reliable and general.

60. In the tournament, individual strategies had no control over the rate of copy error, but it is easy to see how, had mutant strategies been allowed that could reduce copy error, these strategies would have thrived, for the same reason that copying thrives: each time copy error occurs, strategies receive a randomly chosen behavior (or in some of our simulations, no new behavior), while accurate copying taps into those high-return behaviors selected for performance by the demonstrators.

61. Lyons et al. 2011.

62. Van Bergen et al. 2004.

63. The most compelling example is found in the caching behavior of scrub jays (Clayton and Dickinson 1998, 2010; see also Suddendorf and Corballis 2007).

64. Fisher 1930.

65. Boyd and Richerson 1985, Cavalli-Sforza and Feldman 1981.

CHAPTER 4: A TALE OF TWO FISHES

1. Indeed, the whole notion of a "genetic program" has come into scientific disrepute. For developmental biologists, the metaphor of genotype as "program" or "blueprint" has proved inconsistent with the dynamic, reciprocal nature of development and inheritance emerging from their studies (see, e.g., Gottlieb 1992, Keller 2010, and Pigliucci and Müller 2010). Developing organisms both modify gene expression and modify developmental environments, generating feedback in the processes of ontogeny that invoke organism-environment relationships and nongenetic inheritance as causes of species-typical, invariant, phenotypes typically seen as "genetically determined" by evolutionary biologists (Oyama 1985, Gottlieb 1992, Oyama et al. 2001, Gilbert 2003, Jablonka and Lamb 2005, Keller 2010, Bateson and Gluckman 2011, and Uller 2012).

2. See Brown and Laland 2003, 2006; and Laland et al. 2011 for reviews of social learning in fishes.

3. See http://www.fishbase.org/home.htm.

4. Our studies of primates are observational studies conducted in zoos or in the wild, and we have never kept primates in our laboratory.

5. For my laboratory's fish-study derived insights into innovation, see Laland and Reader 1999a, 1999b. For our fish-based insights into social learning, see Day et al. 2001; Brown and Laland 2002; Reader et al. 2003; Brown et al. 2003; Kendal et al. 2004; Kendal, Coolen et al. 2005; Croft et al. 2005; Webster, Adams, and Laland 2008; Webster and Laland 2008, 2011, 2012, 2013; and Duffy et al. 2009. For our studies of diffusion of innovations in fishes, see Reader and Laland 2000, Swaney et al. 2001, Morell et al. 2008, Atton et al. 2012, and Webster et al. 2013. For our studies of traditionality in fishes, see Laland and Williams 1997, 1998; and Stanley et al. 2008.

6. The now prevalent approach of using a specific form of social learning in a specific animal as a model system for exploring some more general issues concerning social learning processes was pioneered by Jeff Galef, the Canadian psychologist whose rat work is described in chapter 2. Galef spent over 30 years investigating the transmission of dietary preferences among rats via cues on their breath, and the system proved highly productive, leading to countless valuable insights that were published in over a hundred papers.

7. Templeton and Giraldeau 1996.

8. *Gasterosteus aculeatus.*

9. *Pungitius pungitius.*

10. In our more recent public-information use experiments, conducted by Mike Webster, the demonstrators and food patches were placed in entirely separate tanks, adjacent to the test tank, to rule out any possibility of residual olfactory cues interfering with the findings. This refinement in the apparatus did not change the experimental findings—ninespines but not threespines exhibited public-information use.

11. Milinski et al. 1990, Frommen et al. 2007, Atton et al. 2012, Webster et al. 2013.

12. Boyd and Richerson 1985.

13. The number of dorsal spines actually varies considerably in *Pungitius pungitius*, ranging from 7–12, but with most individuals having 9 or 10. The same species is also known as the tenspined stickleback in some locations.

14. Hoogland et al. 1957.

15. A theoretical framework for studying such phenomena originated in economics. It considers a scenario where individuals copy others' choices without regard to their personal information, thereby leading to informational cascades where everyone ends up doing the same thing irrespective of whether it is the best thing to do (Bikhchandani et al. 1992, 1998; Giraldeau et al. 2002).

16. Giraldeau et al. 2002.

17. Coolen et al. 2005.

18. The argument being made here is that the experimental subjects swam toward the former location of the larger shoal as a means to find food, even though the demonstrator fish were no longer present, and when present had not been feeding.

This is thought to be plausible, because fish have been shown to learn to associate aggregations of fish with opportunities to feed (Brown and Laland 2003).

19. Van Bergen et al. 2004.

20. We doubt this is simply due to forgetting, as sticklebacks have been shown to retain patch preferences for periods longer than a week (Milinski 1994).

21. Schlag 1998.

22. Kendal et al. 2009.

23. Pike et al. 2010.

24. Schlag 1998.

25. Schlag (1998) termed his rule "proportional observation," and his analysis demonstrated that this rule is an optimal social learning strategy, which will take populations to the fitness maximizing behavior.

26. The "ratcheting up" that occurs through the use of this hill-climbing algorithm in sticklebacks differs from human cumulative culture in that it is *bounded*. The algorithm potentially allows animals to approach the optimal behavior in an environment, but at that juncture cumulative knowledge gain would stop. This stands in contrast to human cumulative culture, in which new behavior or products opens up further possibilities for innovation in an unbounded manner.

27. Webster and Laland 2011.

28. Krause and Ruxton 2002.

29. Wingfield et al. 2001, Kambo and Galea 2006.

30. Kavaliers et al. 2001.

31. The absence of sex differences in nonreproductive fish, and the intermediate level of copying exhibited by them, confirms that it is reproductive state, not sex, that favors divergent sex-specific adaptive social learning strategies (Webster and Laland 2011).

32. *Culaea inconstans.*

33. *Apeltes quadracus* and *Spinachia spinachia.*

34. See, for instance, Emery and Clayton 2004.

35. Laland 2004.

36. The primary source of this idea is Boyd and Richerson's 1985 book *Culture and the Evolutionary Process*. This discussed "transmission biases" that affect the likelihood of exploiting socially available knowledge.

37. See, for instance, Boyd and Richerson 1985, Giraldeau and Caraco 2000, and Schlag 1998.

38. Coolen et al. 2003.

39. Van Bergen et al. 2004.

40. Pike and Laland 2010.

41. Hoppitt and Laland 2013.

42. Grüter et al. 2010, Grüter and Ratnieks 2011.

43. Webster and Laland 2008.

44. Mason 1988.

45. Horner et al. 2010.

46. See Hoppitt and Laland 2013 for a summary of these data.

47. Morgan et al. 2012.

48. Wood et al. 2012 reached comparable conclusions in a study with children.

49. For instance, there are no data consistent with sticklebacks pursuing a random copying strategy, nor is there evidence found for *copy if dissatisfied* (Pike et al. 2010).

50. Laland 2004, Rendell, Fogarty, Hoppitt, et al. 2011; Hoppitt and Laland 2013.

51. Boyd and Richerson 1985, Henrich and McElreath 2003, Hoppitt and Laland 2013.

52. Coolen et al. 2003, Van Bergen et al. 2004.

53. Coolen et al. 2005.

54. Ibid.

55. Pike and Laland 2010.

56. Henrich and Boyd 1998.

57. Laland 2004. There are good evolutionary reasons for anticipating hierarchical organization of behavior (Dawkins 1976), and hierarchical control has been reported for a great deal of human and animal behavior (Byrne et al. 1998).

58. Both of these views can be found in the literature (e.g., Boyd and Richerson 1985, Heyes 2012, Henrich 2015).

59. This stance is explicitly in accordance with behavioral ecologists' use of the *phenotypic gambit* (Grafen 1984). It does not matter whether animals adopt such strategies as a consequence of evolved psychological mechanisms, learning, culture, or some combination of processes. Strategies can still fruitfully be studied as if the simplest genetic system controlled them (Laland 2004). This pragmatic stance has proven extremely productive in the field, but experimental investigations of the underlying learning mechanisms are also important to a complete understanding of the phenomenon.

60. Shettleworth 2001.

61. Mineka and Cook 1988.

62. Marler and Peters 1989.

63. Sealey 2010.

64. Thornton and McAuliffe 2006.

65. Rutz et al. 2010.

66. For instance, in canaries, aggression between males has been found to preclude social transmission (Cadieu et al. 2010).

CHAPTER 5: THE ROOTS OF CREATIVITY

1. Fisher and Hinde 1949, Hinde and Fisher 1951, Martinez del Rio 1993.

2. Hinde and Fisher 1972.

3. Sherry and Galef Jr. 1984, 1990; Kothbauer-Hellman 1990.

4. Lefebvre 1995.

5. Lefebvre 1995.

6. Byrne 2003, Reader and Laland 2003a, Lefebvre et al. 2004, Casanova et al. 2008.

7. Russon 2003.

8. Young 1987.

9. Eaton 1976.

10. J. M. Brown 1985.

11. Goodall 1986.

12. Hosey et al. 1997.

13. Schönholzer 1958.

14. Morand-Ferron et al. 2004. These authors report that there are now several species of birds that have picked up the dunking habit.

15. Reader and Laland 2003a.

16. Kummer and Goodall 1985; Lefebvre et al. 1997; Reader and Laland 2001, 2002, 2003b; Biro et al. 2003.

17. Sol 2003.

18. Sol and Lefebvre 2000, Sol et al. 2002.

19. Greenberg and Mettke-Hofman 2001.

20. Laland and Reader 2009.

21. Boinski 1988.

22. Kummer and Goodall 1985.

23. Thorndike described this as follows: "The Law of Effect is that, other things being equal, the greater the satisfyingness of the state of affairs which accompanies or follows a given response to a certain situation, the more likely that response is to be made to that situation in the future" (1898, p. 103). In modern parlance, we would simply say that actions that are followed by a positive outcome are likely to be repeated, while those followed by a negative outcome will be eliminated.

24. This learning process is nowadays known as "operant" or "instrumental conditioning."

25. Morgan 1912.

26. Thorpe 1956, Cambefort 1981, Lefebvre et al. 2004.

27. Zentall and Galef Jr. 1988, Heyes and Galef Jr. 1996, Box and Gibson 1999, Galef Jr. and Giraldeau 2001, Shettleworth 2001, Fragaszy and Perry 2003.

28. Kummer and Goodall 1985.

29. Kummer & Goodall 1985, p. 213.

30. Hamadryas baboons (*Papio hamadryas*). See Kummer and Kurt 1965.

31. Laland and Reader 1999a, 1999b; Laland 2004; Sol, Lefebvre, et al. 2005; Gajdon et al. 2006.

32. Kummer and Goodall 1985, p. 213. This article is also reproduced in Reader and Laland 2003a, where the quote is found on page 234.

33. Reader and Laland 2003a.

34. Reader and Laland 2003b.

35. See Reader and Laland 2003b for discussion.

36. Nihei 1995.

37. See http://www.snopes.com/photos/animals/carwash.asp.

38. Reader and Laland 2003b.

39. This stance contrasts with some definitions of human innovation, which refer to acquisition of a novel act by any route as innovation, and the initial inception as "invention" (Rogers 1995).

40. Reader and Laland 2003b.

41. Simon and I chose to adopt a broad definition of animal innovation deliberately (Reader and Laland 2003b). Our definition, for instance, made no distinction between totally novel behavior and modifications of existing behavior, as had been the practice among researchers studying birdsong learning (Slater and Lachlan 2003). Some researchers argued that we should reserve the term "innovation" for qualitatively new or cognitively demanding tasks or processes. However, we felt that it would have been a mistake to insist that an innovator must express a previously unobserved motor pattern, demonstrate an unusual cognitive ability, or devise a sophisticated product since this might undermine the collection of data. Moreover, it is open to question whether such criteria could be applied objectively. Virtually all animal innovators have likely used motor actions already in their repertoire (Hinde and Fisher 1951), while subjective judgments of cognitive ability or complexity are infamously vulnerable to anthropocentric prejudice. As I described in chapter 2, when Imo the Japanese macaque first invented sweet potato washing, researchers were greatly impressed, and she was heralded as a "genius" and "gifted." However, subsequent study established that food washing is a common feature in the behavior of several macaque species. Armed with this background knowledge, Imo's invention still qualifies as an innovation by my definition, but would have been mischaracterized had the definition stipulated that some level of complex cognition must be involved. The key characteristic of animal innovation is the introduction of a novel behavior pattern into a population's repertoire. See Ramsey et al. (2007) for an alternative view.

42. Née Rachel Day.

43. Bateson and Martin 2013.

44. Day 2003, Kendal et al. 2005.

45. This finding is replicated in other studies (Reader and Laland 2001, 2009; Laland and Reader 2010).

46. A recent example is provided by the diffusion of foraging information in sticklebacks (Atton et al. 2012).

47. Laland and Janik 2006, Laland and Galef 2009.

48. Day 2003, Kendal et al. 2009.

49. A recent study reveals strikingly similar findings in humans, with innovation increasing with age and experience, and imitation showing a corresponding decrease (Carr et al. 2015).

50. Studies of other primates have found that adults acquire information more efficiently and can recognize and classify objects more quickly than nonadults (Menzel and Menzel 1979, and Kendal et al. 2015).

51. *Leontopithecus* (lion tamarins), *Saguinus* (tamarins), *Callithrix* (marmosets). Day et al. 2003.

52. Gibson 1986.
53. King 1986.
54. Rowe 1996.
55. Ibid.
56. Dunbar 1995.
57. Gibson 1986, King 1986.
58. *Sturnus vulgaris* (starlings).
59. Boogert et al. 2008.
60. If social learning underlies the diffusion of innovations, we would expect individuals acquiring the behavior later in the diffusion to exhibit shorter learning times, given that they will have more demonstrators than the individuals that acquire the behavior early. This proved to be the case. Contact latency and solving duration were negatively correlated, consistent with social learning underlying the spread of solving.
61. Aplin et al. 2012, Aplin et al. 2014.
62. Network based diffusion analysis was initially developed by Franz and Nunn 2009, but was extended by Hoppitt et al. 2010a, 2010b; Hoppitt and Laland 2011; and Nightingale et al. 2015. Statistical packages allowing implementation of the methods are available from my website http://lalandlab.st-andrews.ac.uk/freeware .html.
63. Webster et al. 2013, Allen et al. 2013, Claidiere et al. 2013, Aplin et al. 2014.
64. *Poecilia reticulate.*
65. The mazes were designed so that at each partition, fish were initially required to swim away from the smell of the food in order to reach it. This was important, since it meant that the task could not be solved merely by swimming up an odor gradient.
66. Reader and Laland (2000) found a significant positive correlation between the time taken for guppies to complete a maze for the first time and the number of trials that it took to learn the maze according to a trials-to-criterion measure (e.g., completing the maze within a set time window, with no wrong turns, on three consecutive trials), which legitimized our use of the first solver as a proxy measure of innovation.
67. Laland and Reader 1999a.
68. In addition, in order to be confident that it was the quest to find food that generates these patterns in who swims the maze first, we replicated the experiments with no food in the tanks. When we did so, all of the reported differences disappeared, confirming that the original patterns were specifically related to the foraging task.
69. Oikawa and Itazawa 1992, Pedersen 1997.
70. Reader and Laland 2000.
71. Laland and Reader 1999b.
72. All fish received some food, no health problems to any fish arose from this feeding regime, and the entire procedure was overseen by the university vet to ensure that no animal welfare concerns arose.

73. Bateman 1948, Trivers 1972.
74. Kokko and Monaghan 2001, Kokko and Jennions 2008.
75. Brown et al. 2009.
76. Bateman 1948, Trivers 1972, Davies 1991.
77. Reznick and Yang 1993, Sargent and Gross 1993.
78. Ibid.
79. Farr and Herrnkind 1974.
80. Magurran and Seghers 1994.
81. Lefebvre et al. 1997.
82. Wilson 1985. Wilson also called this hypothesis "behavioral drive," but I prefer the term cultural drive, since it is more descriptive of Wilson's proposed mechanism.
83. See also Wyles et al. 1983.
84. Measured per avian parvorder (a taxon above superfamily and below infraorder).
85. Various factors can bias the number of reports of innovation associated with a species. For instance, in theory, researchers might study some species more than others, some birds might live in locations where their behavior is more likely to be observed, or journals might be more willing to publish articles on some species than others. It is important that such factors are controlled for in comparative analyses, so they do not distort the conclusions.
86. Lefebvre et al. 2004.
87. Sol and Lefebvre 2000; Sol et al. 2002; Sol, Duncan, et al. 2005; Sol 2003.
88. Ibid.
89. Sol, Lefebvre, et al. 2005.
90. Again, consistent with Wilson's hypothesis, both innovation rate and brain size have recently been shown to correlate with avian species (and subspecies) richness, suggesting that evolutionary rates are accelerated in large-brained, innovative taxa, as the behavioral drive hypothesis predicts (Nicolakakis et al. 2003; Sol 2003; and Sol, Stirling, et al. 2005); Lefebvre et al. (2016) reached a similar conclusion in Darwin's finches, a different group of birds.
91. Reader and Laland 2001, 2002. Nearly half of the examples were innovations in foraging, but innovations also appeared in aggressive, communicative, display, grooming, sexual, and play behaviors, and a number of other contexts.
92. Reader and Laland 2001.
93. Jolly 1966, Humphrey 1976, Byrne and Whiten 1988.
94. Reader and Laland 2001, Van Bergen 2004.
95. *Daubentonia madagascariensis* (aye-aye), *Galagos* (busy babies).
96. Reader and Laland 2002.
97. These relationships between rates of innovation or social learning and brain size have now been replicated several times, using a variety of different methods and brain measures, and have proven robust. See Reader et al. 2011, and Navarette et al. 2016 for details.
98. DeVoogd et al. 1993.
99. Krebs et al. 1989, Hampton et al. 1995, Jacobs and Spencer 1994.

100. Barton 1999, Joffe and Dunbar 1997, Harvey and Krebs 1990, De Winter and Oxnard 2001.
101. The same study (Reader and Laland 2002) found that tool use also fit this pattern.
102. Lefebvre et al. 1997, Reader and Laland 2002.
103. We found that the frequency of tool use also covaried with brain size, innovation, and social learning, as might be expected if innovative primates used tools to solve many of their problems (Reader and Laland 2002). A couple of years later Cambridge PhD student Yfke van Bergen found that rates of extractive foraging—that is, removing concealed or embedded food from a substrate, such as eating nuts or honey—also covaried with brain size and innovation in primates (Van Bergen 2004).
104. Leadbeater and Chittka 2007b.
105. Wisenden et al. 1997.
106. Dunbar and Shultz 2007a, 2007b.

CHAPTER 6: THE EVOLUTION OF INTELLIGENCE

1. The cells of multicellular organisms contain mitochondria, the organelles that convert chemical energy from food into a form of energy that the cells can use. Mitochondria possess a small amount of DNA. Mitochondrial DNA is inherited maternally, with no recombination. Therefore, assuming it accumulates mutations at a roughly constant rate, the number of differences between individuals will covary with the time since they shared a common ancestor, and knowledge of the mutation rate allows such ancestors to be dated. See Cann et al. 1987; note mitochondrial Eve is not the common ancestor of all humans, but rather the ancestor of all human mtDNA. Cann et al. extracted mitochondrial DNA from 186 people around the world and used it to construct a genealogy. All of the oldest branches of the tree terminated in the mtDNA of modern Africans, and the African branches exhibited more variation. Since the mitochondria in any population are descended from a single woman in the past, this woman was given the name "Eve"—but she was obviously not the only woman living at the time, nor was she the first female human. The authors' finding suggests Eve was an African who lived about 200 kya (CI = 100–400; kya = 1,000 years ago). Vigilant et al. (1991), also from Wilson's lab, refined this estimate to 166–249 kya.
2. The Out of Africa model specifies that between 100 and 200 kya, anatomically modern *H. sapiens* arose in Africa and later spread around the world, replacing the other archaic populations with little or no gene flow.
3. This was a Human Frontier Science Programme postdoctoral fellowship.
4. Wilson's cultural drive hypothesis was first published in 1983 (Wyles et al. 1983), and later elaborated on in Wilson 1985. However, it was spelled out most clearly in Wilson 1991.
5. Wyles et al. 1983; Wilson 1985, 1991.
6. Wilson 1991, p. 335.

7. Large animals typically have large organs, and small animals, small organs. For instance, being much larger than humans, whales possess larger brains than we do, while cows have bigger brains than most monkeys. When body size and brain size are plotted against each other, a positive correlation is found. Many researchers studying brain evolution maintain that the actual size of an animal's brain relative to the brain size one would expect for an animal of that size is a better indicator of the animal's intelligence than the absolute size of the animal's brain. However, this remains a contentious issue. One complication is that animals' brains vary in their neuronal density, with primates possessing both large relative brain sizes and unusually high neuronal densities.

8. Wyles et al. 1983.

9. Wilson 1985, p. 157.

10. Odling-Smee et al. 2003.

11. Lynch 1990 examined the rates of change of cranial morphology in mammals, and concluded that the rate of morphological divergence of almost all lineages, including the great apes, is consistent with the expectations of a neutral evolution model (i.e. the analysis provides little evidence that selection affects evolutionary rates). The only data point for which an elevated rate of long-term evolution was observed is for humans. Lynch suggested that this could be an artifact of temporal scaling, but in recent years this conclusion has been refuted by extensive genetic data showing a massive increase in the rate of molecular evolution in humans (e.g., Hawks et al. 2007, Cochran and Harpending 2009). At the time of writing, the idea that "behavioral drive" or "cultural drive" has accelerated evolutionary rates in mammals (hominins aside) and birds is supported by little concrete evidence, and it may be that the mechanism Wilson envisaged has solely affected rates of evolution in hominins, or even just humans. In contrast to the impact on evolution rates, the hypothesis that behavioral/cultural drive is causally responsible for the evolution of large brains and complex cognition is now supported by extensive data (Reader and Laland 2002, Reader et al. 2011, and Navarette et al. 2016). Throughout this book, my references to cultural drive refer to this latter effect.

12. Deary 2001.

13. MacPhail 1982.

14. Lefebvre et al. 1997.

15. Reader and Laland 2002.

16. See Kaufman and Kaufman 2015 for a recent overview of this literature. See also Reader et al. 2016 and articles therein.

17. Chittka and Niven 2009.

18. Rendell et al. 2010.

19. Brass and Heyes 2005, Heyes 2009.

20. Kohler 1925, Tomasello 1999.

21. The claim that animals cannot follow the majority behavior if they cannot compute that behavior is not true in all circumstances. For instance, following simple rules, groups of animals can also accurately and quickly come to a consensus and

NOTES

travel in the direction preferred by the majority, despite the fact that individuals are likely to have no explicit knowledge of whether they are in the majority or minority, or even if there are any other informed individuals at all (Couzin et al. 2005, Couzin 2009).

22. Evolutionary models have shown that it is almost always adaptive to do what the majority are doing, because this "conformist transmission" is a very effective means for individuals to acquire the locally optimal behavior (Boyd and Richerson 1985, and Henrich and Boyd 1998).

23. A caveat here is that if the frequency of performance of a behavior correlates positively with the frequency of individuals that exhibit that behavior, then the former may prove to be a useful proxy for the latter, particularly because the former may well be easier for animals to compute. Indeed, a recent study found that in great tits, these two variables were correlated (Aplin et al. 2015).

24. Whiten and Van Schaik 2007, Van Schaik and Burkart 2011.

25. Van Schaik et al. 1999.

26. This advantage arises through the animals increasing the likelihood of their relatives reproducing (i.e., increasing their inclusive fitness); see Hamilton 1964.

27. Fragaszy and Perry 2003, Lonsdorf 2006, Biro et al. 2006, Fragaszy 2012, Schuppli et al. 2016.

28. Goodall 1986, Lonsdorf 2006, Biro et al. 2006.

29. Reader 2000.

30. See previous chapter.

31. Goodall 1986, Lonsdorf 2006, Biro et al. 2006.

32. Boyd and Richerson 1985.

33. Kaplan et al. 2000, Kaplan and Robson 2002.

34. As discussed later, there are grounds for anticipating that cultural drive might also have operated in other animal taxa—for instance, in whales, parrots, and corvids.

35. Cheney and Seyfarth 1988, Reader and Lefebvre 2001.

36. Cosmides and Tooby 1987, Carruthers 2006, Roberts 2007.

37. MacPhail and Bolhuis 2001, Byrne 1995.

38. Dunbar and Shultz 2007a, 2007b.

39. Tomasello and Call 1997, Herrmann et al. 2007.

40. For instance, primate genera differed in their performance in laboratory tests of cognition, with great apes typically outperforming other primates (Tomasello and Call 1997, Deaner et al. 2006). Moreover, in two primate species, cotton-top tamarins and common chimpanzees, individual performance had been found to covary across multiple laboratory tasks (Banerjee et al. 2009, Herrmann et al. 2010). These and related data from other mammals and birds (Emery and Clayton 2004, Matzel et al. 2003, Kolata et al. 2008) can be viewed as consistent with the hypothesis that a single general cognitive factor underlies laboratory performance.

41. Reader and Laland 2002, Deaner et al. 2000, Byrne 1992.

42. Lefebvre et al. 1997.

43. Lefebvre et al. 1997, 2004; Lefebvre 2010.

44. Byrne and Whiten 1988, Whiten and Byrne 1997.
45. Byrne and Whiten 1990.
46. Byrne and Corp 2004.
47. Byrne and Whiten 1988, Whiten and Byrne 1997.
48. Kudo and Dunbar 2001, Nunn and Van Schaik 2002, Kappeler and Heyman 1996.
49. Dunbar 1992.
50. Ibid.
51. Diet breadth data were compiled by allocating prey to 13 categories. These were (1) invertebrate prey; (2) vertebrate prey; (3) fruit, fungus, and honey; (4) seeds and nuts; (5) exudates; (6) flowers; (7) nectar / pollen; (8) roots, tubers, bulbs, and truffles; (9) leaves, shoots, stems, herbs, and buds; (10) wood and bamboo; (11) bark; (12) pith; and (13) lichen, with each species given a score between 1 and 13 to specify diet breadth.
52. Dunbar 1995; Dunbar and Shultz 2007a, 2007b.
53. We also carried out factor analyses, which gave equivalent results to the PCAs (see electronic supplementary material of Reader et al. 2011, table S1).
54. All corrected for research effort (the fact that some species are studied more than others, and hence will have higher reports).
55. Reader et al. 2011 called this single dominate component *primate* g_{s1}. The "*s*" subscript denotes that *primate g* is an across-species construct, while the "*1*" in the subscript specifies that it is the first of multiple measures of *primate g*, computed in different ways.
56. See table 1 in Reader et al. 2011.
57. All 10 pairwise correlations between the five cognitive measures were strongly significant ($p < 0.001$). See table 3 in Reader et al. 2011.
58. We called this principal component *primate* g_{s2}. The dominant eigenvalue = 3.77, and was significantly higher than all other components. The eight-variable PCA also extracted a second component, on which diet breadth, percentage of fruit in diet, and group size loaded. See table 2 in Reader et al. 2011. We found that *primate* g_{s2} covaried strongly with *primate* g_{s1}.
59. If the relationship between different measures of primate cognitive ability reflects an evolutionary history of coevolution, then the *primate* g_s component should remain when phylogeny is taken into account using independent contrasts for each measure. Repeated using *CAIC* (Purvis and Rambaut 1995), the five-variable PCA revealed a single component on which all cognitive measures loaded (n = 57, d.f. = 14, χ^2 = 70.20, p < 0.0001, variance contribution = 0.53, extractive foraging, social learning, and tool-use loadings = 0.76–0.84, innovation and tactical deception loadings = 0.66 and 0.55, respectively). Further analyses can be found in Reader et al. 2011.
60. The finding is no artifact. We took several approaches to address potential confounding variables. The observed associations were not caused by data points that qualified simultaneously for more than one measure, since these were removed. Nor are the associations an artifact of the covariance of each individual measure

with brain volume, body mass, or correlated error variance in research effort; analyses using residuals of each cognitive measure from multiple regressions that included (i) relative brain volume, (ii) body mass, (iii) that did not correct for research effort, or (iv) that deployed five independent measures of research effort gave equivalent results (See Reader et al. 2011, electronic supplementary material). To account for the possibility that observers may be more willing to ascribe behavioral flexibility to the great apes than to other primates, we repeated the analysis with great apes removed, finding equivalent results (Reader et al. 2011, electronic supplementary material). To address the concern that there may be error in individual species data, particularly in the case of species that have not been well studied, we repeated the PCA using the same procedures at the genera level. Once again, we found a single dominant component ($\chi^2 = 116.28$, $p < 0.0001$, variance contribution = 75%). The same pattern was observed when the analysis also incorporated Deaner et al.'s (2006) genus-level composite index of performance in laboratory tests of cognition ($\chi^2 = 131.12$, $p < 0.0001$, variance contribution = 73%). Deaner et al.'s reduced model measure (inversed so that high scores represent high performance) loads heavily on g (loading = 0.73). This supports the argument that laboratory performance is reliant on general intelligence and demonstrates the covariation of six cognitive measures across extant primate genera.

61. Deary 2001, p. 17.

62. The strong correlation between distinct measures of primate cognitive performance that Reader et al. 2011 found, was, of course, strikingly evocative of the correlations in performance on different IQ tests observed in humans. The most likely explanation for this correspondence is that the g factor reported in humans reflects underlying general processes that evolved in our common ancestors and are thus shared by our primate relatives. Naturally, care is warranted in interpreting this finding, since here the observed associations occur across, rather than within, species.

63. This corresponds to each species' loadings on the principal component.

64. *Cacajao calvus.*

65. Principal Component Analysis (PCA) can be used to calculate factor scores for each component it extracts, providing a composite score for the variables loading on a component. Reader et al. 2011 calculated g_{sl} factor scores to provide a composite g_s measure for each species, which can be interpreted as a measure of comparative general intelligence. On average, the Hominoidea (excluding humans) outscored other taxa, but no significant mean differences in g_{sl} scores were found between Cercopithecoidea, Ceboidea, and Prosimii.

66. For instance, neocortex ratio (neocortex size relative to rest of brain size) or executive brain ratio (neocortex and striatum relative to brainstem) give similar results.

67. The associations between g_{sl} and various measures of brain size are given in table 4 of Reader et al. 2011. Subsequent to conducting these analyses we have collected

further brain data, which we used when conducting the analyses again, but with more powerful statistical methods. The findings have proven very robust (Street 2014, Navarette et al. 2016).

68. Gray and Thompson 2004, Van der Maas et al. 2006, Deary 2000, Schoenemann 2006.

69. Deaner et al. 2006.

70. Animals are classified hierarchically, with closely related species in genera, closely related genera in families, closely related families in orders, and so forth. Primates occupy the taxonomic rank of an order, and comprise multiple families (including the great apes, the Old World monkeys, the capuchin and squirrel monkeys, etc.). Each primate genera can be regarded as the major subdivision of one of these primate families, typically consisting of more than one species.

71. Deaner et al. 2006.

72. Riddell and Corl 1977.

73. Lefebvre 2010, Healy and Rowe 2007, Dechmann and Safi 2009.

74. Reader and Laland 2002; Van Schaik and Burkart 2011; Dunbar 1995; Byrne and Whiten 1988; Whiten and Byrne 1997; Harvey and Krebs 1990; Barton 2006; Deaner et al. 2000; Dunbar and Shultz 2007a, 2007b; Clutton-Brock and Harvey 1980.

75. Reader et al. 2011.

76. This was done with the use of MACCLADE v. 4.08.

77. *Cebus*.

78. *Papio*.

79. *Macaca*.

80. Hominoidea. It is important to note that while humans are located within the superfamily with the highest g_s scores, the analysis conflicts with a *Scala naturae* conception, suggesting instead that convergent selection may have repeatedly favored intelligence in distant primate lineages. This interpretation involves fewer evolutionary events, and is thus more parsimonious than all alternative scenarios, such as that high general intelligence evolved once in the common ancestors of apes and Old World monkeys and was then repeatedly lost. Some variance among genera is probably attributable to measurement error, and error variance is expected to be highest in the least-studied taxa.

81. To address the concern that the data for little-studied species may be unreliable, Reader et al. 2011 repeated the phylogenetic reconstruction with the less well-studied species removed, finding similar results. Reader et al.'s 2011 results are consistent with findings that great apes' social learning abilities are superior to other primates (Van Schaik and Burkart 2011, Whiten 2011) and are broadly consistent with Van Schaik and Burkart's 2011 meta-analysis of taxonomic differences in primate cognition; however, the Reader et al. analysis provides no evidence for a difference between prosimians and monkeys.

82. Cosmides and Tooby 1987, Carruthers 2006.

83. Fernandes et al. 2014.

84. *Cheirogaleus medius.*
85. Striedter 2005, Montgomery et al. 2010.
86. Finlay and Darlington 1995, Rilling and Insel 1999, Striedter 2005.
87. Barton and Venditti 2013.
88. Striedter 2005.
89. Ibid.
90. Finlay and Darlington 1995, Shultz and Dunbar 2006.
91. This statement also holds irrespective of whether brain evolution is better characterized by variation in a single dimension of size or by the independent evolution of component parts (Finlay and Darlington 1995, Barton and Harvey 2000, Deaner et al. 2000, and Dunbar and Shultz 2007b).
92. Stephan et al. 1981.
93. Healy and Rowe 2007.
94. Sally Street was cosupervised by Gillian Brown.
95. This work was funded by grants from the John Templeton Foundation and an ERC Advanced Grant.
96. These specimens derived from recently deceased animals in zoos, which had died for other reasons. No animals were sacrificed for the study.
97. The analyses entailed a combination of phylogenetically controlled multiple regression models, multivariate Bayesian phylogenetic Poisson models, and phylogenetically controlled causal graph analysis.
98. Strictly speaking, what these analyses can achieve is to identify variables that are unlikely to be causally involved, and thereby allow us to reject alternative hypotheses. For instance, the analyses do not prove that selection for longer, slower lifespans was a primary cause of the evolution of large primate brains and high intelligence. However, a cluster of hypotheses in which longevity is causally involved in the evolution of intelligence remain credible, while alternative hypotheses stressing other factors as the primary agents of selection can be dismissed as inconsistent with the data.
99. Street 2014, Navarette et al. 2016. At the time of writing, this work is ongoing.
100. Kaplan et al. 2000.
101. Whiten et al. 1999, Van Schaik et al. 2003, Perry et al. 2003, Hoppitt and Laland 2013.
102. Kaplan et al. 2000.
103. Causal-graph analyses support these conclusions, with the best-supported graphs in phylogenetic exploratory path analyses linking technical innovation directly to brain size and social learning, and nontechnical innovation to brain size via diet and life-history measures (Navarette et al. 2016).
104. Kaplan et al. 2000. Mathematical theory suggests that a longer juvenile period can also be favored if there are important gains in productive ability with body size—for instance, through juvenile social learning from adults (Kaplan and Robson 2002).
105. Dunbar 1995; Dunbar and Schultz 2007a, 2007b.
106. Reader et al. 2011.

107. Group size can be viewed as a marker or indicator variable for this social intelligence, with large brains favored among primates because of the computational demands of living in large, complex societies. Recently Dunbar and Shultz refined their argument slightly to suggest that what favors large brains is selection for the cognitive skills needed to manage intense forms of pair bonding, with primate societies based on bonded relationships that are only found in pair bonds in other taxa (Dunbar and Schultz 2007b).

108. Here various measures of slow life history (particularly maximum longevity and duration of the period of juvenile dependency) can be viewed as a marker or indicator variable for this cultural intelligence, with large brains and longer lifespans favored in specific primate lineages because of the computational demands of acquiring complex skills (e.g., tool use) through social learning.

109. Henrich 2004a; Powell et al. 2009.

110. Striedter 2005.

111. Ibid.

112. I say dubious because there are multiple other differences among primates, whales, and elephants, both with respect to their anatomy and the environments that they encounter (e.g., terrestrial vs. aquatic), that make such comparisons difficult to interpret. For this reason, most comparative analyses of brain evolution focus on smaller taxonomic groups, such as the mammalian orders.

113. Several researchers have emphasized the importance of the size of the prefrontal cortex or frontal lobes (Deaner et al. 2006, Dunbar 2011).

114. Byrne 1995, Dunbar 2011, Striedter 2005, MacLean et al. 2014.

115. Striedter 2005.

116. Ibid.

117. Deacon 1990.

118. Rakic 1986, Purves 1988.

119. By the same reasoning, disproportionately large brain regions should attract some new inputs, and disproportionately small regions should lose some old inputs.

120. Heffner and Masterton 1975, 1983.

121. Deacon 1990.

122. Finlay and Darlington 1995, Barton and Venditti 2013.

123. Fernandes et al. 2014.

124. Barton 2012.

125. Navarette et al. 2016.

126. Barton and Capellini 2011. The relationship between brain size and longevity in mammals is well established, but has generally been interpreted as indicative of a constraint on, rather than a cause of, brain evolution, including in primates (Dunbar and Shultz 2007a). Our analyses of this issue are ongoing and are not entirely unequivocal. Hence the hypothesis that the observed relationship between brain size and longevity in primates signifies developmental constraint rather than cultural drive remains a possibility, even though it is currently not the best-supported explanation.

127. Glickstein and Doron 2008, Wolpert et al. 1998.
128. On nut cracking in chimpanzees, see Marshall-Pescini and Whiten 2008. On nettle processing in gorillas, see Byrne et al. 2011.
129. Byrne 1997.
130. Hunt and Gray 2003, Rendell and Whitehead 2001, Emery and Clayton 2004, Bugnyar and Kortschal 2002.
131. Interestingly, episodic-like memory has recently been demonstrated in rats and chimpanzees (Corballis 2013, and Clayton and Dickinson 2010), species not renowned for their food caching, but famed for their social learning. This fits with Allan Wilson's argument.
132. Emery and Clayton 2004.
133. Ibid. Note, such comparisons between distantly related species are dubious (see n. 112, above).
134. Emery and Clayton 2004.
135. Heyes and Saggerson 2002, Pepperberg 1988, Moore 1996.
136. Rendell and Whitehead 2001, 2015; Krutzen et al. 2005; Allen et al. 2013.
137. Wilson 1985.
138. Moas: Dinornithiformes. Elephant birds: Aepyornithidae.
139. A fascinating case study here is the Hawaiian crow (*Corvus hawaiiensis*), which shows remarkable tool-using capabilities and yet lies on the verge of extinction (Rutz et al. 2016). This species would appear to be highly adept at using tools to probe for invertebrate prey, but seemingly fails to show the behavioral flexibility to adjust to changed environmental conditions.
140. Wilson 1985, Boyd and Richerson 1985, Pagel 2012.

CHAPTER 7: HIGH FIDELITY

1. Joseph Henrich has recently written a compelling book on this topic (*The Secret of Our Success*, 2015) that answers this question in a manner very consistent with the arguments presented here; however, his discussion focuses largely on humans.
2. Henrich 2004a, Powell et al. 2009.
3. Enquist et al. 2010.
4. Cultural parents, as the term implies, are the cultural equivalents of biological parents. They are transmitters of information. For instance, if an individual acquires cultural knowledge from two biological parents and one schoolteacher, then they have three cultural parents.
5. For instance, the Enquist et al. 2010 analysis seemed to imply that one-to-one transmission was infeasible, though the social learning literature was replete with claims to the contrary, including daughters learning from their mothers, or sons from their fathers.
6. Tomasello 1994.
7. Laland and Hoppitt 2003, Laland et al. 2009.
8. Hoppitt and Laland 2008, 2013.

9. Laland et al. 1993.
10. Whiten and Custance 1996, Whiten 1998.
11. Whiten et al. 1999.
12. Darwin 1871.
13. Whiten et al. 1999.
14. This too would serve to promote more culture in humans rather than other animal groups, since the former are typically more populous. The findings reinforce earlier theoretical studies that imply demographic factors, such as how large and well connected a population is, could make a big difference to how much culture it would exhibit (Henrich 2004a, Powell et al. 2009).
15. Boesch 2003 (chimpanzees), Hunt and Gray 2003 (New Caledonian crows).
16. Tomasello 1994, Tennie et al. 2009, Galef Jr. 1992, Heyes 1993.
17. Galef Jr. 1992, Galef 2009, Heyes 1993, Whiten and Erdal 2012, Dawkins 1976. Dawkins emphasizes the importance of fidelity to effective replicators.
18. Lewis and Laland 2012.
19. Such analyses, which are a commonly used tool within behavioral and evolutionary biology, iterate processes or recursive equations through the use of a computer program to investigate how systems are expected to evolve under diverse conditions.
20. Lewis and Laland 2012.
21. An increase in novel invention has been shown to increase the number of independent cultural traits held by an individual and a population (Lehmann et al. 2011).
22. Enquist et al. 2011; Boyd and Richerson 1985; Reader and Laland 2003a, 2003b.
23. Lehmann et al. 2011, Enquist et al. 2011, Eriksson et al. 2007, Strimling et al. 2009, Van der Post and Hogeweg 2009.
24. Enquist et al. 2011.
25. Lewis and Laland 2012.
26. Our analysis built on the modeling framework developed by Enquist et al. 2011, which focuses on the cultural traits present in a population, rather than considering the specific traits held by individuals, or individual-level processes. This allows us to look at the cultural development of a population based on general cultural rates, and explore the complex dependencies among cultural processes; we then can go beyond the simple accumulation of a single one-dimensional improvement or of multiple unrefined traits.
27. After Enquist et al. 2011. For simplicity, we assumed that there were up to 10 seed traits in a population. To begin, each cultural group was initialized with two cultural seed traits, selected at random.
28. We carried out two types of simulation. For the first, the four events were constrained such that $\rho_1 + \rho_2 + \rho_3 + \rho_4 = 1$. In this case, one of these four events must happen at the next time step, which corresponds to the assumption that the time between events is variable. This analysis allowed us to avoid any assumptions about underlying demographic rates (births, deaths, migrations) and to interpret

the results in several ways with respect to time. Additionally, this approach is computationally simple and therefore allowed exploration of many parameter sets. Loss rate ranged from 0.2 to 0.7 in increments of 0.1; the remaining parameters were varied in increments of 0.1, with no parameter allowed to be zero (because we were particularly interested in how the processes interact and not in the effects of major evolutionary changes that bring forth the processes). For each parameter set, 10 independent, replicate cultural groups were simulated. In the second set of simulations, rates varied independently, allowing additional analysis of parameter interactions, but the approach was more computationally expensive and so was limited to exploring more course parameter space. The results of the two approaches were in broad agreement.

29. We also assumed that each cultural seed had a "utility," which was a quantitative measure of its usefulness to the user, and that on average, less useful traits would be lost before more useful traits. Traits generated through modifications or combinations were allotted similar utilities as the originals, while the probability that a particular trait was lost was inversely proportional to the trait's utility. The latter assumption is consistent with the findings from several social learning experiments that have shown humans and other animals regularly copy the most successful individuals, or copy in proportion to the demonstrator's payoff (Morgan et al. 2012, Mesoudi 2008, Kendal et al. 2009, and Pike et al. 2010). We devised several measures of cumulative culture, including the "number of traits" in a population, "mean trait complexity" (number of elements comprising a trait), "number of lineages," "mean lineage complexity," and "maximum utility." Together, these complexity and utility measures enabled us to describe the cumulative qualities of the cultures of each simulated population. For further methodological details, see Lewis and Laland 2012.

30. In our analyses, typically loss rates <0.5 were required for cumulative culture. See figure 1, Lewis and Laland 2012.

31. See figure 2 in Lewis and Laland 2012.

32. We also found that loss rate has the largest coefficient of individual regressions relating the various cultural rates to the measure of cumulative culture (extracted by PCA). In the individual regression, loss rate explains 75% of the variance in this measure, compared with 30% explained by combination when considered independently, and with lower percentages for the other processes. These findings draw from a constrained analysis, where we considered the loss rate relative to the other, creative, processes. In this analysis, the three other (creative) processes occurred with probability $1 - \rho_4$, so it might be argued that the cumulative nature of the culture could equally be due to an increase in overall creativity. However, we also conducted simulations where the parameters were allowed to vary independently, which confirmed that loss rate is genuinely the most important of the rates rather than an artifact. The four strongest effects in a linear regression analysis of the rates against a measure of cumulative culture were all either interactions that included loss rate or the loss rate alone. Thus, the independent parameter

simulations strongly support our finding that loss rate is the most important factor in determining the cumulative nature of culture.

33. Tomasello 1994, 1999; Tennie et al. 2009.
34. Hoppitt et al. 2008.
35. Bickerton 2009.
36. Whiten 1998, Dorrance and Zentall 2002, Saggerson et al. 2005.
37. Herrmann et al. 2007, Whiten et al. 2009.
38. Henrich and Boyd 2002.
39. Simonton 1995.
40. Petroski 1992.
41. Basalla 1988.
42. This study also supports the argument (e.g., Henrich 2004a, Powell et al. 2009) that large population sizes facilitate cumulative knowledge gain (Lewis and Laland 2012).
43. Fogarty et al. 2011.
44. Csibra and Gergely 2006, Csibra 2007. It may be more accurate to write that human teaching *draws on* key psychological adaptations, rather than that it is one. While cases of animal teaching are probably legitimately regarded as adaptations, human teaching may be better regarded as a domain-general competence (Premack 2007), although one that is likely heavily reliant on evolved capabilities that are possibly adaptations. Such capabilities may include the motivation to teach and be taught, the facility to comprehend the knowledge state of the pupil, and the ability to produce and attend to infant-directed speech.
45. Boyd and Richerson 1985, Tomasello 1994, Fehr and Fischbacher 2003, Csibra and Gergely 2006, Csibra 2007.
46. Danchin et al. 2004.
47. Hoppitt and Laland (2013) discuss how the term "demonstrator" misleadingly implies an active role for the individual being copied; they note that other terms exist, such as "model," that are more neutral, and hence that might be preferred were this the only consideration. There are also problems with the term "observer." The authors note, however, that alternative terms bring their own potential for confusion, and resolve that on balance there are no clearly superior terms to "demonstrator" and "observer."
48. Pearson 1989.
49. Tim Caro is now at UC Davis, and Marc Hauser is at Risk-Eraser, LLC, West Falmouth, UK. Caro and Hauser 1992.
50. Caro and Hauser 1992, p. 153.
51. Caro 1980a, 1980b, 1995.
52. Cheetah (*Acinonyx jubatus*). Domestic cat (*Felis silvestris catus*).
53. Franks and Richardson 2006, Leadbeater et al. 2006, Hoppitt et al. 2008.
54. *Suricata suricatta.*
55. Thornton 2007.
56. Thornton and McAuliffe 2006.

57. Thornton and McAuliffe are now at the University of Exeter and Yale University, respectively.
58. In Thornton & McAuliffe's experiments, the scorpions were rendered stingless, so that inexperienced pups would not be hurt. The researchers went on to record instances of pups being "psuedo-stung" (stung with a stingless tail), and found that all pups trained solely on dead scorpions would have been killed by stings from their prey, whereas only one pup trained on live prey was pseudo-stung.
59. Clutton-Brock et al. 1999.
60. Moglich and Holldobler 1974, Moglich 1978, Pratt et al. 2005.
61. Here the study species is *Temnothorax albipennis*, but many groups of ants engage in tandem running, so teaching may be widespread in ants.
62. Such following occurs in other taxa too—for instance fishes. See Laland and Williams 1997.
63. Franks and Richardson 2006.
64. Leadbeater and Chittka 2007b.
65. Observations of other species of ants that are moving their nest site support this interpretation (Moglich 1978). Some species use both tandem running and carrying to transport fellow workers to new nest sites. During the first phase of the move, the number of tandem runners stays constant, but the number of carriers increases, indicating that the number of workers with knowledge of the route is increasing. However, once the population at the new nest site reaches a critical mass, tandem running ceases, and the remaining workers are carried. This suggests that the function of tandem running during nest emigration is to ensure that sufficient workers know the location of the new nest site to ensure that the move occurs efficiently.
66. Franks and Richardson 2006, Leadbeater et al. 2006, Thornton and McAuliffe 2006, Raihani and Ridley 2008, Rapaport and Brown 2008, Colombelli-Negrel et al. 2012; see Hoppitt et al. 2008, and Rapaport and Brown 2008 for reviews.
67. Csibra and Gergely 2006, Hoppitt et al. 2008.
68. Hoppitt et al. 2008.
69. Leadbeater et al. 2006, Csibra 2007, Premack 2007.
70. Cavalli-Sforza and Feldman 1981, Boyd and Richerson 1985, Feldman and Zhivotovsky 1992, Rendell et al. 2010.
71. Kirby et al. 2007, 2008; Boyd et al. 2010.
72. Boyd and Richerson 1985, Peck and Feldman 1986, Boyd et al. 2003, Fehr and Fischbacher 2003, Gintis 2003.
73. West et al. 2007.
74. Sachs et al. 2004, Lehmann and Keller 2006, West et al. 2007.
75. Fogarty et al. 2011.
76. This was formalized by William Hamilton (1964) as $c < br$, where c is the reproductive cost to the helper, b is the reproductive benefit to the recipient, and r is the degree of relatedness between them; relatedness is defined as the probability that they share an altruistic gene.

77. A real cumulative culture scenario would have multiple bouts of refinement, rather than a single one, but our model suffices to establish the nature of the effects that cumulative culture has.

78. The analysis also found that the fitness advantage of teaching over nonteaching increased with the fidelity of teaching, and did so more sharply in a cumulative compared with a noncumulative setting.

79. A second learning opportunity in the noncumulative culture setting (when $w1 \geq w1w2$) does not increase a teacher's fitness. Thus, the difference between the cumulative and noncumulative model is not explained by the fact that there are two learning opportunities in the cumulative setting.

80. This is shown in figure 2 of Fogarty et al. 2011.

81. Franks and Richardson 2006, Leadbeater et al. 2006, Thornton and McAuliffe 2006, Hoppitt et al. 2008.

82. Strong conclusions on the frequency of teaching are difficult in the absence of experimental confirmation of teaching in a number of cases (e.g., cats), where circumstantial evidence for teaching exists. Fogarty et al. (2011) present a formal analysis that establishes the precise conditions under which teaching will evolve in each genetic system. This suggests that there are likely to be circumstances in which teaching evolves more readily among haplodiploid workers than in diploids—other factors being equal—because of the possibility of higher levels of relatedness between workers in relatively monogamous haplodiploid colonies (Cornwallis et al. 2010).

83. Lukas and Clutton-Brock 2012.

84. Hrdy 1999.

85. Thornton and McAuliffe 2006.

86. Langen 2000.

87. Cooperative-breeding species also often exhibit high levels of relatedness (Cornwallis et al. 2010), which may be another factor that enhances the likelihood of teaching evolving among them.

88. Thornton and Raihani 2008.

89. The obvious counterexample are felids, for which teaching of young by mothers may perhaps be favored because hunting skills, or the opportunities to gain them, are difficult to acquire through asocial or inadvertent social learning (corresponding to low A and S, but high T in our model).

90. Hoppitt et al. 2008, Thornton and Raihani 2008, Laland and Hoppitt 2003.

91. In the analysis, the fidelity of human teaching corresponds to the magnitude of T. On the high fidelity of human teaching relative to other animals, see Tomasello 1994, Csibra and Gergely 2006, and Csibra 2007.

92. Tomasello and Call 1997, Premack 2007.

93. Draper 1976, Whiten et al. 2003, Lewis 2007, McDonald 2007.

94. Csibra and Gergely 2006.

95. Tehrani and Riede 2008, Hewlett et al. 2011, Garfield et al. 2016.

96. Dean was cosupervised by Rachel Kendal, a primatologist at Durham University, UK.
97. Dean et al. 2012.
98. Tomasello 1994, Galef Jr. 1992, Heyes 1993, Tennie et al. 2009.
99. Tennie et al. 2009, Tomasello 1999, Laland 2004, Marshall-Pescini and Whiten 2008.
100. Large social networks may enhance cultural diversity and promote cumulative culture (Henrich 2004a, Hill et al. 2011), but we did not consider this hypothesis because it presupposes the existence of the necessary cognitive capabilities.
101. Giraldeau and Lefebvre 1987.
102. Coussi-Korbel and Fragaszy 1995.
103. Reader and Laland 2001, Biro et al. 2003.
104. Hrubesch et al. 2008.
105. At least, no such study that was capable of evaluating all of the plausible hypotheses had simultaneously tested humans and other animals using the same apparatus.
106. Cumulative culture had been investigated through historical analysis (Basalla 1988) in psychology laboratories (Caldwell and Millen 2008), and through experimentation in chimpanzees (Marshall-Pescini and Whiten 2008).
107. A first experiment included two conditions. One was an "open" condition, where groups could gain access to all stages; and the other was a "scaffolded" condition, where guards prevented access to the manipulanda associated with higher stages until performance at the lower stage reached the criterion. In a second experiment, conducted only with chimpanzees, one female from each of four additional groups was isolated from her group and trained to use the puzzle box to stage 3. The use of trained females of differing status as demonstrators allowed us to investigate how social rank affected the spread of solutions.
108. The Michale E. Keeling Center for Comparative Medicine and Research, University of Texas.
109. The Centre de Primatologie, Strasbourg.
110. Chimpanzees and capuchins were selected because the evidence for cultural traditions is as strong in these species as in any nonhuman, which maximized the chances of observing cumulative cultural learning. Moreover, as our closest relative, chimpanzees provide an appropriate comparator to humans, with the performance of capuchins aiding interpretation of any chimpanzee-human differences. Children are widely used in comparative studies to help tease out the effects of culture, as adults have been greatly enculturated by society.
111. A further 4 chimpanzees reached stage 2; however, each group witnessed multiple solvers at stage 1.
112. Wood et al. 1976.
113. There was a positive, rather than the predicted negative, correlation between the amount of scrounging an individual falls victim to and performance in capuchins, chimpanzees, and children; and there was no sign that scrounging hindered performance. Dominant children and chimpanzees did not monopolize the puzzle

box, and although there was a positive correlation between rank and puzzle box use among capuchins in 2007, this was not repeated in 2008. When manipulating the box, low rankers did not receive less attention than high rankers, nor was there any evidence for satisficing or conservatism, with individuals continuing to manipulate the dials and buttons after they had found the solution to stage 1. In the open condition, where they received rewards at all stages, both chimpanzees and children manipulated the puzzle box slightly more, rather than less, than individuals in the scaffolded condition, despite the latter being unrewarded at the previous stage(s). Although we did not find a significant difference between the proportions of rewards that chimpanzees scrounged at each stage, they expressed clear and strong preferences for the three foods in pilot work, and olfactory holes in the doors allowed these to be detected in the apparatus prior to their extraction. Moreover, many of the chimpanzees performed failed attempts to access the foods by "termiting" (inserting stalks through the olfactory holes), and all 29 cases involved an attempt to reach the highest-stage food that was available. In the children and capuchins, more low-stage than high-stage rewards were scrounged, which reflects a greater motivation to retain high-grade rewards.

114. Whiten et al. 1999, Perry et al. 2003.
115. Whiten et al. 2007, Dindo et al. 2008.
116. For that view, see Tennie et al. 2009, and Tomasello 1999. All alternative causal hypotheses can be rejected as highly improbable. It is not clear why success in solving the task should cause children to imitate, be taught by, or receive rewards from others, nor how an unspecified third variable might account for our within-species data. For instance, although it is possible that the relationship between imitation and performance reflects the child's cognitive ability, this explanation cannot account for the relationships of both teaching and prosociality with performance, because in both cases the donor (of knowledge or reward) is a different individual from the learner.
117. Tomasello 1994, 1999; Tennie et al, 2009.

CHAPTER 8: WHY WE ALONE HAVE LANGUAGE

1. These well-known phrases appear in a speech in Shakespeare's *Hamlet* (Act II, Scene 2): "What a piece of work is a man! How noble in reason, how infinite in faculty! In form and moving how express and admirable! In action how like an Angel! In apprehension how like a god! The beauty of the world! The paragon of animals!"
2. Vygotsky (1934) 1986.
3. Chomsky 1968.
4. For excellent overviews, see Fitch 2010, and Hurford 2014.
5. Washburn and Lancaster 1968.
6. Miller 2001.
7. Dunbar 1998.

8. Deacon 2003a.
9. Falk 2004.
10. Power 1998.
11. Greenfield 1991.
12. Burling 1993.
13. Hauser et al. 2014.
14. Fitch 2010, Hurford 2014, Bolhuis et al. 2010.
15. Bickerton 2009, Hurford 2014.
16. Hauser 1996.
17. Wheeler and Fischer 2012, 2015.
18. But see Watson et al. 2014.
19. Pika et al. 2005.
20. Ibid.
21. Fitch 2010, Hurford 2014.
22. Hauser 1996, Bickerton 2009, Fitch 2010, Hauser et al. 2014, Hurford 2014.
23. Yang 2013, Truswell 2015.
24. Terrace 1979.
25. Bickerton 2009.
26. Szamado and Szathmary 2006.
27. Bickerton 2009.
28. Odling-Smee and Laland 2009.
29. Grafen 1990; but see Kotiaho 2001.
30. Searcy and Nowicki 2005.
31. Maynard-Smith 1991.
32. For instance, see Smith et al. 2003.
33. Szamado and Szathmary 2006.
34. By protolanguage, I mean strings of words with no structural relationships or syntax.
35. A symbol is an object or a concept that represents or suggests another idea, belief, action, or material entity.
36. Szamado and Szathmary 2006.
37. Hurford 1999.
38. Szamado and Szathmary 2006.
39. Gray and Atkinson 2003, Pagel et al. 2007, Smith and Kirby 2008, Kirby et al. 2008.
40. Janik and Slater 1997, Fitch 2010, Hurford 2014.
41. Bergman and Feldman 1995, Boyd and Richerson 1985, Feldman et al. 1996, Stephens 1991.
42. For instance, see Rendell et al. 2010.
43. Laland 2016.
44. This claim is supported by both ethnographic (e.g., Hewlett et al. 2011, Tehrani and Riede 2008) and experimental (e.g., Dean et al. 2012) data.
45. Stringer and Andrews 2005.
46. Hrdy 1999.

47. See also Isler and Van Schaik 2012.

48. Anton 2014.

49. Stringer and Andrews 2005, Anton 2014.

50. Fitch 2004 makes a compelling argument along these lines.

51. Trivers 1974.

52. Tomasello 1999, Gergely and Csibra 2005, Gergely et al. 2007, Csibra 2010.

53. Walden and Ogan 1988.

54. Tomasello 1999, Gergely and Csibra 2005, Gergely et al. 2007, Csibra 2010.

55. Tomasello 1999. Some of our own experimental data also demonstrates this. For instance, in one study, Dean et al. (2012) found that all 23 observed instances of teaching that involved verbal instruction (e.g., "push that button," "slide that door") combined with gesture and movement that effectively grounded the vocalization.

56. Stringer and Andrews 2005.

57. Whiten et al. 1999, Van Schaik et al. 2003.

58. Oldowan tools are named after the Olduvai Gorge in Tanzania, where the tools were first discovered, and date from c. 2.5–1.2 mya. Oldowan artifacts have been recovered from several localities in eastern, central, and southern Africa, the oldest of which is a site at Gona, Ethiopia. Oldowan technology is typified by what are known as "flakes" and "choppers." Microscopic surface analysis of the flakes struck from cores has shown that these flakes were used as tools for cutting plants and butchering animals. Choppers are stone cores with flakes removed from part of the surface, creating a sharpened edge that was also used for cutting, chopping, and scraping. Acheulean stone tools, named after the site of St. Acheul in France, where artifacts from this tradition were first discovered, date from c. 1.7–0.2 mya, and are widely thought to be a technological advance on Oldowan tools. They have been found over an immense area of the Old World, from southern Africa to northern Europe and from western Europe to the Indian subcontinent. Acheulean technology is best characterized by its distinctive pear-shaped and symmetrical stone hand axes.

59. McBrearty and Brooks 2000, D'Errico and Stringer 2011.

60. Laland et al. 2000.

61. Galef Jr. 1988, Laland et al. 1993.

62. Odling-Smee et al. 2003.

63. Laland et al. 2000.

64. Odling-Smee et al. 2003.

65. Ibid.

66. Odling-Smee et al. 1996, 2003; Smith 2007a, 2007b; Kendal et al. 2011.

67. Laland et al. 2010.

68. Laland et al. 2000.

69. Ibid.

70. Watson et al. 2014. However, this claim remains contentious (see Fischer et al. 2015).

71. Pika et al. 2005, Janik and Slater 1997, Zuberbuhler 2005.

72. Fitch 2010, Hurford 2014.

73. Enquist and Ghirlanda 2007, Enquist et al. 2011.

74. Tomasello 1994.

75. Whiten et al. 1999, Reader 2000.

76. Csibra and Gergely 2011.

77. While this claim holds at the time of writing, it is only honest to acknowledge that similar claims have been made in the past for other hypotheses with respect to other, or more restricted, sets of criteria (e.g., Bickerton 2009). Hence, the fact that a particular hypothesis meets all the designated criteria is no guarantee that it is correct, or that it will stand the test of time as new sets of criteria are devised.

78. Odling-Smee and Laland 2009.

79. Here a distinction between an "adaptation" (a character favored by natural selection for a particular role or function) and an "adaptive trait" (a character that increases biological fitness) is relevant. To my knowledge, no one (Chomsky included) denies that communication through language is an adaptive trait. The debate concerns whether the original function of the psychological and neural mechanisms that allow language was communication.

80. Noam Chomsky has argued that the language faculty, specifically the property of discrete infinity or recursion that plays a central role in his theory of Universal Grammar, may have evolved as a spandrel: in this view, Chomsky initially pointed to language being a result of increased brain size and increasing neural complexity, as well as the enhancements in human thinking these afford. However, he provides no definitive answers as to what factors led to the brain attaining the size and complexity of which discrete infinity is a consequence. A recent account is given in Hauser et al. 2002.

81. See, for instance, Pinker and Jackendoff 2005.

82. Fitch 2004; see also Nowicki and Searcy 2014, and Smit 2014.

83. Sterelny 2012a, 2012b.

84. Fitch 2005, Ridley 2011.

85. Pagel 2012.

86. Nowak and Highfield 2011.

87. Fehr and Gachter 2002, Fehr and Fischbacher 2003, Boyd and Richerson 1985, Henrich 2015.

88. Pagel 2012.

89. Ibid.

90. Deacon 1997; see also Bickerton 2009.

91. Hauser 1996, Bickerton 2009.

92. Hauser 1996, Bickerton 2009, Hauser et al. 2014,

93. Derek Bickerton and Terrence Deacon have both argued that the continued use and manipulation of symbols exerted selection on the human mind and favored an evolved structure for language learning, including the learning of syntax. See Deacon 1997, and Bickerton 2009. Other researchers have proposed theories of

language evolution that stress both biological and cultural evolution, but without greatly emphasizing this selective feedback; notably, Arbib (2012).

94. Rilling et al. 2008, Schenker et al. 2010.
95. Deacon 1997, 2003b. The Baldwin effect is named after psychologist James Mark Baldwin, who proposed a specific mechanism for adaptive evolution in 1896 (Baldwin 1896, 1902). Baldwin argued that animals could adjust to environmental conditions phenotypically—for instance, through learning, which would not only help them to survive but create a situation in which natural selection would favor an unlearned tendency to perform the same behavior. Although largely ignored by evolutionary biologists for almost a century, the idea is experiencing a renaissance. For instance, Baldwin effect arguments have influenced the field of research exploring the role of developmental plasticity in evolution (e.g., West-Eberhard 2003).
96. Bickerton 2009.
97. Hauser 1996, Bickerton 2009.
98. Falk 2004; I prefer the term "infant-directed speech," because many of us fathers also adjust our speech for infants.
99. Deacon 1997; Fitch 2004, 2005; Falk 2004.
100. Thiessen et al. 2005, Fitch 2010.
101. Pinker 1995.
102. Waterson 1978, Thiessen et al. 2005, Fitch 2010; Huttenlocher et al. 2002.
103. Falk (2004) claims that infant-directed speech occurs worldwide, but this claim is contested. For instance, Masataka 2003 (p. 137) writes: "Evidence for cross-cultural universality of the production of motherese is less convincing and actually very controversial."
104. Bloom 1997, 2000; Fitch 2010.
105. Bickerton 2009.
106. Chomsky 1965.
107. Fitch 2010.
108. Brighton et al. 2005.
109. This argument was made most prominently by Chomsky (1980), who emphasized "the poverty of the stimulus," or the fact that data from which language must be learned lack direct evidence for a number of features of grammar that children must learn.
110. Deacon 1997.
111. Smith and Kirby 2008, Kirby et al. 2007.
112. Smith and Kirby 2008.
113. Kirby et al. 2008.
114. Smith and Kirby 2008; Kirby et al. 2007, 2008, 2015.
115. Byrne 2016.
116. Wollstonecroft 2011.
117. Byrne and Russon 1998, Byrne 2016.
118. Whalen et al. 2015.

119. Hoppitt and Laland 2013.
120. Heyes 2012.
121. Bolhuis et al. 2014.
122. Morgan et al. 2015.
123. Delagnes and Roche 2005.
124. Toth 1987.
125. Roche et al. 1999, Callahan 1979.
126. Schick and Toth 2006, Braun et al. 2009.
127. Hovers 2012.
128. Blumenschine 1986, Potts 2013.
129. Callahan 1979.
130. Potts 2013.
131. Schick and Toth 2006.
132. Stout et al. 2000, Uomini and Meyer 2013.
133. A recent review of Acheulean toolmaking found that reduction strategies (i.e., the manner in which flakes are produced) were highly consistent across individuals (Shipton et al. 2009). The authors suggest "true imitation" (i.e., reproducing the motor pattern of another individual through observational learning) is the minimal form of social transmission that could produce such consistency. Furthermore, an unpublished experimental study found that "demonstrative gestures" were sufficient for the cooperative procurement and initial reduction of bedrock slabs (Petraglia et al. 2005). Only two studies have directly investigated the ability of contemporary adult humans to make tools following different means of social transmission, both comparing the efficacy of speech with symbolic gestural communication. One investigated the acquisition of Levallois technology (Ohnuma et al. 1997), which is a complex technology prevalent from 300–30 kya; the study reported no differences between the conditions. However, the measure of performance was a binary (yes / no) assessment by the experimenter, leaving the possibility that more subtle differences existed but were undetected. The second (Putt et al. 2014) investigated bifacial knapping, a technique associated with Acheulean technology. Although the tools produced in both conditions showed similar shape, symmetry, and quality, the two groups used different techniques, with verbally taught participants more accurately replicating the technique of the instructor (even though they lacked the skill to enact it effectively). As verbal and gestural communication are both symbolic forms of communication, further differences may yet emerge if a wider range of social transmission mechanisms, including imitation, emulation, and subtle forms of pedagogy, are considered. This is particularly relevant to the manufacture of Oldowan technology, where the debate over the underlying transmission mechanisms is at its fiercest.
134. Whether Oldowan stone toolmaking has implications for the evolution of human language and teaching is hotly debated. See, for instance, Gibson and Ingold 1993, and Ambrose 2001.

135. Wynn et al. 2011.

136. Hovers 2012.

137. Bickerton 2009.

138. Morgan et al. 2015.

139. Participants were assigned to conditions at random and blinding was not possible.

140. This involved using a novel metric we developed. It assessed quality, taking into account flake mass, cutting edge length, and diameter. See "Supplementary Methods" in Morgan et al. 2015 for details.

141. It also improved performance relative to the imitation/emulation and basic teaching conditions.

142. Improvements are specified relative to the reverse engineering condition.

143. The magnitude of this was 430%.

144. Analyses of the utterances by participants in the verbal teaching condition showed that the rate of decline varied with topic.

145. Morgan et al. (2015) suggest that the steady decline of performance to baseline levels, even in conditions that allow teaching and language, merely reflects the short learning time employed in this study. Previous transmission chain studies have established that periods of individual practice can bolster the stability of socially transmitted knowledge (see Hoppitt and Laland 2013 for a review of this literature). This suggests that with more time to learn, with bouts of teaching and language integrated with periods of individual practice, the benefits of teaching and language would likely have been preserved for longer.

146. The short learning period in our experiment (five minutes long) was clearly unrealistic compared with the length of time that Oldowan hominins likely had available for learning. This is particularly true given available data showing that precise control of conchoidal fracture can take decades to acquire (Nonaka et al. 2010) and anthropological data showing that knapping skills are acquired across an apprenticeship lasting several years (Stout 2002). However, a short learning period is sufficient to examine the relative rates of transmission, which was the focus of this work. Naturally, we cannot rule out the possibility that with a longer learning period, performance across conditions would have converged. But given that knapping skills are known to take years to develop fully, we suspect that increasing the time spent learning would initially only increase the differences in performance across conditions, with any convergence only occurring after extensive learning. Given their magnitude, the observed differences in performance between conditions would likely translate into significant fitness differences in the shorter term.

147. For example, if verbal teaching provided transmission benefits, but simpler forms of teaching did not, then the coevolutionary process would not be able to account for the evolution of these simpler forms of teaching. Likewise, if the transmission of toolmaking benefitted from simple teaching, but gained no further benefit from verbal teaching, then the coevolutionary process would stop with simpler forms of teaching and could not explain the evolution of verbal teaching.

148. That the low level of performance with imitation/emulation and reverse engineering is stable along chains (and that performance with teaching and language eventually decreases to this level) suggests a baseline level of performance reliant on little transmitted knowledge. It is possible that this baseline level could also be achieved through intuition and individual trial-and-error learning. We suggest that the decline of performance with teaching and language to this baseline merely reflects the short learning time employed in this study. Previous transmission chain studies have established that periods of individual practice can bolster the stability of socially transmitted knowledge (Hoppitt and Laland 2013). This suggests that with more time to learn, with bouts of teaching and language integrated with periods of individual practice, the benefits of teaching and language would likely have been preserved for longer.

149. Stout et al. 2000, 2010; Uomini 2009.

150. After the experiment, we tested our subjects' knowledge, asking them a series of questions about toolmaking to determine what they had learned.

151. The platform angle is the angle between the struck surface (the "platform") and the underhanging surface below. For successful knapping this must be below 90 degrees, and ideally around 70 degrees. The angle is measured in the plane generated by rotating the plane of the platform 90 degrees around an axis that runs through (1) the point of impact, and (2) the closest point on the edge.

152. Hovers 2012, Ambrose 2001, De la Torre 2011, Stout et al. 2010.

153. Beyene et al. 2013, Lepre et al. 2011.

154. Schick and Toth 2006.

155. Evans 2016.

156. This raises the supplementary question: If the selective advantage was present, why did it take 700,000 years for more complex communication to evolve? The most likely explanation is that this did indeed evolve during the Oldowan, but the appearance of more advanced tools may have additionally been contingent on the evolution of other aspects of cognition, such as technical comprehension or the hierarchical planning of actions, as well as demographic factors.

157. Bickerton 2009, Donald 1991, Corballis 1993.

158. Bickerton 2009.

159. Tomasello 2008.

160. Wallace 1869; see also "The Limits of Natural Selection as Applied to Man (S165: 1869/1870)" on the Alfred Russel Wallace Page (http://people.wku.edu/charles .smith/wallace/S165.htm).

CHAPTER 9: GENE-CULTURE COEVOLUTION

1. Morgan et al. 2015.

2. Other signals of recent selection include high-frequency alleles in linkage disequilibrium and unusually long haplotypes of low diversity.

3. Feldman and Cavalli-Sforza 1976, Cavalli-Sforza and Feldman 1981, Boyd and Richerson 1985, Rogers 1988, Feldman and Laland 1996, Enquist et al. 2007, Richerson and Boyd 2005.

4. Feldman and Cavalli-Sforza 1976, Cavalli-Sforza and Feldman 1981, Boyd and Richerson 1985, Rogers 1988, Feldman and Laland 1996, Enquist et al. 2007, Richerson and Boyd 2005.

5. Laland et al. 1995, Cavalli-Sforza and Feldman 1973, Otto et al. 1995.

6. Feldman and Laland 1996, Richerson and Boyd 2005, Richerson et al. 2010.

7. Corballis 1991.

8. Ibid.

9. In principle, selective regimes such as heterozygote advantage (Annett 1985, McManus 1985) or frequency-dependent selection (Faurie and Raymond 2005) could preserve variation in handedness.

10. Morgan and Corballis 1978.

11. This data is based on a meta-analysis of 14 twin studies in McManus 1985.

12. While isolated studies (e.g., Warren et al. 2006) have reported positive heritabilities for some handedness measures, the overall picture across multiple studies remains that handedness, at least as measured in the vast majority of questionnaire and performance studies, does not exhibit strong heritability (McManus 1985, Neale 1988, and Su et al. 2005).

13. Corballis 1991.

14. Harris 1980, Corballis 1991.

15. Hardyck et al. 1976, Hung et al. 1985, Teng et al. 1976.

16. Laland et al. 1995.

17. Previous studies had established that if there was a genetic influence on handedness, it operated through dextralizing and chance alleles, rather than right-biased and left-biased alleles (Annett 1985, McManus 1985).

18. Bishop 1990.

19. In principle, genetic variation affecting handedness could be preserved through forms of selection such as frequency dependent selection or heterozygote advantage. However, there is little compelling evidence that such selection is widespread (Laland et al. 1995).

20. The hypothesis that human populations are currently evolving toward an equilibrium in which dextralizing alleles are fixed is inconsistent with data revealing a decreasing trend in right handedness in the United States and Australia over the last century (Corballis 1991). This data is generally interpreted as reflecting a relaxation in the social pressure to conform to a right-handed standard.

21. The method used is maximum likelihood. See Laland et al. 1995.

22. The one study with a poor fit was the earliest, which has been criticized for its methodology. In early studies of handedness, researchers were unaware of potential biases generated by advertising for experimental subjects to take part in a handedness investigation. Such advertisements distort the subject pool, since

left-handers are disproportionately interested in the topic. For further details, see Laland et al. 1995.

23. Laland et al. 1995.
24. Toth 1985, Uomini 2011.
25. Toth 1985, Uomini 2011.
26. Bradshaw 1991, Hopkins and Cantalupo 2004.
27. Boyd and Richerson 1985, Feldman and Laland 1996, Kumm et al. 1994, Laland et al. 1995.
28. Holden and Mace 1997, Burger et al. 2007.
29. Peng et al. 2012.
30. Bersaglieri et al. 2004.
31. Voight et al. 2006.
32. Bersaglieri et al. 2004.
33. Feldman and Laland 1996; Perreault 2012.
34. Feldman and Cavalli-Sforza 1989.
35. Boyd and Richerson 1985, Feldman and Laland 1996, Feldman and Cavalli-Sforza 1989, Laland 1994, Laland et al. 1995.
36. Hawks et al. 2007. Other factors may have played a role in the speeding up of human evolution, such as the increased number of new mutations in the larger populations that have been facilitated by agriculture. See Cochran and Harpending 2009.
37. Gleibermann 1973, Williamson et al. 2007.
38. Helgason et al. 2007.
39. Myles, Hradetzky, et al. 2007; Neel 1962.
40. O'Brien and Laland 2012.
41. Posnansky 1969.
42. Livingstone 1958.
43. Evans and Wellems 2002.
44. Kappe et al. 2010.
45. Hawley et al. 1987.
46. Roberts and Buikstra 2003, Barnes 2005.
47. Barnes et al. 2011.
48. Ibid.
49. Bellamy et al. 1998, Li et al. 2006.
50. Govoni and Gros 1998.
51. Barnes et al. 2011.
52. Aidoo et al. 2002.
53. Weatherall et al. 2006.
54. Durham 1991.
55. Lovejoy 1989.
56. Piel et al. 2010.
57. Jonxis 1965.
58. O'Brien and Laland 2012.

59. Agbai 1986, Houston 1973.
60. O'Brien and Laland 2012.
61. Rendell, Fogarty, and Laland 2011.
62. Rendell, Fogarty, and Laland 2011 labelled this process "runaway niche construction."
63. O'Brien and Laland 2012.
64. Such signals include high-frequency alleles in linkage disequilibrium, unusually long haplotypes of low diversity, and an excess of rare variants.
65. Voight et al. 2006; Wang et al. 2006; Sabeti et al. 2006, 2007; Nielsen et al. 2007; Tang et al. 2007.
66. Stefansson et al. 2005, Nguyen et al. 2006, Prabhakar et al. 2008, Quach et al. 2009.
67. Laland et al. 2010.
68. Ibid.
69. Perry et al. 2007. This is a good example of gene-culture coevolution in which changes in human diet favor copies of a gene.
70. Voight et al. 2006, Richards et al. 2003.
71. Voight et al. 2006, Williamson et al. 2007, Han et al. 2007.
72. Haygood 2007.
73. Hunemeier, Amorim et al. 2012; Acuna-Alonzo et al. 2010; Magalon et al. 2008; Sabbagh et al. 2011.
74. Kelley and Swanson 2008.
75. Soranzo et al. 2005.
76. Armelagos 2014.
77. Striedter 2005, Schick and Toth 2006, Stringer and Andrews 2005.
78. Armelagos 2014.
79. Ibid.
80. Remick et al. 2009, Armelagos 2014.
81. Aunger 1992, 1994a, 1994b; Armelagos 2014.
82. Ibid.
83. Williamson et al. 2007.
84. Wang et al. 2006.
85. Olalde et al. 2014.
86. Williamson et al. 2007, López Herráez et al. 2009.
87. Izagirre et al. 2006; Lao et al. 2007, Myles, Somel, et al. 2007.
88. Olalde et al. 2014.
89. Voight et al. 2006, Myles et al. 2008.
90. Voight et al. 2006, Wang et al. 2006. Genes expressed in hair follicles include ectodysplasin A receptor (*EDAR*) and *EDA2R*. Those expressed in eye and hair color include solute carrier family 24, member 4 (*SLC24A4*), KIT ligand (*KITLG*), tyrosinase (*TYR*), and oculocutaneous albinism II (*OCA2*). Genes expressed in freckles include *6p25.3* and melanocortin 1 receptor (*MC1R*).
91. Sexual selection is a mode of natural selection in which some individuals outreproduce others in a population because they are better at securing (more or

higher quality) mates (Darwin 1871). Sexual selection can lead to the evolution of costly and viability reducing traits, such as elaborate plumage or large antlers, because of the advantage that these afford in attracting or fighting for mates.

92. Laland 1994.

93. Laurent et al. 2012.

94. Jones et al. 2007, Little et al. 2008.

95. Laland 1994, Ihara et al. 2003.

96. Stedman et al. 2004.

97. Williamson et al. 2007 previously identified several genes involved in nervous system development as candidates of recent selective sweeps. These include abnormal spindle, microcephaly associated (*ASPM*) and microcephalin 1 (*MCPH1*). Other genes involved in nervous system development, synapse formation, nerve growth factors, and both neuron and dendritic spine generation are identified by de Magalhaes and Matsuda 2012. See also Hill and Walsh 2005.

98. de Magalhaes and Matsuda 2012.

99. Laland et al. 2010. Examples include cyclin dependent kinase 5 regulatory subunit-associated protein 2 (*CDK5RAP2*), centromere protein J (*CENPJ*), and γ-aminobutyric acid A receptor, subunit α4 (*GABRA4*). Other genes implicated in human brain evolution and identified through comparative genomics include microcephalin 1 (*MCPH1*), asp homologue, microcephaly associated (*Drosophila*) (*ASPM*), CDK5 regulatory subunit associated protein 2 (*CDK5RAP2*), solute carrier family 2 (facilitated glucose transporter), member 1 (*SLC2A1*), *SLC2A4*, neuroblastoma breakpoint family (*NBPF*) genes, growth arrest and DNA-damage-inducible gamma (*GADD45G*), ret finger protein-like 1, 2, and 3 (*RFPL1, RFPL2* and *RFPL3*), and genes associated with neuronal functionality (dopamine receptor D5 [*DRD5*]), glutamate receptor, ionotropic, NMDA 3A (*GRIN3A*), *GRIN3B*, and SLIT-ROBO Rho GTPase activating protein 2 (*SRGAP2*). For further details see Somel et al. 2013.

100. These include a deletion of one enhancer of the growth arrest and DNA-damage-inducible gamma (*GADD45G*) gene, which is postulated to have led to brain size expansion and is shared with Neanderthals. Likewise, a human-specific duplication of a truncated version of the gene SLIT-ROBO Rho GTPase activating protein 2 (*SRGAP2*) was dated to approximately 2.4 mya. This duplication is predicted to increase dendritic spine density in the cortex and could have enhanced signal processing in hominin brains. See Somel et al. 2013 for further detail.

101. Uddin et al. 2004, Caceres et al. 2003.

102. Somel et al. 2014. The promoter regions of many neural genes have also experienced positive selection during human evolution (Haygood et al. 2007).

103. Somel et al. 2014.

104. Hill and Walsh 2005, Sakai et al. 2011, Blazek et al. 2011.

105. Striedter 2005, Marino 2006, Sherwood et al. 2006.

106. Blazek et al. 2011.

107. Relative to chimpanzees, studies have identified increases in (1) glia to neuron ratios, (2) spacing between neurons, (3) astrocyte complexity, and (4) neuropile density in the human brain. Some of these changes, however, may be a result of the overall brain size increase on the human evolutionary lineage. See Marino 2006, Sherwood et al. 2006, Semendeferi et al. 2011, Oberheim et al. 2009, and Spocter et al. 2012.

108. Blazek et al. 2011.

109. Deacon 1997, 2003a; Milbrath 2013.

110. For example, forkhead box P2 (*FOXP2*), *ASPM*, Microcephalin (*MCPH1*), protocadherin 11 X-linked (*PCDH11X*), and *PCDH11Y* are all linked to language or speech. Somel et al. 2013.

111. Fisher et al. 1998.

112. Enard, Przeworski, et al. 2002.

113. Enard et al. 2009.

114. Liao and Zhang 2008.

115. Nasidze et al. 2006.

116. Kayser et al. 2006.

117. Oota et al. 2001, Cordaux et al. 2004.

118. Hunemeier, Gómez-Valdés, et al. 2012.

119. Stearns et al. 2010.

120. Ibid.

121. Ibid.

122. Ibid, p. 621.

123. Wilson 1975.

124. Wilson 1978. p. 167.

125. Laland and Brown 2011.

126. Laland et al. 2000.

127. Odling-Smee et al. 2003.

128. Laland et al. 2000.

129. Laland 1994; Rendell, Fogarty, and Laland 2011.

130. Cavalli-Sforza and Feldman 1981, Boyd and Richerson 1985, Richerson and Boyd 2005, Henrich 2015.

131. Laland et al. 2010.

132. Striedter 2005.

133. Rutz et al. 2010, Kruetzen et al. 2014.

134. Whitehead 1998, Baird & Whitehead 2000.

135. Morris 1967, Buss 1999.

136. Such structures are often "extended phenotypes" (Dawkins 1982); these are biological adaptations, such as nests, burrows, and webs, expressed outside the body of the constructor.

137. Odling-Smee et al. 2003.

138. Dawkins 1982, Odling-Smee et al. 2003.

139. Laland and Brown 2006.
140. Bird et al. 2013.
141. Boivin et al. 2016; see Diamond 2006 for the evidence that over-exploitation led to collapse.
142. Laland and Brown 2006.

CHAPTER 10: THE DAWN OF CIVILIZATION

1. This assertion, while almost certainly correct, has not to my knowledge been conclusively proven. One complication in its demonstration is that researchers know a lot more about recent evolution on short time scales than they know about what happened deeper in the past. As biological evolution on short time scales is faster than that observed at longer intervals (Gingerich 1983), this would create an illusion of accelerating evolution in the recent past, even if no such acceleration had occurred. Perreault (2012) has shown that the same considerations apply to cultural evolution. What has been established is that cultural evolution is faster than biological evolution, when controlling for the correlation between rate and time intervals, and the generation time of the species (Perreault 2012), and that many aspects of knowledge and technology show a hyperexponential increase over time (Enquist et al. 2008, Enquist et al. 2011). As described in the previous chapter, several theoretical studies have concluded that gene-culture coevolution is typically faster than conventional biological evolution, in part because cultural evolution occurs at faster rates than biological evolution. There is also genetic evidence that human evolution has accelerated during the last 40,000 years (Hawks et al. 2007). Collectively, these findings render my claim highly likely to be correct.

2. Lynch 1990 reports this finding, but draws attention to potential sources of bias and expresses the concern that it may be an artifact. More recent genetic data suggest that the accelerated rates of evolution observed in our species are probably genuine (e.g., Hawks et al. 2007).

3. Hawks et al. 2007, Cochran and Harpending 2009. The more recent acceleration in human genetic evolution is likely also to reflect increases in global population size (Cochran and Harpending 2009). Human numbers have increased from less than a million, 50,000 years ago, to several billion today. That represents several orders of magnitude more novel mutations being introduced into the population, and upon which selection can act. However, as that population growth was made possible by culture, ultimately the acceleration can be regarded as an example of gene-culture coevolution.

4. Vitousek et al. 1997, Waters et al. 2016, Boivin et al. 2016.

5. There is some evidence that human numbers were modest until somewhere in the realm of 50 kya (e.g., Li and Durbin 2011). This might be interpreted as implying that the energetic cost of our brain was such that it was barely able to "pay its way" until more recent times.

6. Kaplan et al. 2000; Kaplan and Robson 2002.

7. Vitousek et al. 1997.

8. Cochran and Harpending 2009. Chance fluctuations in genotype frequencies typically dictate evolutionary dynamics in small populations, but as populations get larger, natural selection starts to dominate as it becomes less probable that a beneficial allele will be lost by chance.

9. Henrich 2004a, Powell et al. 2009, Lewis and Laland 2012.

10. Rendell, Boyd, et al. 2011, Laland and Rendell 2013.

11. Ibid., Boyd and Richerson 1985, Henrich 2015.

12. Enquist et al. 2010; Lewis and Laland 2012.

13. Crittenden 2011.

14. Ibid.

15. Marlowe 2001.

16. Ibid.

17. Indicatoridae spp.

18. Crittenden 2011.

19. Boyd and Richerson 1985, Henrich and McElreath 2003, Richerson & Boyd 2005, Henrich and Henrich 2007.

20. Richerson et al. 1996.

21. For instance, for the !Kung Bushmen, plant foods supply 60–80% of calories (Lee 1979).

22. Richerson et al. 1996, Hill et al. 2009.

23. Wollstonecroft 2011.

24. Basgall 1987.

25. Richerson et al. 1996.

26. Ibid.

27. Ibid.

28. Blurton Jones 1986, 1987.

29. Lee and Daly 1999.

30. Richerson et al. 1996.

31. Lee and Daly 1999.

32. Collard et al. 2012.

33. Lee and Daly 1999.

34. As archaeological data has accumulated, increasing evidence for cultural change in the behavior of hunter-gatherer communities—in dietary shifts, for instance—is being detected (see, for example, Bettinger et al. 2015). Nonetheless, the general point holds that rates of change in technology appear modest relative to agricultural societies.

35. Armelagos et al. 1991.

36. Smith 1998, 2007a, 2007b.

37. Smith 1998, Richerson et al. 2001, Zeder et al. 2006, Bar-Yosef and Price 2011.

38. Winterhalder and Kennett 2006, Childe 1936, Wright 1977, Richerson et al. 2001.

39. Richerson et al. 2001.

40. Barlow 2002.
41. Smith 2007b, p. 196.
42. Smith 2007a, 2007b.
43. Laland and O'Brien 2010, O'Brien and Laland 2012.
44. Richerson 2013.
45. Ibid.
46. Ibid.
47. Actually, this assertion is not quite accurate, as leaf-cutter ants (*Atta* and *Acromyrmex*) have evolved a capability to harvest a fungus (*Lepiotaceae*), which could be regarded as a form of agriculture. The ants' cultivation, however, is clearly an adaptive specialization rather than a flexible, learned, and socially transmitted general capability to domesticate another species. Another interesting challenge to the claim that agriculture is unique to humans is provided by the spotted bowerbird (*Ptilonorhynchus maculates*). These birds have been found to grow plants with berries close to their bowers, which are then used to decorate the bowers as part of the mating process; however, whether the cultivation is deliberate remains to be established (Madden et al. 2012).
48. Laland and O'Brien 2010.
49. Laland and O'Brien 2010, O'Brien and Laland 2012.
50. Laland and O'Brien 2010, Zeder 2016.
51. Archaeologists and anthropologists sometimes refer to this as "traditional ecological knowledge."
52. Zeder 2016.
53. Zeder 2012, 2016.
54. Beja-Pereira et al. 2003, Zeder 2016.
55. Zeder 2016.
56. Smith 1998, 2007a, 2007b; Zeder 2016.
57. Winterhalder and Kennett 2006, p. 3.
58. Smith 1998, 2007a, 2007b.
59. Smith 2001.
60. Diamond 1997. For instance, Blumler 1992 found that, in contrast to Mediterranean regions, California had no large-seeded annual grasses; as a consequence, no agriculture could develop until the import of maize.
61. Diamond 1997.
62. Feldman and Kislev 2007.
63. The DNA of living organisms is found on chromosomes, which typically are found in paired sets. Polyploid organisms are those containing more than two paired sets of chromosomes. A hybrid is the offspring resulting from the cross-breeding of two distinct organisms or strains.
64. Bronowski 1973.
65. Badr et al. 2000. The Fertile Crescent is a crescent-shaped region of the Middle East that curves from northern Egypt through to the Persian Gulf, encompassing regions of present-day Israel, Jordan, Lebanon, Syria, and Iraq.

66. Matsuoka et al. 2002.
67. *Lagenaria siceraria*.
68. Erickson et al. 2005.
69. Lu et al. 2009.
70. Fuller 2011. See also http://archaeology.about.com/od/domestications/a/rice .htm.
71. Armelagos et al. 1991, O'Brien and Laland 2012.
72. Cohen and Armelagos 1984, Cohen and Crane-Kramer 2007.
73. O'Brien and Laland 2012.
74. Gibbons 2009.
75. O'Brien and Laland 2012.
76. Armelagos and Harper 2005, Pearce-Duvet 2006.
77. O'Brien and Laland 2012.
78. Gibbons 2009.
79. Mahatma Gandhi famously quipped: "What do I think of Western civilization? I think it would be a very good idea." The quote beautifully illustrates some of baggage that comes with the use of the term "civilization." For instance, the term implies that Western societies are in some respects superior or more sophisticated than other societies. My use of the term in the title of this chapter rejects such connotations, and refers solely to large-scale structured societies with extensive division of labor. See O'Brien and Laland 2012.
80. Smith 1998, 2007a, 2007b.
81. Cohen 1989.
82. Diamond 1997.
83. Cohen 1989, Rowley-Conwy and Layton 2011.
84. Smith 1998, 2007a, 2007b; Laland and O'Brien 2010.
85. Laland and O'Brien 2010.
86. Odling Smee et al. 2003, Smith 2007a.
87. Armelagos et al. 1991.
88. Tellier 2009.
89. Gignoux et al. 2011.
90. Caselli et al. 2005.
91. Of course, a major demographic effect of the industrial revolution has been the demographic transition to below-replacement fertility in many countries, which severs the link between innovation and population growth.
92. Armelagos et al. 1991.
93. The oldest city is thought to be Uruk, in Mesopotamia. It was first settled circa 4500 BCE (http://www.ancient.eu/city/).
94. Smith 2009.
95. Boyd and Richerson 1985; Richerson et al. 2014.
96. Reader and Laland 2003b.
97. Bronowski 1973.
98. Smith 1776, p. 195.

99. Petroski 1992.
100. Bronowski 1973. See also http://www.ploughmen.co.uk/about-us/history-of-the -plough.
101. Farming practices that rely on rainfall for water, known as "rainfed agriculture," provide much of the food consumed by poor communities in developing countries, including regions of Africa, Latin America, and southeast Asia. Early agricultural experiments may have also occurred in some of these regions. However, in the absence of effective irrigation systems, levels of productivity remain low.
102. Bronowski 1973. See http://www.ploughmen.co.uk/about-us/history-of-the -plough.
103. Ibid.
104. Anthony 2007.
105. Bronowski 1973.
106. Some Mediterranean regions were also reliant on rainfed agriculture.
107. Bronowski 1973.
108. Ibid.
109. Ibid.
110. Bushnell 1957.
111. See http://www.ancientegyptonline.co.uk/renenutet.html.
112. Kramer 1964.
113. Grant and Hazel 2002. I am indebted to Thomas Laland-Brown for drawing this point to my attention.
114. Grant and Hazel 2002.
115. Ibid.
116. Bronowski 1973, Clarke and Crisp 1983.
117. Lane 2016.
118. Bronowski 1973.
119. Ibid.
120. Anthony and Brown 2000, Anthony 2007.
121. Bronowski 1973.
122. See https://www.britannica.com/technology/spoked-wheel.
123. The assertion that the pace of cultural evolution is accelerating is supported by an analysis of 154 inventions created between 1800 and 1960, from microwave ovens to electroencephalographs (Michel et al. 2011). More recent innovations took far less time to become widely adopted.
124. Lane 2016, Beinhocker 2006.
125. Petroski 1992.
126. Virality refers to the tendency of a piece of information to be circulated rapidly across the internet (i.e., to go "viral").
127. Bijker 1995.
128. Ibid.

129. See http://content.time.com/time/specials/packages/completelist/0,29569, 1991915,00.html and http://thestubble.com/6-weird-inventions-couldve -invented.html.

130. See http://www.farnhamconsulting.com/A31%20Watson.htm.

131. Futuyma 1998.

132. Feldman and Cavalli-Sforza 1981.

133. See also Henrich 2004a and Powell et al. 2009.

134. Laland and Rendell 2013.

135. Derex and Boyd 2015.

136. Henrich 2004a, Powell et al. 2009.

137. Mesoudi and O'Brien 2008, Derex and Boyd 2015.

138. Richerson and Boyd 2005, Derex and Boyd 2015, Henrich 2015.

139. Henrich 2004a, Powell et al. 2009.

140. Henrich 2004a.

141. Ibid.

142. Kline and Boyd 2010.

143. Laland and Rendell 2013.

144. Boyd and Richerson 1985, Laland and Brown 2006.

145. Rendell, Boyd, et al. 2011.

146. Hawkesworth 2014; see also http://nzetc.victoria.ac.nz/tm/scholarly/tei-Haw Acco-t1-g1-t1-body-d3-d4.html, and http://www.tourism.net.nz/new-zealand /about-new-zealand/regions/bay-of-plenty/history.html.

147. Blunden 2003; see also http://nzetc.victoria.ac.nz/tm/scholarly/tei-HawAcco-t1 -g1-t1-body-d3-d4.html.

148. Blunden 2003.

149. On the language evolution, see Gray and Jordan 2000. For the mitochondrial DNA evidence, see Trejaut et al. 2005.

150. Donald 1991.

151. Evison et al. 2008.

152. Laland and Rendell 2013.

153. See https://www.google.co.uk/search?client=safari&rls=en&q=how+many +bytes+in+a+book&ie=UTF-8&oe=UTF-8&gfe_rd=cr&ei=U5uiVp GKLcn5 -gb8jLjYDw#q=how+much+information+on+the+internet (dated 23 January 2013).

154. See also Smith 1998, 2007a, 2007b; and Zeder 2012, 2016 for complementary arguments.

155. Jaradat 2007.

156. Boivin et al. 2016.

157. Ibid.

158. Ibid.

159. Ibid.

160. See http://www.bbc.co.uk/news/business-30875633 (dated 19 January 2015).

161. The ratio between incomes of the richest and poorest nations widened from 3:1 in 1820 to 70:1 in 2000 (http://www.rgs.org/OurWork/Schools/Teaching+re sources/Key+Stage+3+resources/Who+wants+to+be+a+billionaire/Is+it+ok +for+the+rich+to+keep+getting+richer.htm). This report claims that by making an annual contribution of just 1% of their wealth, the world's 200 richest people could provide primary education for every child in the world.
162. Lane 2016.

CHAPTER 11: FOUNDATIONS OF COOPERATION

1. Hill et al. 2011.
2. Guillermo 2005.
3. Shaw 2003.
4. Fehr and Fischbacher 2003; Henrich 2015.
5. West et al. 2007.
6. Fogarty et al. 2011.
7. Boyd and Richerson 1985, 1988; Henrich 2004b.
8. West et al. 2011.
9. West et al. 2011, p. 255.
10. Fogarty et al. 2011.
11. Fogarty et al. 2011.
12. Kline et al. 2013.
13. Caro and Hauser 1992.
14. Thornton and McAuliffe 2006.
15. Caro and Hauser 1992.
16. Nicol and Pope 1996.
17. Castro and Toro 2004 make a similar argument.
18. Fehr and Fischbacher 2003.
19. Sterelny 2012a.
20. Fehr and Fischbacher 2003.
21. Fehr and Fischbacher 2003; Tomasello 2010.
22. Sterelny 2012a.
23. This tendency to reward cooperators and punish noncooperators is known as "strong reciprocity." See Fehr and Gachter 2002, and Fehr and Fischbacher 2003.
24. Tomasello 2010.
25. Tomasello 1999, 2008, 2010.
26. Sterelny 2012a.
27. Ibid.
28. Ibid.
29. Brosnan et al. 2012.
30. Sterelny 2012a, de Waal 2001.
31. Sterelny 2012a.
32. Bronowski 1973, Clarke and Crisp 1983.

33. Boyd and Richerson 1985; Fehr and Fischbacher 2003; Gintis 2003; Henrich 2004b, 2015; Henrich and Henrich 2007; Richerson and Boyd 2005; Nowak and Highfield 2011.

34. Nowak and Highfield 2011.

35. Oxpecker: *Buphagus* spp.

36. There are, as yet unpublished, reports and video footage of a group of Balinese long-tailed macaques stealing tourist objects, such as mobile phones and sunglasses; subsequently, they are willing to relinquish the objects in exchange for food. This "trade" appears to have emerged as a cultural tradition. See http://jbleca.webs.com/currentresearch.htm for further details.

37. Stanford et al. 1994.

38. Gilby et al. 2010.

39. Ridley 2011.

40. Ibid.

41. Davies and Bank 2002.

42. Richerson and Boyd 2005.

43. I suspect that some simple institutions could be learned without language.

44. Pagel 2012 makes a similar argument.

45. Trivers 1971.

46. Alexander 1987.

47. Nowak and Sigmund 1998.

48. Nowak and Highfield 2011.

49. Fehr and Gachter 2002, Fehr and Fischbacher 2003.

50. Boyd and Richerson 1985, Richerson and Boyd 2005, Henrich and Boyd 1998.

51. In this respect, experiments by Karen Kinzler and colleagues are of interest (Kinzler et al. 2009). These authors show that when presented with photographs and voice recordings of novel children, 5-year-olds chose to be friends with native speakers of their own language rather than foreign-language or foreign-accented speakers. Also, children chose those of the same race as friends when the target children were silent, but they chose those of other races with a native accent when accent was pitted against race. The results suggest that children preferentially evaluate others along dimensions that distinguished social groups in prehistoric human societies.

52. Richerson and Boyd 2005, Richerson et al. 2014.

53. Boyd and Richerson 1985.

54. Richerson and Henrich 2012.

55. Richerson et al. 2014, Henrich 2004b.

56. Lee and Daly 1999.

57. Richerson et al. 2014.

58. Tomasello 1999.

59. Fehr and Gachter 2002, Fehr and Fischbacher 2003.

60. Frequency-dependent payoffs and multiple stable equilibria are probably very common in human social institutions. See, for instance, Cooper 1999.

61. Bell et al. 2009.
62. Boyd and Richerson 1992, Fehr and Gachter 2002, Fehr and Fischbacher 2003, Richerson et al. 2014, Henrich 2004b.
63. Richerson et al. 2014.
64. This idea was championed by Richerson and Boyd (1998), who refer to the evolved psychological mechanisms as "tribal social instincts." However, I find the term "instinct" problematical (see note 114, chapter 2), and hence prefer the alternative term "norm psychology" promoted by Chudek and Henrich (2011). See also Fehr and Fischbacher (2003) and Richerson and Henrich (2012) for related treatments.
65. Chudek & Henrich 2011, p. 218.
66. Chudek & Henrich 2011.
67. Ibid., Fehr and Fischbacher 2003.
68. Richerson and Boyd 1998, Richerson et al. 2014.
69. Richerson and Boyd 1998, Richerson and Henrich 2012.
70. Further support for the hypothesis that humans could have been subject to natural selection favoring docility comes from the findings of the Second Social Learning Strategies Tournament. The tournament found that a cumulative culture setting created a selective context in which the best strategy was simply to play OBSERVE once and then play EXPLOIT thereafter. The greater the buildup of cumulative culture (the more the move REFINE had been played in that environment), the greater the advantage to this simple strategy relative to more sophisticated strategies that exhibit greater social or asocial learning.
71. Tomasello 2010.
72. Tomasello 1999, 2010.
73. Dean et al. 2012.
74. See, for instance, Melis et al. 2006.
75. Sterelny 2012a, 2012b.
76. Van Schaik and Burkart 2011.
77. Laland and Bateson 2001, Pawlby 1977.
78. These phenomena could also have arisen through a cultural evolutionary process. See chapter 8 for discussion and references.
79. Heyes 2012.
80. Byrne 1994, Hoppitt and Laland 2008a.
81. Van Baaren et al. 2009.
82. Chartrand and Van Baaren 2009.
83. Heyes 2012, Chartrand and Van Baaren 2009, Van Baaren et al. 2004.
84. Chartrand and Bargh 1999.
85. Van Swol 2003.
86. On finding time more enjoyable when imitated, see Tanner et al. 2008.
87. Carpenter et al. 2013.
88. Van Baaren et al. 2004.
89. Stel et al. 2010, Heyes 2012.

90. Heyes 2012, Yabar et al. 2006.
91. Heyes 2012.
92. Ibid.
93. Boyd and Richerson 1985.
94. Heyes 2012.
95. Laland and Bateson 2001, Heyes 2005.
96. Heyes 2012, Wen et al. 2016.
97. This is known as "secondary reinforcement," a well-established finding in research into learning processes.
98. Tarr et al. 2014.
99. Heyes 2012.
100. Laland and Bateson 2001.
101. Pawlby 1977.
102. Van Schaik and Burkart (2011) make a similar argument.
103. Rose and Lauder 1996.
104. Gergely and Csibra 2005, Gergely et al. 2007, Csibra 2010.
105. Tennie et al. 2009.
106. Tomasello 1999, 2010.
107. Tomasello et al. 2007.
108. For clear experimental evidence and further discussion on conformity, see Morgan et al. 2012, and Morgan and Laland 2012.
109. Haun et al. 2012. See also Herrman et al. 2013.
110. Call et al. 2004; Call and Tomasello 1998, 2008.
111. For the neural evidence, see Striedter 2005, and Marino 2006. For the genetic evidence, see Laland et al. 2010. Examples of the latter include cyclin dependent kinase 5 regulatory subunit-associated protein 2 (*CDK5RAP2*); centromere protein J (*CENPJ*); and γ-aminobutyric acid A receptor, subunit α4 (*GABRA4*). Other genes implicated in human brain evolution, identified through comparative genomics, include microcephalin 1 (*MCPH1*); asp homologue, microcephaly associated (*Drosophila*) (*ASPM*); CDK5 regulatory subunit associated protein 2 (*CDK5RAP2*); solute carrier family 2 (facilitated glucose transporter), member 1 and 4 (*SLC2A1*, *SLC2A4*); neuroblastoma breakpoint family (*NBPF*) genes; growth arrest and DNA-damage-inducible gamma (*GADD45G*); ret finger protein-like 1, 2, and 3 (*RFPL1*, *RFPL2* and *RFPL3*); and genes associated with neuronal functionality (dopamine receptor D5) (*DRD5*), glutamate receptor, ionotropic, NMDA 3A and 3B (*GRIN3A*, *GRIN3B)*, and SLIT-ROBO Rho GTPase activating protein 2 (*SRGAP2*). For further details, see Somel et al. 2013.
112. Somel et al. 2014.
113. That is, our capability for classical conditioning, operant conditioning, and reversal learning.
114. Bolhuis et al. 2011; see Carey 2009 for illustrations of how such cognitive adaptations emerge.
115. Boyd and Richerson 1985, Henrich 2009, Chudek and Henrich 2011.

CHAPTER 12: THE ARTS

1. Mesoudi et al. 2004, 2006; Mesoudi 2011.
2. This is perhaps because the arts place a particular premium on creativity and originality, and also possibly because evolution has a bad name in some areas of the humanities (see Laland and Brown 2011 for historical details).
3. Morgan 1877; Spencer 1857, (1855) 1870; Tylor 1871.
4. Darwin 1859.
5. A story pervades that the apple logo found on iPhones and Macintosh computers is a tribute to Alan Turning, the father of modern computing, who died by biting into an apple laced with cyanide. While differing opinions abound, this story sadly would appear more likely to be an urban legend than the truth.
6. Turing 1937; Minsky 1967, p. 104.
7. Gould and Vrba 1982.
8. Hoppitt and Laland 2013.
9. Galef Jr. 1988.
10. Strictly, the "most learning" referred to here should read most "instrumental" (or "operant") learning.
11. Pullium and Dunford 1980.
12. For a recent review, see Brass and Heyes 2005.
13. Laland and Bateson 2001.
14. Rizzolatti and Craighero 2004.
15. Ibid.
16. Striedter 2005.
17. Iacoboni et al. 1999.
18. This form of social learning is typically referred to as "observational conditioning."
19. Bronowski 1973.
20. Striedter 2005.
21. Deacon 1997.
22. Striedter 2005; Heffner and Masterton 1975, 1983.
23. Barton 2012.
24. Prints of these are still available for $20 each (https://secure.donationpay.org/chimphaven/chimpart.php).
25. For an accessible animal behaviorist's assessment of elephant painting, see http://www.dailymail.co.uk/sciencetech/article-1151283/Can-jumbo-elephants-really-paint—Intrigued-stories-naturalist-Desmond-Morris-set-truth.html.
26. Tourist camps in Thailand that boast "painting elephants" have attracted criticism from animal rights activists who express concerns that the training regimes may be cruel to the animals. The tourist camps, in response, claim that the elephants are mentally and socially healthy.
27. Footage of painting elephants and chimpanzees can easily be found on YouTube.
28. Stoeger et al. 2012.
29. Plotnik et al. 2006.

30. Consistent with this argument is the finding that rhesus monkeys can be trained to produce mirror-induced, self-directed behavior resembling mirror self-recognition, with appropriate visual-somatosensory training that links visual and somatosensory information (Chang et al. 2015).

31. Striedter 2005.

32. Hugo (1831) 1978, p. 189.

33. The first use of perforated shells as beads is dated to over 100,000 years ago (D'Errico and Stringer 2011, McBrearty and Brooks 2000). The shells frequently have geometrical patterns cut into them, and have been colored with pigments. The use of red ochre as a painting material dates back further. Engraved ostrich shells dated to 60,000 years ago have been found in South Africa (Texier et al. 2010). By around 45,000 to 35,000 years ago, art was widespread (at least in western Europe) and highly consistent, and comprised pierced beads of ivory and shells, etched and carved stones, engraved decorations on bone and antler tools and weapons, and sculpted statues of animals and female figures, which were thought to be fertility symbols. However, the most evocative and striking images of Paleolithic artwork are unquestionably the magnificent cave art paintings discovered in several European countries (Sieveking 1979). Many caves are renowned for their artwork; the oldest include the spectacular paintings found at the Le Chauvet Cave in France, dated to 30,000 years ago. Perhaps the most remarkable collection of cave paintings is at Lascaux in Dordogne, France, where an incredible 2,000 painted images of horses, deer, cattle, bison, humans, and a 5-meter high bull, have been dated to 18,000–12,000 years ago. Also renowned is the beautiful painted ceiling of the cave at Altamira, in northern Spain. This was the first cave art to be discovered, in 1879. The art at Altamira, which has been dated to around 19,000–11,000 years ago, comprises stunning representations of bison, horses, and other large animals, with extraordinary use of colors and shading to indicate depth. The quaint story of its discovery details that the paintings, which are on a low ceiling, were initially missed by the team of archaeologists, but were spotted by one of the team's 8-year-old daughter; she was the only individual small enough to stand erect and still look up at the ceiling (Tattersall 1995).

34. There is the expected regional variation, with particular techniques, styles and materials used in specific locations, indicating that the art expressed particular meanings that were socially learned and shared by the members of the community (Zaidel 2013). The paintings record for posterity what dominated the minds of those peoples, the animals that they lived by and stalked, and the power and potencies that those creatures symbolized. The correspondence between those species that were painted and those that have been independently verified as present is sufficiently tight that ecologists now use paleolithic art to infer species distributions (Yeakel et al. 2014). There is also continuity over time, as the same methods and skills are reproduced throughout the millennia. For instance, the European cave art tradition lasts tens of thousands of years, while the use of pigments, such as red ochre, in rock paintings is still used today (McBrearty and

Brooks 2000). These traditions were passed on from one generation to the next, picking up innovations from numerous creative, avant-garde, or radical individuals along the way, in a continuum that stretches back to the origins of our species, and forward to those exhibits found in today's contemporary art museums. Finally, the observed patterns of change are historically contingent. Like technology, novel art does not spring forth fully formed from the mind of the maker, but rather is a creative reworking of existing artistic forms.

35. The company was Ballet Rambert until 1966, and then Rambert Dance Company until 2013.

36. Laland et al. 2016.

37. Byrne 1999, Laland and Bateson 2001, Heyes 2002, Brass and Heyes 2005.

38. Carpenter 2006.

39. Heyes and Ray 2000, Laland and Bateson 2001.

40. Brown et al. 2006.

41. Some animals' movements, such as the coordinated jumping and wing-flapping courtship of pairs of Japanese cranes, or the communication system of honeybees, possess some dance-like properties, but these are species-specific behavior patterns that have evolved to fulfil quite separate functions.

42. Nettl 2000.

43. Fitch 2011.

44. Patel 2006.

45. In contrast to this hypothesis, I also place emphasis on motor imitation.

46. Doupe 2005, Jarvis 2004.

47. *Cacatua galerita eleonora*.

48. You can see Snowball on YouTube at https://www.youtube.com/watch?v=cJOZp2ZftCw.

49. Moore 1992.

50. Patel et al. 2009.

51. Schachner et al. 2009, Patel et al. 2009, Dalziell et al. 2013.

52. Hoppitt and Laland 2013.

53. Fitch 2013.

54. Hoppitt and Laland 2013.

55. Cook et al. 2013, Fitch 2013.

56. Ibid.

57. Dalziell et al. 2013.

58. Indeed, in a number of both classical and modern dance forms, motor imitation is key. Dancers are required to copy the process but not the product of the movement, and operate under socially constrained rules that depend critically on the technique and style of their particular school (e.g., Martha Graham vs. Merce Cunningham styles).

59. Feenders et al. 2008.

60. Clarke and Crisp 1983.

61. Clarke and Crisp 1983, Dudley 1977.
62. Clarke and Crisp 1983.
63. Laubin and Laubin 1977.
64. Clarke and Crisp 1983.
65. Correction may also occur through manual shaping of the dancer's body by the teacher or, to a lesser degree, through verbal instruction. In some dances, specific steps are given verbal labels, as in ballet in particular, which has its own elaborate glossary of terms, such as *fondu, arabesque, chassé,* and *grand jeté,* each with its own characteristic movements. Except in those cases, however, describing bodily movements with words is typically difficult. Hence, when dance instruction is given verbally, it is often through the use of imagery, where again an ability to relate one's own bodily movements to another object, emotion, or entity is required.
66. I am indebted to Nicky Clayton for drawing my attention to many of these points.
67. Whalen et al. 2015.
68. Lewontin 1970.
69. Plotkin 1994.
70. Kirschner and Gerhart 2005.
71. Darwin 1871, Pagel et al. 2007, Gray and Atkinson 2003, Gray and Jordan 2000, Kirby et al. 2008.
72. Clarke and Crisp 1983.
73. Ibid., Dudley 1977.
74. Clarke and Crisp 1983.
75. Lawson 1964.
76. Clarke and Crisp 1983.
77. Ibid.
78. Ibid.
79. Lawson 1964, Clarke and Crisp 1983.
80. Clarke and Crisp 1983.
81. Lawson 1964, Clarke and Crisp 1983.
82. Clarke and Crisp 1983.
83. Ibid.
84. Lawson 1964, Clarke and Crisp 1983.
85. Clarke and Crisp 1983.
86. Homer. *The Odyssey.* Book I, p. 29.
87. Tarr et al. 2014.
88. Ibid.
89. Tomasello et al. 2005.
90. Tarr et al. 2014.
91. Ibid.
92. Heyes 2012, Chartrand and Van Baaren 2009, Van Baaren et al. 2009.

93. Heyes 2012.
94. Oxford University psychologist and expert on imitation, Celia Heyes, has suggested that certain social dance forms may have spread through a cultural group selection mechanism precisely because they generated these positive, prosocial sentiments (Heyes 2012).
95. Fossil evidence is supplemented by other data sources, most obviously molecular data.
96. The *mambo* also has an interesting and unexpected history, with both African and European roots. In Cuba, it designates a sacred song of the *Congos*, a group of Cubans of Bantu origin. *Mambo* is the Bantu name for a musical instrument that is used in religious rituals. However, the *mambo* can also be traced back to the unlikely source of English country dance, which in the seventeenth century became the *contredanse* at the French court, and later the *contradanza* in Spain. A century later, *contradanza* reached Cuba, where it became known as *danza*. *Danza* was subsequently modified by the arrival of the planters and their slaves who fled from Haiti after it became independent; they added a syncopation called *cinquillo*. By the nineteenth century, freer and more spontaneous dancing by couples had replaced the formality of the *contredanse*, with the new music known as *danzon*; this, in turn, gave rise to *mambo* in the 1930s. See Clarke and Crisp (1983) for further details.
97. See http://www.pbt.org/community-engagement/brief-history-ballet.
98. Clarke and Crisp 1983.
99. Ibid., Dudley 1977.
100. Clarke and Crisp 1983.
101. Ibid.
102. Dudley 1977.
103. McDonagh 1976.
104. Fitch 2016.
105. Steele 2013.
106. Clarke and Crisp 1983.
107. Steele 2013.
108. Kant 2007.
109. Steele 2013.
110. Clarke and Crisp 1983.
111. The role of dance and adornment in premodern societies remains significant even in modern societies, both because so many choreographers have been inspired by visions of "ancient" or "primitive" cultures and ritual dance, and because some of the most famous ballets draw on myths that go back thousands of years; these have not only influenced ballet narratives but also their costumes (Steele 2013).
112. Steele 2013.
113. Ibid.
114. Clarke and Crisp 1983, Steele 2013.
115. Ibid.

REFERENCES

Abbott, A. 2015. Clever fish. *Nature* 521:412–414.

Acuna-Alonzo, V., T. Flores-Dorantes, J. K. Kruit, T. Villarreal-Molina, O. Arellano-Campos, T. Hunemeier, A. Moreno-Estrada, et al. 2010. A functional ABCA1 gene variant is associated with low HDL-cholesterol levels and shows evidence of positive selection in Native Americans. *Human Molecular Genetics* 19:2877–2885.

Agbai, O. 1986. Anti-sickling effect of dietary thiocyanate in prophylactic control of sickle cell anemia. *Journal of the National Medical Association* 78:1053–1056.

Aidoo, M., D. J. Terlouw, M. S. Kolczak, P. D. McElroy, F. O. ter Kuile, S. Kariuki, B. L. Nahlen, et al. 2002. Protective effects of the sickle cell gene against malaria morbidity and mortality. *Lancet* 359:1311–1312.

Alexander, R. D. 1987. *The Biology of Moral Systems*. New York, NY: Aldine de Gruyter.

Allen, J., M. Weinrich, W. Hoppitt, and L. Rendell. 2013. Network-based diffusion analysis reveals cultural transmission of lobtail feeding in humpback whales. *Science* 340:485–488.

Ambrose, S. H. 2001. Paleolithic technology and human evolution. *Science* 291:1748–1753.

Anderson, C. A., and B. J. Bushman. 2001. Effects of violent video games on aggressive behavior, aggressive cognition, aggressive affect, physiological arousal, and pro-social behavior: a meta-analytic review of the scientific literature. *Psychological Science* 12:353–359.

Annett, M. 1985. *Left, Right, Hand and Brain: The Right Shift Theory*. London, UK: Erlbaum.

Anthony, D. W. 2007. *The Horse, the Wheel, and Language: How Bronze-Age Riders from the Eurasian Steppes Shaped the Modern World*. Princeton, NJ: Princeton University Press.

Anthony, D. W., and D. Brown. 2000. Neolithic horse exploitation in the Eurasian steppes: diet, ritual and riding. *Antiquity* 74:75–86.

Anton, S. C. 2014. Evolution of early *Homo*: an integrated biological perspective. *Science* 345:6192, doi:10.1126/science.1236828.

Apesteguia, J., S. Huck, and J. Oechssler. 2007. Imitation-theory and experimental evidence. *Journal of Economic Theory* 136:217–235.

Aplin, L. M., D. R. Farine, J. Morand-Ferron, A. Cockburn, A. Thornton, and B. C. Sheldon. 2014. Experimentally induced innovations lead to persistent culture via conformity in wild birds. *Nature* 518:538–541.

Aplin, L. M., D. R. Farine, J. Morand-Ferron, A. Cockburn, A. Thornton, and B. C. Sheldon. 2015. Counting conformity: evaluating the units of information in frequency-dependent social learning. *Animal Behaviour* 110:E5–E8.

Aplin, L. M., D. R. Farine, J. Morand-Ferron, and B. C. Sheldon. 2012. Social networks predict patch discovery in a wild population of songbirds. *Proceedings of the Royal Society of London B* 279:4199–4205.

Arbib, M. A. 2012. *How the Brain Got Language: The Mirror System Hypothesis*. Oxford, UK: Oxford University Press.

Ardrey, R. 1966. *The Territorial Imperative*. London, UK: Collins.

Armelagos, G. J. 2014. Brain evolution, the determinates of food choice, and the omnivore's dilemma. *Critical Reviews in Food Science and Nutrition* 54:1330–1341.

Armelagos, G. J., A. H. Goodman, and K. H. Jacobs. 1991. The origins of agriculture: population growth during a period of declining health. *Population and Environment* 131:9–22.

Armelagos, G. J., and K. S. Harper. 2005. Genomics at the origins of agriculture, part two. *Evolutionary Anthropology* 14:109–121.

Arnold, C., L. J. Matthews, and C. L. Nunn. 2010. The 10k Trees website: A new online resource for primate phylogeny. *Evolutionary Anthropology* 19:114–118.

Asch, S. E. 1955. Opinions and social pressure. *Scientific American* 193(5):31–35.

Atton, N. 2013. Investigations into Stickleback Social Learning. PhD diss., University of St Andrews.

Atton, N., W. Hoppitt, M. M. Webster, B. G. Galef, and K. N. Laland. 2012. Information flow through threespine stickleback networks without social transmission. *Proceedings of the Royal Society of London B* 279:4272–4278.

Auer, P., N. Cesa-Bianchi, and P. Fischer. 2002. Finite-time analysis of the multi-armed bandit problem. *Machine Learning* 47:235–256.

Aunger, R. 1992. The nutritional consequences of rejecting food in the Ituri Forest of Zaire. *Human Ecology* 30:1–29.

———. 1994a. Are food avoidances maladaptive in the Ituri Forest of Zaire? *Journal of Anthropological Research* 50:277–310.

———. 1994b. Sources of variation in ethnographic interview data: food avoidances in the Ituri Forest, Zaire. *Ethnology* 33:65–99.

Avital, E., and E. Jablonka. 2000. *Animal Traditions*. Cambridge, UK: Cambridge University Press.

Axelrod, R. 1984. *The Evolution of Cooperation*. New York, NY: Basic Books.

Badr, A. M., K. Sch, R. Rabey, H. E. Effgen, S. Ibrahim, H. H. Pozzi, C. Rohde, and W. F. Salamini. 2000. On the origin and domestication history of Barley (*Hordeum vulgare*). *Molecular Biology and Evolution* 17(4):499–510.

Baird, R. W. 2000. The killer whale: foraging specializations and group hunting. In: *Cetacean Societies: Field Studies of Dolphins and Whales*, ed. J. Mann, R. C. Connor, P. L. Tyack, and H. Whitehead. Chicago, IL: University of Chicago Press.

Baird, R. W., and H. Whitehead. 2000. Social organization of mammal-eating killer whales: group stability and dispersal patterns. *Canadian Journal of Zoology* 78:2096–2105.

Baldwin, J. M. 1896. A new factor in evolution. *American Naturalist* 30:441–451.

Baldwin, J. M. 1902. *Development and Evolution*. New York, NY: Macmillan.

Bandura, A., and F. L. Menlove. 1968. Factors determining vicarious extinction of avoidance behavior through symbolic modeling. *Journal of Personality and Social Psychology* 8:99–108.

Bandura, A., D. Ross, and S. A. Ross. 1961. Transmission of aggression through the imitation of aggressive models. *Journal of Abnormal Social Psychology* 63:575–582.

Banerjee, K., C. F. Chabris, V. E. Johnson, J. J. Lee, F. Tsao, and M. D. Hauser. 2009. General intelligence in another primate: individual differences across cognitive task performance in a new world monkey (*Saguinus oedipus*). *PLOS ONE* 4:e5883, doi:org/10.1371/journal.pone.0005883.

Barlow, K. R. 2002. Predicting maize agriculture among the Fremont: an economic comparison of farming and foraging in the American southwest. *American Antiquity* 67:65–88.

Barnard, C. J., and R. M. Sibly. 1981. Producers and scroungers: a general model and its application to captive flocks of house sparrows. *Animal Behaviour* 29:543–550.

Barnes, E. 2005. Diseases and human evolution. Albuquerque, NM: University of New Mexico Press.

Barnes, I., A. Duda, O. G. Pybus, and M. G. Thomas. 2011. Ancient urbanization predicts genetic resistance to tuberculosis. *Evolution* 65:842–848.

Barnett, S. A. 1975. The Rat: A Study in Behavior. Chicago, IL: University of Chicago Press.

Barrett, D. B., G. T. Kurian, and T. M. Johnston. 2001. *World Christian Encyclopedia*. Oxford, UK: Oxford University Press.

Barton, R. A. 1999. The evolutionary ecology of the primate brain. In: *Comparative Primate Socioecology*, ed. P. C. Lee. Cambridge, UK: Cambridge University Press, pp. 167–194.

———. 2006. Primate brain evolution: integrating comparative, neurophysiological, and ethological data. *Evolutionary Anthropology* 15:224–236.

———. 2012. Embodied cognitive evolution and the cerebellum. *Philosophical Transactions of the Royal Society of London B* 367:2097–2107.

Barton, R. A., and I. Capellini. 2011. Maternal investment, life histories, and the costs of brain growth in mammals. *Proceedings of the National Academy of Sciences USA* 108:6169–6174.

Barton, R. A., and P. H. Harvey. 2000. Mosaic evolution of brain structure in mammals. *Nature* 405:1055–1058.

Barton, R. A., and C. Venditti. 2013. Human frontal lobes are not relatively large. *Proceedings of the National Academy of Sciences USA* 110:9001–9006.

Bar-Yosef, O., and T. D. Price, eds. 2011. The origins of agriculture: new data, new ideas. *Current Anthropology* 52, special supplement 4.

Basalla, G. 1988. *The Evolution of Technology*. Cambridge, UK: Cambridge University Press.

Basgall, M. E. 1987. Resource intensification among huntergatherers: acorn economies in prehistoric California. *Research in Economic Anthropology* 9:21–52.

Bateman, A. J. 1948. Intra-sexual selection in Drosophila. *Heredity* 2:349–368.

Bateson, P., and P. Gluckman. 2011. *Plasticity, Robustness, Development and Evolution.* Cambridge, MA: Cambridge University Press.

Bateson, P., and P. Martin. 2000. *Design for a Life: How Behavior and Personality Develop.* New York, NY: Simon & Schuster.

———. 2013. *Play, Playfulness, Creativity and Innovation.* Cambridge, UK: Cambridge University Press.

Beinhocker, E. 2006. *The Origin of Wealth: Evolution Complexity and the Radical Remaking of Economics.* Boston, MA: Harvard Business School Press.

Beja-Pereira, A., G. Luikart, P. R. England, D. G. Bradley, O. C. Jann, G. Bertorelle, A. T. Chamberlain, et al. 2003. Gene–culture co-evolution between cattle milk protein genes and human lactase genes. *Nature Genetics* 35:311–313.

Bell, A. V., P. J. Richerson, and R. McElreath. 2009. Culture rather than genes provides greater scope for the evolution of large-scale human prosociality. *Proceedings of the National Academy of Sciences USA* 106:17671–17674.

Bellamy, R., C. Ruwende, T. Corrah, K. McAdam, H. Whittle, and A. Hill. 1998. Variations in the *Nramp1* gene and susceptibility to tuberculosis in West Africans. *New England Journal of Medicine* 338:640–644.

Bergemann, D., and J. Valimaki. 1996. Learning and strategic pricing. *Econometrica* 64:1125–1149.

Berger, S. M. 1962. Conditioning through vicarious instigation. *Psychological Review* 69:450–466.

Bergman, A., and M. W. Feldman. 1995. On the evolution of learning: representation of a stochastic environment. *Theoretical Population Biology* 48:251–276.

Bersaglieri, T., P. C. Sabeti, N. Patterson, T. Vanderploeg, S. F. Schaffner, J. A. Drake, M. Rhodes, et al. 2004. Genetic signatures of strong recent positive selection at the lactase gene. *American Journal of Human Genetics* 74:1111–1120.

Bettinger, R. L., R. Garvey, and S. Tushingham. 2015. *Hunter-Gatherers: Archaeological and Evolutionary Theory.* New York, NY: Springer.

Beyene, Y., S. Katoh, G. WoldeGabriel, W. K. Hart, K. Uto, M. Sudo, M Kondo, et al. 2013. The characteristics and chronology of the earliest Acheulean at Konso, Ethiopia. *Proceedings of the National Academy of Sciences USA* 110:1584–1591.

Bickerton, A. 2009. *Adam's Tongue.* New York, NY: Hill and Wang.

Biesmeijer, J. C., and T. D. Seeley. 2005. The use of the waggle dance information by honey bees throughout their foraging careers. *Behavioral Ecology and Sociobiology* 59:133–142.

Bijker, W. 1995. *Of Bicycles, Bakelites, and Bulbs: Toward a Theory of Sociotechnical Change.* Cambridge, MA: MIT Press.

Bikhchandani, S., D. Hirshleifer, and I. Welch. 1992. A theory of fads, fashion, custom, and cultural change as informational cascades. *Journal of Political Economy* 100:992–1026.

———. 1998. Learning from the behavior of others: conformity, fads, and informational cascades. *Journal of Economic Perspectives* 12:151–170.

Bird, R. B., N. Taylor, and F. Codding. 2013. Niche construction and dreaming logic: aboriginal patch mosaic burning and varanid lizards (*Varanus gouldii*) in Australia. *Proceedings of the Royal Society of London B*, doi:10.1098/rspb.2013.2297.

Birney, E. 2012. An integrated encyclopedia of DNA elements in the human genome. The ENCODE Project consortium. *Nature* 489:57–74.

Biro, D., N. Inoue-Nakamura, R. Tonooka, G. Yamakoshi, C. Sousa, and T. Matsuzawa. 2003. Cultural innovation and transmission of tool use in wild chimpanzees: evidence from field experiments. *Animal Cognition* 6:213–223.

Biro, D., C. Sousa, and T. Matsuzawa. 2006. Ontogeny and cultural propagation of tool use by wild chimpanzees at Bossou, Guinea: case studies in nut cracking and leaf folding. In: *Cognitive Development in Chimpanzees*, ed. T. Matsuzawa, M. Tomonaga, and M. Tanaka. Tokyo, Japan: Springer, pp. 476–508.

Bishop, D.V.M. 1990. *Handedness and Developmental Disorder*. Hove, UK: Erlbaum.

Blazek, V., J. Bruzek, and M. F. Casanova. 2011. Plausible mechanisms for brain structural and size changes in human evolution. *Collegium Antropologicum* 35:949–955.

Bloom, P. 1997. Intentionality and word learning. *Trends in Cognitive Sciences* 1: 9–12.

———. 2000. *How Children Learn the Meaning of Words*. Cambridge, MA: MIT Press.

Blumenschine, R. J. 1986. *Early Hominid Scavenging Opportunities: Implications of Carcass Availability in the Serengeti and Ngorongoro Ecosystems*. Oxford, UK: Archaeopress.

Blumler, M. A. 1992. Seed Weight and Environment in Mediterranean-type Grasslands in California and Israel. PhD diss., University of California, Berkeley.

Blunden, G. 2003. *Charco Harbour*. Sydney, Australia: Sydney University Press.

Blurton Jones, N. G. 1986. Bushman birth spacing: a test for optimal inter-birth intervals. *Ethology and Sociobiology* 7:91–105.

———. 1987. Bushman birth spacing: direct tests of some simple predictions. *Ethology and Sociobiology* 8:183–204.

Boeda, E., J. Connan, D. Dessort, S. Muhesen, N. Mercier, H. Valladas, and N. Tisnerat. 1996. Bitumen as a hafting material on Middle Paleolithic artefacts. *Nature* 380:336–338.

Boesch, C. 2003. Is culture a golden barrier between human and chimpanzee? *Evolutionary Anthropology* 12:26–32.

Boesch, C., and M. Tomasello. 1998. Chimpanzee and human cultures. *Current Anthropology* 39:591–604.

Boinski, S. 1988. Sex differences in the foraging behavior of squirrel monkeys in a seasonal habitat. *Behavioral Ecology and Sociobiology* 23:177–186.

Boivin, N. L., M. A. Zeder, D. Q. Fuller, A. Crowther, G. Larson, J. M. Erlandson, T. Denham, and M. D. Petraglia. 2016. Ecological consequences of human niche construction: examining long-term anthropogenic shaping of global species distributions. *Proceedings of the National Academy of Sciences USA* 113(23):6388–6396.

Bolhuis, J. J., G. R. Brown, R. C. Richardson, and K. N. Laland. 2011. Darwin in mind: new opportunities for evolutionary psychology. *PLOS Biology* 9(7):e1001109, doi:org/1371/journal.pbio.1001109.

Bolhuis, J. J., K. Okanoya, and C. Scharff. 2010. Twitter evolution: converging mechanisms in birdsong and human speech. *Nature Reviews Neuroscience* 11:747–759.

Bolhuis, J. J., I. Tattersall, N. Chomsky, and R. C. Berwick. 2014. How could language have evolved? *PLOS Biology* 12(8):e1001934, doi:10.1371/journal.pbio.1001934.

Bonner, J. T. 1980. *The Evolution of Culture in Animals.* Princeton, NJ: Princeton University Press.

Boogert, N. J., S. M. Reader, W. Hoppitt, and K. N. Laland. 2008. The origin and spread of innovations in starlings. *Animal Behaviour* 75:1509–1518.

Boulle, P. (1963) 2011. *Planet of the Apes.* New York, NY: Vintage Books.

Box, H. O., and K. R. Gibson, eds. 1999. *Mammalian Social Learning: Comparative and Ecological Perspectives.* Cambridge, UK: Cambridge University Press.

Boyd, R., and P. J. Richerson. 1985. *Culture and the Evolutionary Process.* Chicago, IL: University of Chicago Press.

———. 1988. The evolution of reciprocity in sizable groups. *Journal of Theoretical Biology* 132:337–356.

———. 1992. Punishment allows the evolution of cooperation or anything else in sizable groups. *Ethology and Sociobiology* 133:171–195.

———. 1995. Why does culture increase human adaptability? *Ethology and Sociobiology* 16:125–143.

———. 1996. Why culture is common, but cultural evolution is rare. *Proceedings of the British Academy of Science* 88:77–93.

———. 2005. *The Origin and Evolution of Cultures.* Oxford, UK: Oxford University Press.

Boyd, R., and J. Silk. 2015. *How Humans Evolved,* 7th ed. New York, NY: Norton.

Boyd, R., H. Gintis, and S. Bowles. 2010. Coordinated punishment of defectors sustains cooperation and can proliferate when rare. *Science* 328:617–620.

Boyd, R., H. Gintis, S. Bowles, and P. J. Richerson. 2003. The evolution of altruistic punishment. *Proceedings of the National Academy of Sciences USA* 100:3531–3535.

Boyd, R., P. J. Richerson, and J. Henrich. 2011. The cultural niche: why social learning is essential for human adaptation. *Proceedings of the National Academy of Sciences USA* 108:10918–10925.

Bradshaw, J. L. 1991. Animal asymmetry and human heredity: dextrality, tool use and language in evolution—10 years after Walker (1980). *British Journal of Psychology* 82:39–59.

Brass, M., and C. Heyes. 2005. Imitation: Is cognitive neuroscience solving the correspondence problem? *Trends in Cognitive Sciences* 9:489–495.

Braun, D. R., T. Plummer, P. W. Ditchfield, L. C. Bishop, and J. V. Ferraro. 2009. Oldowan technology and raw material variability at Kanjera South. In: *Interdisciplinary Approaches to Oldowan,* ed. E. Hovers and D. R. Braun. New York, NY: Springer, pp. 99–110.

Brighton, H., S. Kirby, and K. Smith. 2005. Cultural selection for learnability: three principles underlying the view that language adapts to be learnable. In: *Language Origins: Perspectives on Evolution*, ed. M. Tallerman. Oxford, UK: Oxford University Press, pp. 291–309.

Bronowski, J. 1973. *The Ascent of Man*. London, UK: BBC Books.

Brosnan, S. F., B. J. Wilson, and M. J. Beran. 2012. Old world monkeys are more like humans than New World monkeys when playing a coordination game. *Proceedings of the Royal Society of London B*, 279:1522–1530.

Brown, C., and R. L. Day. 2002. The future of stock enhancements: lessons for hatchery practice from conservation biology. *Fish and Fisheries* 3:79–94.

Brown, C., and K. N. Laland. 2001. Social learning and life skills training for hatchery reared fish. *Journal of Fish Biology* 59:471–493.

———. 2002. Social learning of a novel avoidance task in the guppy: conformity and social release. *Animal Behaviour* 64:41–47.

———. 2003. Social learning in fishes: a review. *Fish and Fisheries* 4:280–288.

———. 2006. Social learning in fishes. In: *Fish Cognition and Behaviour*, ed. C. Brown, K. N. Laland, and J. Krause. Oxford, UK: Blackwell, pp. 186–202.

Brown, C., T. Davidson, and K. N. Laland. 2003. Environmental enrichment and prior experience improve foraging behaviour in hatchery-reared Atlantic salmon. *Journal of Fish Biology* 63:187–196.

Brown, G. R., K. N. Laland, and M. Borgerhoff Mulder. 2009. Bateman's principles and human sex roles. *Trends in Ecology & Evolution* 24:297–304.

Brown, G. R., T. Dickins, R. Sear, and K. N. Laland. 2011. Evolutionary accounts of human behavioural diversity. *Philosophical Transactions of the Royal Society of London B* 366:313–324.

Brown, C., A. Markula, and K. N. Laland. 2003. Social learning of prey location in hatchery-reared Atlantic salmon. *Journal of Fish Biology* 63:738–745.

Brown, M. F. 1985. Rooks feeding on human vomit. *British Birds* 78:513.

Brown, R. E. 1985. The rodents II: suborder Myomorpha. In: *Social Odours in Mammals*, ed. R. E. Brown and D. W. Macdonald. Oxford, UK: Clarendon Press, pp. 345–457.

Brown, S., M. J. Martinez, and L. M. Parsons. 2006. The neural basis of human dance. *Cerebral Cortex* 16:1157–1167.

Bshary, R. 2011. Machiavellian intelligence in fishes. In: *Fish Cognition and Behaviour*, ed. C. Brown, K. N. Laland, and J. Krause. Oxford, UK: Blackwell, pp. 277–297.

Bshary, R., W. Wickler, and H. Fricke. 2002. Fish cognition: a primate's eye view. *Animal Cognition* 5:1–13.

Bugnyar, T., and K. Kortschal. 2002. Observational learning and the raiding of food caches in ravens (*Corvus corax*): Is it "tactical deception"? *Animal Behaviour* 64:185–195.

Burger, J., M. Kirchner, B. Bramanti, W. Haak, and M. G. Thomas. 2007. Absence of the lactase-persistence associated allele in early Neolithic Europeans. *Proceedings of the National Academy of Sciences USA* 104:3736–3741.

Burling, R. 1993. Primate calls, human language, and nonverbal communication. *Current Anthropology* 34:25–53.

Bushnell, G.H.S. 1957. *Peru*. London, UK: Thames and Hudson.

Buss, D. M. 1999. *Evolutionary Psychology: The New Science of the Mind*. London, UK: Allyn & Bacon.

Byrne, R. W. 1992. The evolution of intelligence. In: *Behaviour and Evolution*, ed. P.J.B. Slater and T. R. Halliday. Cambridge, UK: Cambridge University Press, pp. 223–265.

———. 1994. The evolution of intelligence. In: *Behaviour and Evolution*, ed. P.J.B. Slater and T. R. Halliday. Cambridge, UK: Cambridge University Press, pp. 223–265.

———. 1995. *The Thinking Ape*. Oxford, UK: Oxford University Press.

———. 1997. The Technical Intelligence hypothesis: An additional evolutionary stimulus to intelligence? In: *Machiavellian Intelligence II: Extensions and Evaluations*, ed. A. Whiten and R. W. Byrne. Cambridge, UK: Cambridge University Press, pp. 289–311.

———. 1999. Imitation without intentionality. Using string parsing to copy the organization of behaviour. *Animal Cognition* 2:63–72.

———. 2003. Novelty in deceit. In: *Animal Innovation*, ed. S. M. Reader and K. N. Laland. Oxford, UK: Oxford University Press.

———. 2016. *Evolving Insight*. Oxford, UK: Oxford University Press.

Byrne, R. W., and N. Corp. 2004. Neocortex size predicts deception rate in primates. *Proceedings of the Royal Society of London B* 271:1693–1699.

Byrne, R. W., and A. E. Russon. 1998. Learning by imitation: a hierarchical approach. *Behavioral and Brain Sciences* 21:667–684.

Byrne, R. W., and A. Whiten. 1988. *Machiavellian Intelligence: Social Expertise and the Evolution of Intellect in Monkeys, Apes and Humans*. Oxford, UK: Oxford University Press.

———. 1990. Tactical deception in primates: the 1990 database. *Primate Report* 27 (entire vol.):1–101.

Byrne, R. W., C. Holbaiter, and M. Klailova. 2011. Local traditions in gorilla manual skill: evidence for observational learning of behavioral organization. *Animal Cognition* 14:683–693.

Caceres, M., J. Lachuer, M. A. Zapala, J. C. Redmond, L. Kudo, D. H. Geschwind, D. J. Lockhart, et al. 2003. Elevated gene expression levels distinguish human from non-human primate brains. *Proceedings of the National Academy of Sciences USA* 100:13030–13035.

Cadieu, N., S. Fruchard, and J.-C. Cadieu. 2010. Innovative individuals are not always the best demonstrators: feeding innovation and social transmission in *Serinus canaria*. *PLOS ONE* 5:e8841, doi:org/10.1371/journal.pone.0008841.

Calarco, J. A., Y. Xing, M. Caceres, J. P. Calarco, X. Xiao, Q. Pan, C. Lee, et al. 2007. Global analysis of alternative splicing differences between humans and chimpanzees. *Genes & Development* 21:2963–2975.

Caldwell, C., and A. Millen. 2008. Experimental models for testing hypotheses about cumulative cultural evolution. *Evolution and Human Behavior* 29:165–171.

Caldwell, M. C., and D. K. Caldwell. 1972. Behavior of marine mammals. In: *Mammals of the Sea*, ed. H. Ridgway. Springfield, IL: C. C. Thomas, pp. 419–465.

Call, J., and M. Tomasello. 1998. Distinguishing intentional from accidental actions in orangutans (*Pongo pygmaeus*), chimpanzees (*Pan troglodytes*), and human children (*Homo sapiens*). *Journal of Comparative Psychology* 112:192–206.

———. 2008. Does the chimpanzee have a theory of mind? 30 years later. *Trends in Cognitive Sciences* 12:187–192.

Call, J., B. Hare, M. Carpenter, and M. Tomasello. 2004. Unwilling or unable? Chimpanzees' understanding of intentional actions. *Developmental Science* 7: 488–498.

Callahan, E. 1979. *The Basics of Biface Knapping in the Eastern Fluted Point Tradition: A Manual for Flintknappers and Lithic Analysts*. Bethlehem, CT: Eastern States Archaeological Federation.

Cambefort, J. P. 1981. A comparative study of culturally transmitted patterns of feeding habits in the chacma baboon (*Papio ursinus*) and the vervet monkey (*Cercopithecus aethiops*). *Folia Primatologica* 36:243–263.

Cann, R. L., M. Stoneking, and A. C. Wilson. 1987. Mitochondrial DNA and human evolution. *Nature* 325:31–36.

Carey, S. 2009. *The Origin of Concepts*. New York, NY: Oxford University Press.

Caro, T. M. 1980a. Predatory behaviour in domestic cat mothers. *Behaviour* 74:128–147.

———. 1980b. Effects of the mother, object play and adult experience on predation in cats. *Behavioral and Neural Biology* 29:29–51.

———. 1995. Short-term costs and correlates of play in cheetahs. *Animal Behaviour* 49:333–345.

Caro, T. M., and M. D. Hauser. 1992. Is there teaching in nonhuman animals? *Quarterly Review of Biology* 67:151–174.

Carpenter, M. 2006. Instrumental, social and shared goals and intentions in imitation. In: *Imitation and the Social Mind: Autism and Typical Development*, ed. S. J. Rogers and J.H.G. Williams. New York, NY: Guilford, pp. 48–70.

Carpenter, M., J. Uebel, and M. Tomasello. 2013. Being mimicked increases prosocial behaviour in 18-month-old infants. *Child Development* 84:1511–1518.

Carr, K., R. L. Kendal, and E. G. Flynn. 2015. Imitate or innovate? Children's innovation is influenced by the efficacy of observed behaviour. *Cognition* 142:322–332.

Carroll, S. B. 2005. *Endless Forms Most Beautiful: The New Science of Evo Devo*. New York, NY: W. W. Norton.

Carruthers, P. 2006. *The Architecture of the Mind: Massive Modularity and the Flexibility of Thought*. Oxford, UK: Oxford University Press.

Casanova, C., R. Mondragon-Ceballos, and P. C. Lee. 2008. Innovative social behavior in chimpanzees (*Pan troglodytes*). *American Journal of Primatology* 70:54–61.

Casar, C., K. Zuberbuehler, R. J. Young, and R. W. Byrne. 2013. Titi monkey call sequences vary with predator location and type. *Biology Letters* 9, 20130535, doi: 10.1098/rsbl.2013.0535.

Caselli, G., J. Vallen, and G. Wunsch. 2005. *Demography: Analysis & Synthesis*. Cambridge, MA: Academic Press.

Castro, L., and M. A. Toro. 2004. The evolution of culture: from primate social learning to human culture. *Proceedings of the National Academy of Sciences USA* 101:10235–10240.

Catchpole, C. K., and P.J.B. Slater. 1995. *Bird Song: Biological Themes and Variations*. Cambridge, UK: Cambridge University Press.

———. 2008. *Bird Song: Biological Themes and Variations*, 2nd ed. Cambridge, UK: Cambridge University Press.

Cavalli-Sforza, L. L., and M. W. Feldman. 1973. Models for cultural inheritance I: group mean and within-group variation. *Theoretical Population Biology* 4:42–55.

———. 1981. *Cultural Transmission and Evolution: A Quantitative Approach*. Princeton, NJ: Princeton University Press.

Chang, L., Q. Fang, S. Zang, M. Poo, and N. Gong. 2015. Mirror-induced self-directed behaviors in rhesus monkeys after visual-somatosensory training. *Current Biology* 25(2):212–217.

Chartrand, T. L., and J. Bargh. 1999. The chameleon effect: the perception-behavior link and social interaction. *Journal of Personality and Social Psychology* 766:893–910.

Chartrand, T. L., and R. van Baaren. 2009. Human mimicry. *Advances in Experimental Social Psychology*. 4108:219–274.

Cheney, D. L., and R. M. Seyfarth. 1988. Social and non-social knowledge in vervet monkeys. In: *Machiavellian Intelligence: Social Expertise and the Evolution of Intellect in Monkeys, Apes and Humans*, ed. R. W. Byrne and A. Whiten. Oxford, UK: Oxford University Press, pp. 255–270.

Childe, V. G. 1936. *Man Makes Himself*. London, UK: Watts.

Chittka, L., and J. Niven. 2009. Are bigger brains better? *Current Biology* 19:R995–R1008.

Chomsky, N. 1965. *Aspects of the Theory of Syntax*. Cambridge, MA: MIT Press.

———. 1968. *Language and Mind*. Cambridge, UK: Cambridge University Press.

———. 1980. *Rules and Representations*. New York, NY: Columbia University Press.

Chudek, M., and J. Henrich. 2011. Culture-gene coevolution, norm psychology and the emergence of human prosociality. *Trends in Cognitive Sciences* 155:218–226.

Claidiere, N., E.J.E. Messer, W. Hoppitt, and A. Whiten. 2013. Diffusion dynamics of socially learned foraging techniques in squirrel monkeys. *Current Biology* 23:1251–1255.

Clarke, M., and C. Crisp. 1983. *The History of Dance*, 5th ed. New York, NY: Crown.

Clayton, N. S., and A. Dickinson. 1998. Episodic-like memory during cache recovery by scrub jays. *Nature* 395:272–274.

———. 2010. Mental time travel: Can animals recall the past and plan for the future? In: *The New Encyclopedia of Animal Behaviour*, Vol. 2, ed. M. D. Breed ad J. Moore. Oxford, UK: Academic Press, pp. 438–442.

Cloutier, S., R. C. Newberry, K. Honda, and J. R. Alldredge. 2002. Cannibalistic behaviour spread by social learning. *Animal Behaviour* 63:1153–1162.

Clutton-Brock, T. H., and P. H. Harvey. 1980. Primates, brain and ecology. *Journal of Zoology*, London 190:309–323.

Clutton-Brock, T. H., D. Gaynor, G. M. MacIlrath, A.D.C. MacColl, R. Kansky, P. Chadwick, M. Manser, et al. 1999. Predation, group size and mortality in a cooperative mongoose (*Suricata suricatta*). *Journal of Animal Ecology* 68:672–683.

Cochran, G., and H. Harpending. 2009. *The 10,000 Year Explosion: How Civilization Accelerated Human Evolution*. New York, NY: Basic Books.

Cohen, M. N. 1989. *Health and the Rise of Civilizations*. New Haven, CT: Yale University Press.

Cohen, M. N., and G. Armelagos. 1984. *Paleopathology at the Origins of Agriculture*. Orlando, FL: Academic.

Cohen, M. N., and G.M.M. Crane-Kramer. 2007. *Ancient Health: Skeletal Indicators of Agricultural and Economic Intensification*. Gainesville, FL: University of Florida Press.

Collard, M., B. Buchanan, A. Ruttle, and M. J. O'Brien. 2012. Niche construction and the toolkits of hunter–gatherers and food producers. *Biological Theory* 6:251–259.

Colombelli-Négrel, D., M. E. Hauber, J. Robertson, F. J. Sulloway, H. Hoi, M. Griggio, and S. Kleindorfer. 2012. Embryonic learning of vocal passwords in superb fairy-wrens reveals intruder cuckoo nestlings. *Current Biology* 22:2155–2160.

Cook, P., A. Rouse, M. Wilson, and C. Reichmuth. 2013. A California sea lion (*Zalophus californianus*) can keep the beat: motor entrainment to rhythmic auditory stimuli in a non-vocal mimic. *Journal of Comparative Psychology* 127(4):412–427.

Coolen, I., R. L. Day, and K. N. Laland. 2003. Species difference in adaptive use of public information in sticklebacks. *Proceedings of the Royal Society of London B* 270:2413–2419.

Coolen, I., A. J. Ward, P.J.B. Hart, and K. N. Laland. 2005. Foraging nine-spined sticklebacks prefer to rely on public information over simpler social cues. *Behavioral Ecology* 16:865–870.

Cooper, R. W. 1999. *Coordination Games: Complementarities and Macroeconomics*. Cambridge, UK, and New York, NY: Cambridge University Press.

Corballis, M. C. 1991. *The Lopsided Ape: Evolution of the Generative Mind*. Oxford, UK: Oxford University Press.

———. 1993. *The Lopsided Ape: Evolution of the Generative Mind*. Oxford, UK: Oxford University Press.

———. 2013. Mental time travel: a case for evolutionary continuity. *Trends in Cognitive Sciences* 17:5–6, doi:10.1016/j.tics.2012.10.009.

Cordaux, R., R. Aunger, G. Bentley, I. Nasidze, S. M. Sirajuddin, and M. Stoneking. 2004. Independent origins of Indian caste and tribal paternal lineages. *Current Biology* 14:231–235.

Cornwallis, C., S. West, K. Davis, and A. Griffin. 2010. Promiscuity and the evolutionary transition to complex societies. *Nature* 466:969–972.

Cosmides, L., and J. Tooby. 1987. From evolution to behavior: evolutionary psychology as the missing link. In: *The Latest on the Best: Essays on Evolution and Optimality*, ed. J. Dupre. Cambridge, MA: MIT Press, pp. 277–306.

Coussi-Korbel, S., and D. Fragaszy. 1995. On the relation between social dynamics and social learning. *Animal Behaviour* 50:1441–1453.

Couzin, I. D. 2009. Collective cognition in animal groups. *Trends in Cognitive Sciences* 13(1):36–43.

Couzin, I. D., J. Krause, N. R. Franks, and S. A. Levin. 2005. Effective leadership and decision-making in animal groups on the move. *Nature* 433:513–516.

Cowie, A. 2014. Experimental Studies of Social Foraging in Budgerigars (*Melopsittacus undulates*). PhD diss., University of St Andrews.

Crittenden, A. N. 2011. The importance of honey consumption in human evolution. *Food and Foodways: Explorations in the History and Culture of Human Nourishment* 19:257–273.

Croft, D. P., R. James, A.J.W. Ward, M. S. Botham, D. Mawdsley, and J. Krause. 2005. Assortative interactions and social networks in fish. *Oecologia* 143:211–219.

Csibra, G. 2007. Teachers in the wild. *Trends in Cognitive Sciences* 11:95–96.

———. 2010. Recognizing communicative intentions in infancy. *Mind & Language* 25:141–168.

Csibra, G., and G. Gergely. 2006. Social learning and social cognition: the case for pedagogy. In: *Processes of Change in Brain and Cognitive Development*, ed. Y. Munakata and M. H. Johnson. Oxford, UK: Oxford University Press, pp. 249–274.

———. 2011. Natural pedagogy as evolutionary adaptation. *Philosophical Transactions of the Royal Society B* 366:1149–1157.

Curio, E. 1988. Cultural transmission of enemy recognition by birds. In: *Social Learning: Psychological and Biological Perspectives*, ed. B. G. Galef and T. R. Zentall. Hillsdale, NJ: Erlbaum, pp. 75–97.

Curio, E., U. Ernst, and W. Vieth. 1978. Cultural transmission of enemy recognition: one function of mobbing. *Science* 202:899–901.

Currie D. J., and J. T. Fritz. 1993. Global patterns of animal abundance and species energy use. *Oikos* 67:56–68.

Dalziell, A. H., R. A. Peters, A. Cockburn, A. D. Dorland, A. C. Maisey, and R. D. Magrath. 2013. Dance choreography is coordinated with song repertoire in a complex avian display. *Current Biology* 23:1132–1135.

Danchin, E., L.-A. Giraldeau, T. J. Valone, and R. H. Wagner. 2004. Public information: from nosy neighbours to cultural evolution. *Science* 305:487–491.

Darwin C. R. 1986. Letter from Charles Darwin to the Gardener's Chronicle, [Aug. 16, 1841]. In: *The Correspondence of Charles Darwin*, Vol. 2. Cambridge, UK: Cambridge University Press, pp. 1837–1843. Darwin Correspondence Project, "Letter no. 607," accessed on 12 July 2016, http://www.darwinproject.ac.uk/DCP-LETT-607.

———. (1859) 1968. *On the Origin of Species by Means of Natural Selection, or the Preservation of Favoured Races in the Struggle for Life*. London: John Murray. First edition reprint. London, UK: Penguin Books, London.

———. (1871) 1981. *The Descent of Man and Selection in Relation to Sex*. London: John Murray. First edition reprint. Princeton, NJ: Princeton University Press.

———. 1872. *The Expression of the Emotions in Man and Animals*. London: John Murray.

Davies, G., and J. H. Bank. 2002. *A History of Money: From Ancient Times to the Present Day*. Cardiff, UK: University of Wales Press.

Davies, N. B. 1991. *Dunnock Behaviour and Social Evolution*. Oxford, UK: Oxford University Press.

Dawkins, R. 1976. *The Selfish Gene*. Oxford, UK: Oxford University Press.

———. 1982. *The Extended Phenotype*. Oxford, UK: Oxford University Press.

Dawkins, M. 2012. *Why Animals Matter*. Oxford, UK: Oxford University Press.

Day, R., T. MacDonald, C. Brown, K. N. Laland, and S. M. Reader. 2001. Interactions between shoal size and conformity in guppy social foraging. *Animal Behaviour* 62:917–925.

Day, R. L. 2003. Innovation and Social Learning in Monkeys and Fish: Empirical Findings and Their Application to Reintroduction Techniques. PhD diss., University of Cambridge.

Day, R. L., R. L. Coe, J. R. Kendal, and K. N. Laland. 2003. Neophilia, innovation and social learning: a study of intergeneric differences in callitrichid monkeys. *Animal Behaviour* 65:559–571.

Deacon, T. W. 1990. Fallacies of progression in theories of brain-size evolution. *International Journal of Primatology* 11:193–236.

———. 1997. The *Symbolic Species: The Coevolution of Language and the Brain*. New York, NY: Norton.

———. 2003a. *The Symbolic Species*, 2nd ed. London, UK: Penguin Books.

———. 2003b. Multilevel selection in a complex adaptive system: the problem of language origins. In: *Evolution and Learning: The Baldwin Effect Reconsidered*, ed. B. Weber and D. Depew. Cambridge, MA: MIT Press, pp. 81–106.

Dean, L. G., R. L. Kendal, S. J. Schapiro, B. Thierry, and K. N. Laland. 2012. Identification of the social and cognitive processes underlying human cumulative culture. *Science* 335:1114–1118.

Dean, L. G., G. Vale, K. N. Laland, E. Flynn, and R. L. Kendal. 2014. Human cumulative culture: a comparative perspective. *Biological Reviews* 89:284–301.

Deaner, R. O., C. L. Nunn, and C. P. van Schaik. 2000. Comparative tests of primate cognition: different scaling methods produce different results. *Brain, Behavior and Evolution* 55:44–52.

Deaner, R. O., C. van Schaik, and V. Johnson. 2006. Do some taxa have better domain-general cognition than others? A meta-analysis of nonhuman primate studies. *Evolutionary Psychology* 4:149–196.

Deary, I. J. 2000. *Looking Down on Human Intelligence: From Psychometrics to the Brain*. Oxford, UK: Oxford University Press.

———. 2001. *Intelligence: A Very Short Introduction*. Oxford, UK: Oxford University Press.

Dechmann, D.K.N., and K. Safi. 2009. Comparative studies of brain evolution: a critical insight from the chiroptera. *Biological Review* 84:161–172.

Delagnes, A., and H. Roche. 2005. Late Pliocene hominid knapping skills: the case of Lokalalei 2C, West Turkana, Kenya. *Journal of Human Evolution* 48:435–472.

De la Torre, I. 2011. The origins of stone tool technology in Africa: a historical perspective. *Philosophical Transactions of the Royal Society of London B* 366:1028–1037.

de Magalhaes, J. P., and A. Matsuda. 2012. Genome-wide patterns of genetic distances reveal candidate loci contributing to human population-specific traits. *Annals of Human Genetics* 76:142–158.

Dennett, D. C. 1983. Intentional systems in cognitive ethology: the 'Panglossian paradigm' defended. *Behavioral and Brain Sciences* 6:343–355.

Derex, M., and R. Boyd. 2015. The foundations of the human cultural niche. *Nature Communications* 6:8398, doi:10.1038/ncomms9398.

d'Errico, F., and C. Stringer. 2011. Evolution, revolution or saltation scenario for the emergence of modern cultures? *Philosophical Transactions of the Royal Society of London B* 366:1060–1069.

DeVoogd, T. J., J. R. Krebs, S. D. Healy, and A. Purvis. 1993. Relations between song repertoire size and the volume of brain nuclei related to song: comparative evolutionary analysis among oscine birds. *Proceedings of the Royal Society of London B* 254:75–82.

de Waal, F. 1990. *Peacemaking among Primates*. Cambridge, MA: Harvard University Press.

———. 1996. *Good Natured: The Origins of Right and Wrong in Humans and Other Animals*. Cambridge, MA: Harvard University Press.

———. 2001. *The Ape and the Sushi Master*. London, UK: Penguin Books.

———. 2007. *Chimpanzee Politics: Power and Sex among Apes*. Baltimore, MD: Johns Hopkins University Press.

———. 2010. *The Age of Empathy: Nature's Lessons for a Kinder Society*. London, UK: Souvenir Press.

de Winter, W., and C. E. Oxnard. 2001. Evolutionary radiations and convergences in the structural organization of mammalian brains. *Nature* 409:710–714.

Diamond, J. 1991. *The Rise and Fall of the Third Chimpanzee*. London, UK: Vintage.

———. 1997. *Guns, Germs, and Steel: The Fates of Human Societies*. London, UK: Jonathon Cape.

———. 2006. *Collapse: How Societies Choose to Fail or Succeed*. London, UK: Penguin Books.

Dindo, M., B. Thierry, and A. Whiten. 2008. Social diffusion of novel foraging methods in brown capuchin monkeys (*Cebus apella*). *Proceedings of the Royal Society of London B* 275:187–193.

Donald, M. 1991. *Origins of the Modern Mind: Three Stages in the Evolution of Culture and Cognition*. Cambridge, MA: Harvard University Press.

Dornhaus, A., and L. Chittka. 1999. Evolutionary origins of bee dances. *Nature* 401:38.

Dorrance, B. R., and T. R. Zentall. 2002. Imitation of conditional discriminations in pigeons (*Columba livia*). *Journal of Comparative Psychology* 116:277–285.

Doupe, A. J., D. J. Perkel, A. Reiner, and E. A. Stern. 2005. Birdbrains could teach basal ganglia research a new song. *Trends in Neuroscience* 28:353–363.

Draper, P. 1976. Social and economic constraints on child life among the !Kung. In: *Kalahari Hunter-Gatherers: Studies of the !Kung San and Their Neighbors*, ed. R. B. Lee and I. DeVore. Cambridge, MA: Harvard University Press, pp. 199–217.

Dudley, J. 1977. The early life of an American modern dancer. In: *The Encyclopedia of Dance and Ballet*, ed. M. Clarke and D. Vaughan. London, UK: Pitman.

Duffy, G. A., T. W. Pike, and K. N. Laland. 2009. Size-dependent directed social learning in nine-spined sticklebacks. *Animal Behaviour* 78:371–375.

Dugatkin, L. A. 1992. Sexual selection and imitation: females copy the mate choice of others. *American Naturalist* 139:1384–1389.

Dugatkin, L. A., and J. Godin. 1992. Reversal of female mate choice by copying in the guppy (*Poecilia reticulata*). *Proceedings of the Royal Society of London B* 249:179–184.

Dunbar, R.I.M. 1992. Neocortex size as a constraint on group size in primates. *Journal of Human Evolution* 20:469–493.

———. 1995. Neocortex size and group size in primates: a test of the hypothesis. *Journal of Human Evolution* 28:287–296.

———. 1998. Theory of mind and the evolution of language. In: *Approaches to the Evolution of Language*, ed. J. R. Hurford, M. Studdert-Kennedy, and C. Knight. Cambridge, UK: Cambridge University Press, pp. 92–110.

———. 2011. Evolutionary basis of the social brain. In: *Oxford Handbook of Social Neuroscience*, ed. J. Decety and J. Cacioppo. Oxford, UK: Oxford University Press, pp. 28–38.

Dunbar, R.I.M., and S. Shultz. 2007a. Understanding primate brain evolution. *Philosophical Transactions of the Royal Society of London B* 362:649–658.

———. 2007b. Evolution in the social brain. *Science* 317:1344–1347.

Durham, W. H. 1991. *Coevolution: Genes, Culture, and Human Diversity*. Stanford, CA: Stanford University Press.

Eaton, G. G. 1976. The social order of Japanese macaques. *Scientific American* 234: 96–106.

Elgar, M. A., and B. J. Crespi. 1992. Ecology and evolution of cannibalism. In: *Cannibalism. Ecology and Evolution among Diverse Taxa*, ed. M. A. Elgar and B. J. Crespi. Oxford, UK: Oxford University Press, pp. 1–12.

Emery, N. J. 2004. Are corvids 'feathered apes'? Cognitive evolution in crows, jays, rooks and jackdaws. In: *Comparative Analysis of Minds*, ed. S. Watanabe. Tokyo, Japan: Keio University Press, pp. 181–213.

Emery, N. J., and N. S. Clayton. 2004. The mentality of crows: convergent evolution of intelligence in corvids and apes. *Science* 306:1903–1907.

Enard, W., S. Gehre, K. Hammerschmidt, S. M. Holter, T. Blass, M. Somel, M. K. Bruckner, et al. 2009. A humanized version of Foxp2 affects cortico-basal ganglia circuits in mice. *Cell* 137:961–971.

Enard, W., P. Khaitovich, J. Klose, S. Zollner, F. Heissiq, P. Giavalisco, K. Nieselt-Struwe, et al. 2002. Intra- and interspecific variation in primate gene expression patterns. *Science* 296:340–343.

Enard, W., M. Przeworski, S. E. Fisher, C. Lai, V. Wiebe, T. Kitana, A. P. Monaco, et al. 2002. Molecular evolution of FOXP2, a gene involved in speech and language. *Nature* 418:869–872.

Enquist, M., and S. Ghirlanda. 2007. Evolution of social learning does not explain the origin of human cumulative culture. *Journal of Theoretical Biology* 1479:449–454.

Enquist, M., K. Eriksson, and S. Ghirlanda. 2007. Critical social learning: a solution to Rogers' paradox of nonadaptive culture. *American Anthropologist* 109:727–734.

Enquist, M., S. Ghirlanda, and K. Eriksson. 2011. Modelling the evolution and diversity of cumulative culture. *Philosophical Transactions of the Royal Society of London B* 366:412–423.

Enquist, M., S. Ghirlanda, A. Jarrick, and C.-A. Wachtmeister. 2008. Why does human culture increase exponentially? *Theoretical Population Biology* 74:46–55.

Enquist, M., P. Strimling, K. Eriksson, K. N. Laland, and J. Sjostrand. 2010. One cultural parent makes no culture. *Animal Behaviour* 79:1353–1362.

Erickson, D. L., B. D. Smith, A. C. Clarke, D. H. Sandweiss, and N. Tuross. 2005. An Asian origin for a 10,000-year-old domesticated plant in the Americas. *Proceedings of the National Academy of Sciences USA* 102(51):18315–18320.

Eriksson, K., M. Enquist, and S. Ghirlanda. 2007. Critical points in current theory of conformist social learning. *Journal of Evolutionary Psychology* 5:67–87.

Evans, A. G., and T. E. Wellems. 2002. Coevolutionary genetics of *Plasmodium malaria* parasites and their human hosts. *Integrative and Comparative Biology* 42:401–407.

Evans, C. 2016. Empirical Investigations of Social Learning, Cooperation, and Their Role in the Evolution of Complex Culture. PhD diss., University of St Andrews.

Evison, S. F., O. Petchey, A. Beckerman, and F. W. Ratnieks. 2008. Combined use of pheromone trails and visual landmarks by the common garden ant (*Lasius niger*). *Behavioral Ecology and Sociobiology* 63:261–267.

Falk, D. 2004. Prelinguistic evolution in early hominins: whence motherese? *Behavioral and Brain Sciences* 27:491–503.

Farr, J. A., and W. F. Herrnkind. 1974. A quantitative analysis of social interaction of the guppy (*Poecilia reticulata*) as a function of population density. *Animal Behaviour* 22:582–591.

Faurie, C., and M. Raymond. 2005. Handedness, homicide and negative frequency-dependent selection. *Proceedings of the Royal Society of London B* 272:25–28.

Feenders, G., M. Liedvogel, M. Rivas, M. Zapka, H. Horita, E. Hara, K. Wada, et al. 2008. Molecular mapping of movement-associated areas in the avian brain: a motor theory for vocal learning origin. *PLOS ONE* 3:e1768, doi:org/10.1371/journal.pone.0001768.

Fehr, E., and U. Fischbacher. 2003. The nature of human altruism. *Nature* 425:785–791.

Fehr, E., and S. Gachter. 2002. Altruistic punishment in humans. *Nature* 415:137–140.

Feldman, M., and M. E. Kislev. 2007. A century of wheat research—from wild emmer discovery to genome analysis. *Israel Journal of Plant Sciences* 55(3–4):207–221.

Feldman, M. W., and L. L. Cavalli-Sforza. 1976. Cultural and biological evolutionary processes, selection for a trait under complex transmission. *Theoretical Population Biology* 9:238–259.

Feldman, M. W., and L. L. Cavalli-Sforza. 1981. *Cultural Evolution: A Quantitative Approach.* Princeton, NJ: Princeton University Press.

———. 1989. On the theory of evolution under genetic and cultural transmission with application to the lactose absorption problem. In: *Mathematical Evolutionary Theory*, ed. M. W. Feldman. Princeton, NJ: Princeton University Press, pp. 145–173.

Feldman, M. W., and K. N. Laland. 1996. Gene-culture co-evolutionary theory. *Trends in Ecology & Evolution* 11:453–457.

Feldman, M. W., and L. A. Zhivotovsky. 1992. Gene-culture coevolution: toward a general theory of vertical transmission. *Proceedings of the National Academy of Sciences USA* 89:935–938.

Feldman, M. W., K. Aoki, and J. Kumm. 1996. Individual versus social learning: evolutionary analysis in a fluctuating environment. *Anthropological Science* 104:209–232.

Fernandes, H.B.F., M. A. Woodley, and J. te Nijenhuis. 2014. Differences in cognitive abilities among primates are concentrated of *G*: Phenotypic and phylogenetic comparisons with two meta-analytical databases. *Intelligence* 46:311–322.

Finlay, B. L., and R. B. Darlington. 1995. Linked regularities in the development and evolution of mammalian brains. *Science* 268:1578–1584.

Fischer, J., B. C. Wheeler, and J. P. Higham. 2015. Is there any evidence for vocal learning in chimpanzee food calls? *Current Biology* 25:1028–1029.

Fisher, J., and R. A. Hinde. 1949. The opening of milk bottles by birds. *British Birds* 42:347–357.

Fisher, R. A. 1930. *The Genetical Theory of Natural Selection.* Oxford, UK: Clarendon Press.

Fisher, S. E., F. Vargha-Khadem, K. E. Watkins, A. P. Monaco, and M. E. Pembrey. 1998. Localisation of a gene implicated in a severe speech and language disorder. *Nature Genetics* 18:168–170.

Fitch, W. T. 2004. Kin selection and "mother tongues": a neglected component in language evolution. In: *Evolution of Communication Systems: A Comparative Approach*, ed. D. Kimbrough Oller and U. Griebel. Cambridge, MA: MIT Press, pp. 275–296.

———. 2005. The evolution of language: a comparative review. *Biology and Philosophy* 20:193–230.

———. 2010. *The Evolution of Language.* Cambridge, UK: Cambridge University Press.

———. 2011. The biology and evolution of rhythm: unraveling a paradox. In: *Language and Music as Cognitive Systems*, ed. P. Rebuschat, M. Rohrmeier, J. Hawkins, and I. Cross. Oxford, UK: Oxford University Press, pp. 73–95.

————. 2013. Rhythmic cognition in humans and animals: distinguishing meter and pulse perception. *Frontiers Systems Neuroscience* 7:1–16.

————. 2016. Dance, music, meter and groove: a forgotten partnership. *Frontiers in Human Neuroscience*, doi:org/10.3389/fn-hum.2016.00064.

Flynn, E. 2008. Investigating children as cultural magnets: do young children transmit redundant information along diffusion chains? *Philosophical Transactions of the Royal Society B* 363(1509):3541–3551.

Fogarty, L., P. Strimling, and K. N. Laland. 2011. The evolution of teaching. *Evolution* 65:2760–2770.

Forsgren, E., A. Karlsson, and C. Kvarnemo. 1996. Female sand gobies gain direct benefits by choosing males with eggs in their nests. *Behavioral Ecology and Socio-biology* 39:91–96.

Fortna, A., Y. Kim, E. MacLaren, K. Marshall, G. Hahn, L. Meltesen, M. Brenton, et al. 2004. Lineage-specific gene duplication and loss in human and great ape evolution. *PLOS Biology* 2(7):e207, doi:10.1371/journal.pbio.0020207.

Fragaszy, D. M. 2012. Community resources for learning: how capuchin monkeys construct technical traditions. *Biological Theory* 6:231–240, doi:10.1007/s13752 –012–0032–8.

Fragaszy, D. M., and S. Perry, eds. 2003. *The Biology of Traditions: Models and Evidence.* Cambridge, UK: Cambridge University Press.

Franks, N. R., and T. Richardson. 2006. Teaching in tandem-running ants. *Nature* 439:153.

Franz, M., and C. L. Nunn. 2009. Network-based diffusion analysis: a new method for detecting social learning. *Proceedings of the Royal Society of London B* 276:1829–1836.

Frazer, K. A., X. Chen, D. A. Hinds, P. V. Pant, N. Patil, and D. R. Cox. 2002. Genomic DNA insertions and deletions occur frequently between humans and nonhuman primates. *Genome Research* 13:341–346.

Frommen, J. G., C. Luz, and T.C.M. Bakker. 2007. Nutritional state influences shoaling preference for familiars. *Zoology* 110:369–376.

Fuller, D. 2011. Pathways to Asian civilizations: tracing the origins and spread of rice and rice cultures. *Rice* 43:78–92.

Futuyma, D. J. 1998. *Evolutionary Biology*, 3rd ed. Sunderland, MA: Sinauer.

Gagneux, P., C. Wills, U. Gerloff, D. Tautz, P. A. Morin, C. Boesch, B. Fruth, et al. 1999. Mitochondrial sequences show diverse evolutionary histories of African hominoids. *Proceedings of the National Academy of Sciences USA* 96:5077–5082.

Gajdon, G. K., N. Fijn, and L. Huber. 2006. Limited spread of innovation in a wild parrot, the Kea (*Nestor notabilis*). *Animal Cognition* 9:173–181.

Galef, B. G., Jr. 1988. Imitation in animals: history, definition and interpretation of the data from the psychological laboratory. In: *Social learning: Psychological and Biological Perspectives*, ed. B. G. Galef Jr. and T. R. Zentall. Hillsdale, NJ: Erlbaum, pp. 3–28.

————. 1992. The question of animal culture. *Human Nature* 3:157–178.

————. 2003. Traditional behaviors of brown and black rats *R. norvegicus* and *R. rattus*. In: *The Biology of Traditions: Models and Evidence*, ed. S. Perry and D. Fragaszy. Chicago, IL: University of Chicago Press, pp. 159–186.

Galef, B. G. 2009. Culture in animals? In: *The Question of Animal Culture*, ed. K. N. Laland and B. G. Galef. Cambridge, MA: Harvard University Press, pp. 222–246.

Galef, B. G., Jr., and C. Allen. 1995. A new model system for studying animal tradition. *Animal Behaviour* 50:705–717.

Galef, B. G., Jr., and M. Beck. 1985. Aversive and attractive marking of toxic and safe foods by Norway rats. *Behavioral and Neural Biology* 43:298–310.

Galef, B. G., Jr., and L. L. Buckley. 1996. Use of foraging trails by Norway rats. *Animal Behaviour* 51:765–771.

Galef, B. G., Jr., and M. M. Clark. 1971a. Social factors in the poison avoidance and feeding behavior of wild and domesticated rat pups. *Journal of Comparative Physiology and Psychology* 78:341–357.

————. 1971b. Parent–offspring interactions determine time and place of first ingestion of solid food by wild rat pups. *Psychonomic Science* 25:15–16.

Galef, B. G., Jr., and L. Giraldeau. 2001. Social influences on foraging in vertebrates: causal mechanisms and adaptive functions. *Animal Behaviour* 61:3–15.

Galef, B. G., Jr., and L. Heiber. 1976. The role of residual olfactory cues in the determination of feeding site selection and exploration patterns of domestic rats. *Journal of Comparative Physiology and Psychology* 90:727–739.

Galef, B. G., Jr., and P. W. Henderson. 1972. Mother's milk: a determinant of the feeding preferences of weaning rat pups. *Journal of Comparative Physiology and Psychology* 78:213–219.

Galef, B. G., and K. N. Laland. 2009. *The Question of Animal Culture*. Cambridge, MA: Harvard University Press.

Galef, B. G., Jr., and S. W. Wigmore. 1983. Transfer of information concerning distant foods: a laboratory investigation of the 'information-centre' hypothesis. *Animal Behaviour* 31:748–758.

Galef, B. G., Jr., D. J. Kennett, and S. W. Wigmore. 1984. Transfer of information concerning distant foods in rats: a robust phenomenon. *Animal Learning and Behavior* 12:292–296.

Galef, B. G., Jr., J. R. Mason, G. Preti, and N. J. Bean. 1988. Carbon disulfide: A semiochemical mediating socially–induced diet choice in rats. *Physiology and Behavior* 42:119–124.

Galef, B. G., Jr., B. Rudolf, E. E. Whiskin, E. Choleris, M. Mainardi, and P. Valsecchi. 1998. Familiarity and relatedness: effects on social learning about foods by Norway rats and Mongolian gerbils. *Animal Learning & Behavior* 26:448–454.

Gagneux, P., C. Wills, U. Gerloff, D. Tautz, P. A. Morin, C. Boesch, B. Fruth, et al. 1999. Mitochondrial sequences show diverse evolutionary histories of African hominoids. *Proceedings of the National Academy of Sciences USA* 96:5077–5082.

Gardner, R. A., and B. T. Gardner. 1969. Teaching sign language to a chimpanzee. *Science* 165:664–672.

Garfield, Z. H., M. J. Garfield, and B.S. Hewlett. 2016. A cross-cultural analysis of hunter-gatherer social learning. In: *Social Learning and Innovation in Contemporary Huntergatherers: Evolutionary and Ethnographic Perspectives*, ed. H. Terashima and B. S. Hewlett. Tokyo, Japan: Springer.

Garland, E. C., A. W. Goldizen, M. L. Rekdahl, R. Constantine, C. Garrigue, N. D. Hauser, M. M. Poole, et al. 2011. Dynamic horizontal cultural transmission of humpback whale song at the Ocean Basin scale. *Current Biology* 21:687–691.

Gergely, G., and G. Csibra. 2005. The social construction of the cultural mind: imitative learning as a mechanism of human pedagogy. *Interaction Studies* 6:463–481.

Gergely, G., K. Egyed, and I. Kiraly. 2007. On pedagogy. *Developmental Science* 10:139–146.

Gerull, F. C., and R. M. Rapee. 2002. Mother knows best: effects of maternal modelling on the acquisition of fear and avoidance behaviour in toddlers. *Behaviour Research and Therapy* 40:279–287.

Gibbons, A. 2009. Civilization's cost: the decline and fall of human health. *Science* 324:588.

Gibson, K. R. 1986. Cognition, brain size and the extraction of embedded food resources. In: *Primate Ontogeny, Cognition and Social Behavior*, ed. J. G. Else and P. C. Lee. Cambridge, UK: Cambridge University Press, pp. 93–103.

Gibson, K. R., and T. Ingold. 1993. *Tools, Language and Cognition in Human Evolution*. Cambridge, UK: Cambridge University Press.

Gignoux, C. R., B. M. Henn, and J. L. Mountain. 2011. Rapid, global demographic expansions after the origins of agriculture. *Proceedings of the National Academy of Sciences USA* 108:6044–6049.

Gilbert, S. F. 2003. The morphogenesis of evolutionary developmental biology. International *Journal of Developmental Biology* 47:467–477.

Gilby, I. C., M. E. Thompson, J. D. Ruane, and R. Wrangham. 2010. No evidence of short-term exchange of meat for sex among chimpanzees. *Journal of Human Evolution* 59:44–53.

Gingerich, P. D. 1983. Rates of evolution: effects of time and temporal scaling. *Science* 222:159–161.

Gintis, H. 2003. The hitchhiker's guide to altruism: gene-culture coevolution, and the internalization of norms. *Journal of Theoretical Biology* 220:407–418.

Giraldeau, L.-A., and T. Caraco. 2000. *Social Foraging Theory*. Princeton, NJ: Princeton University Press.

Giraldeau, L.-A., and L. Lefebvre. 1987. Scrounging prevents cultural transmission of food-finding behaviour in pigeons. *Animal Behaviour* 35:387–394.

Giraldeau, L.-A., T. J. Valone, and J. J. Templeton. 2002. Potential disadvantages of using socially acquired information. *Philosophical Transactions of the Royal Society of London B* 357:1559–1566.

Glazko, G., V. Veeramachaneni, M. Nei, and W. Makalowski. 2005. Eighty percent of proteins are different between humans and chimpanzees. *Gene* 346:215–219.

Gleibermann, L. 1973. Blood pressure and dietary salt in human populations. *Ecology of Food and Nutrition* 2:143–155.

Glickstein, M., and K. Doron. 2008. Cerebellum: connections and functions. *Cerebellum* 7:589–594.

Godin, J., E. Herman, and L. A. Dugatkin. 2005. Social influences on female mate choice in the guppy, *Poecilia reticulata*: generalized and repeatable trait-copying behaviour. *Animal Behaviour* 69:999–1005.

Goldsmidt, T., T.C.M. Bakker, and E. Feuth-de Brujin. 1993. Selective choice in copying of female sticklebacks. *Animal Behaviour* 45:541–547.

Goodall, J. 1986. *The Chimpanzees of Gombe: Patterns of Behavior*. Cambridge MA: Harvard University Press.

Gottlieb, G. 1992. *Individual Development and Evolution: The Genesis of Novel Behavior*. New York, NY: Oxford University Press.

Gould, S. J., and E. Vrba. 1982. Exaptation: a missing term in the science of form. *Paleobiology* 8:4–15.

Govoni, G., and P. Gros. 1998. MacrophageNRAMP1 and its role in resistance to microbial infections. *Inflammation Research.* 47:277–284.

Grafen, A. 1984. Natural selection, kin selection and group selection. In: *Behavioural Ecology: An Evolutionary Approach*, 2nd ed., ed. J. R. Krebs and N. B. Davies. Oxford, UK: Blackwell Scientific, pp. 62–84.

———. 1990. Biological signals as handicaps. *Journal of Theoretical Biology* 144: 517–546.

Grant, M., and J. Hazel. 2002. *Who's Who in Classical Mythology*. London, UK: Routledge.

Gray, J. R., and P. M. Thompson. 2004. Neurobiology of intelligence: science and ethics. *Nature Reviews Neuroscience* 5:471–482.

Gray, R. D., and Q. D. Atkinson. 2003. Language tree divergence times support the Anatolian theory of Indo-European origin. *Nature* 426:435–439.

Gray, R. D., and F. M. Jordan. 2000. Language trees support the express-train sequence of Austronesian expansion. *Nature* 405:1052—1055.

Greenberg, R., and C. Mettke-Hofman. 2001. Ecological aspects of neophobia and exploration in birds. *Current Ornithology* 16:119–178.

Greenfield, P. M. 1991. Language, tools and brain: the ontogeny and phylogeny of hierarchically organized sequential behaviour. *Behavior and Brain Science* 14:531–595.

Gross, R., A. I. Houston, E. J. Collins, J. M. McNamara, F. X. Dechaume-Moncharmont, and N. R. Franks. 2008. Simple learning rules to cope with changing environments. *Journal of the Royal Society, Interface* 5:1193–1202.

Grüter, C., and F. L. W. Ratnieks. 2011. Honeybee foragers increase the use of waggle dance information when private information becomes unrewarding. *Animal Behaviour* 81:949–954.

Grüter, C., E. Leadbeater, and F. L. W. Ratnieks. 2010. Social learning: the importance of copying others. *Current Biology* 20:R683–R685.

Guillermo, A. 2005. *The Uruk World System: The Dynamics of Expansion of Early Meso-potamian Civilization*, 2nd ed. Chicago, IL: University of Chicago Press.

Hahn, M., J. P. Demuth, and S.-G. Han. 2007. Accelerated rate of gene gain and loss in primates. *Genetics* 177:1941–1949.

Halgin, R. P., and S. Whitbourne. 2006. *Abnormal Psychology with MindMap II. CD-ROM and PowerWeb*. New York, NY: McGraw-Hill.

Hamilton, W. 1964. The genetical evolution of social behaviour: I. *Journal of Theoretical Biology* 7:1–16.

Hampton, R. R., D. F. Sherry, M. Khurgel, and G. Ivy. 1995. *Brain Behavior and Evolution* 45:54–61.

Han, Y., S. Gu, H. Oota, M. V. Osier, A. J. Pakstis, W. C. Speed, J. R. Kidd, et al. 2007. Evidence of positive selection on a class I ADH locus. *American Journal of Human Genetics* 80:441–456.

Hansen, B. T., L. E. Johannessen, and T. Slagsvold. 2008. Imprinted species recognition lasts for life in free-living great tits and blue tits. *Animal Behaviour* 75:921–927.

Hardyck, C., L. Petriovich, and R. Goldman. 1976. Left handedness and cognitive deficit. *Cortex* 12:266–278.

Harris, L. J. 1980. Left handedness: early theories, facts and fancies. In: *Neuropsychology of Left Handedness*, ed. L. J. Herron. London, UK: Academic Press, pp. 3–78.

Harvey, P. H., and J. R. Krebs. 1990. Comparing brains. *Science* 249:140–146.

Harvey, P. H., and M. D. Pagel. 1991. *The Comparative Method in Evolutionary Biology*. Oxford, UK: Oxford University Press.

Hatfield, G. 2016. René Descartes. In: *The Stanford Encyclopedia of Philosophy* (Summer 2016 Edition), ed. E. N. Zalta, http://plato.stanford.edu/archives/sum2016/entries/descartes/.

Haun, D.B.M., Y. Rekers, and M. Tomasello. 2012. Majority-biased transmission in chimpanzees and human children, but not orangutans. *Current Biology* 22: 727–731.

Hauser, M. D. 1996. *The Evolution of Communication*. Cambridge, MA: MIT Press.

Hauser, M. D., N. Chomsky, and W. T. Fitch. 2002. The faculty of language: what is it, who has it, and how did it evolve? *Science* 298:1569–1579.

Hauser, M. D., C. Yang, R. C. Berwick, I. Tattersall, M. J. Ryan, J. Watumull, N. Chomsky, et al. 2014. The mystery of language evolution. *Frontiers in Psychology* 5:401–412.

Hawks, J., E. T. Wang, G. M. Cochran, H. C. Harpending, and R. K. Moyzis. 2007. Recent acceleration of human adaptive evolution. *Proceedings of the National Academy of Sciences USA* 104:20753–20758.

Hawkesworth, J. (1773) 2014. *An Account of the Voyages Undertaken by the Order of His Present Majesty for Making Discoveries in the Southern Hemisphere*. Cambridge, U.K.: Cambridge University Press.

Hawley, W. A., P. Reiter, R. S. Copeland, C. B. Pumpuni, and G. B. Craig Jr. 1987. *Aedes albopictus* in North America: probable introduction in used tires from northern Asia. *Science* 236:1114–1116.

Hayden, B. 1993. The cultural capacities of Neandertals: a review and re-evaluation. *Journal of Human Evolution* 24:113–146.

Haygood, R., O. Fedrigo, B. Hanson, K. D. Yokoyama, and G. A. Wray. 2007. Promoter regions of many neural- and nutrition-related genes have experienced positive selection during human evolution. *Nature Genetics* 39:1140–1144.

Healy, S. D., and C. Rowe. 2007. A critique of comparative studies of brain size. *Proceedings of the Royal Society of London B* 274:453–464.

Heffner, R. S., and R. B. Masterton. 1975. Variation in the form of the pyramidal tract and its relationship to digital dexterity. *Brain, Behavior and Evolution* 12:161–200.

———. 1983. The role of the corticospinal tract in the evolution of human digital dexterity. *Brain, Behavior and Evolution* 23:165–183.

Helfman G. S., and E. T. Schultz. 1984. Social transmission of behavioural traditions in a coral reef fish. *Animal Behaviour* 32:379–384.

Helgason, A., S. Palsson, G. Thorleifsson, S.F.A. Grant, V. Emilsson, S. Gunnarsdottir, A. Adeyemo, et al. 2007. Refining the impact of TCF7L2 gene variants on type 2 diabetes and adaptive evolution. *Nature Genetics* 39:218–225.

Henrich, J. 2004a. Demography and cultural evolution: why adaptive cultural processes produced maladaptive losses in Tasmania. *American Antiquity* 69:197–221.

———. 2004b. Cultural group selection, coevolutionary processes and large-scale cooperation. *Journal of Economic Behavior and Organization* 53:3–35.

———. 2009. The evolution of costly displays, cooperation and religion: credibility enhancing displays and their implications for cultural evolution. *Evolution of Human Behavior* 30:244–260.

———. 2015. *The Secret of our Success*. Princeton, NJ: Princeton University Press.

Henrich, J., and R. Boyd. 1998. The evolution of conformist transmission and between-group differences. *Evolution of Human Behavior* 19:215–242.

———. 2002. On modeling cognition and culture: why cultural evolution does not require replication of representations. *Journal of Cognitive Culture* 2:87–112.

Henrich, J., and R. Henrich. 2007. *Why Humans Cooperate: A Cultural and Evolutionary Explanation*. Oxford, UK: Oxford University Press.

Henrich, J., and R. McElreath. 2003. The evolution of cultural evolution. *Evolutionary Anthropology* 12:123–135.

Henrich, J., R. Boyd, S. Bowles, C. Camerer, E. Fehr, H. Gintis, and R. McElreath. 2001. In search of Homo economicus: behavioral experiments in 15 small-scale societies. *American Economic Review* 91:73–7.

Hepper, P. 1988. Adaptive fetal learning: prenatal exposure to garlic affects postnatal preferences. *Animal Behaviour* 36:935–936.

Herrmann, E., J. Call, M. V. Hernandez-Lloreda, B. Hare, and M. Tomasello. 2007. Humans have evolved specialized skills of social cognition: the cultural intelligence hypothesis. *Science* 317:1360–1366.

Herrmann, E., M. V. Hernandez-Lloreda, J. Call, B. Hare, and M. Tomasello. 2010. The structure of individual differences in the cognitive abilities of children and chimpanzees. *Psychological Science* 21:102–110.

Herrman, P. A., C. H. Legare, P. L. Harris, and H. Whitehouse. 2013. Stick to the script: the effect of witnessing multiple actors on children's imitation. *Cognition* 129:536–543.

Hewlett, B. S., and C. J. Roulette. 2016. Teaching in hunter-gatherer infancy. *Royal Society Open Science* 3:150403.

Hewlett, B. S., H. N. Fouts, A. H. Boyette, and B. L. Hewlett. 2011. Social learning among Congo Basin hunter-gatherers. *Philosophical Transactions of the Royal Society of London B* 366:1168–1178.

Heyes, C. M. 1993. Imitation, culture and cognition. Animal Behaviour 46:999–1010.

———. 1998. Theory of mind in nonhuman primates. *Behavioral and Brain Sciences* 21:101–114.

———. 2002. Transformational and associative theories of imitation. In: *Imitation in Animals and Artefacts*, ed. K. Dautenhahn and C. L. Nehaniv. Cambridge, MA: MIT Press, pp. 501–524.

———. 2005. Imitation by association. In: *Perspectives on Imitation: From Mirror Neurons to Memes*, ed. S. Hurley and N. Chater. Cambridge, MA: MIT Press, pp. 157–176.

———. 2009. Evolution, development and intentional control of imitation. *Philosophical Transactions of the Royal Society of London B* 364:2293–2298.

———. 2012. What can imitation do for cooperation? In: *Signalling, Commitment & Cooperation*, ed. B. Calcott, R. Joyce, and K. Sterelny. Cambridge, MA: MIT Press.

Heyes, C. M., and B. G. Galef Jr. 1996. *Social Learning in Animals: The Roots of Culture.* Cambridge, MA: Academic Press.

Heyes, C. M., and Ray, E. D. 2000. What is the significance of imitation in animals? *Advances in the Study of Behavior* 29:215–245.

Heyes, C. M., and A. Saggerson. 2002. Testing for imitative and non-imitative social learning in the budgerigar using a two-object /two-action test. *Animal Behaviour* 64:851–859.

Hill, K., M. Barton, and A. M. Hurtado. 2009. The emergence of human uniqueness: characters underlying behavioral modernity. *Evolutionary Anthropology* 18:174–187.

Hill, K. R., R. S. Walker, M. Bozicevic, J. Eder, T. Headland, B. Hewlett, A. M. Hurtado, et al. 2011. Co-residence patterns in hunter-gatherer societies show unique human social structure. *Science* 331:1286–1289.

Hill, R. S., and C. A. Walsh. 2005. Molecular insights into human brain evolution. *Nature* 437:64–67, doi:10.1038/nature04103.

Hinde, R. A., and J. Fisher. 1951. Further observations on the opening of milk bottles by birds. *British Birds* 44:393–396.

———. 1972. Some comments on the republication of two papers on the opening of milk bottles by birds. In: *Function and Evolution of Behavior: An Historical Sample from the Pen of Ethologists*, ed. P. H. Klopfer and J. P. Hailman. Boston, MA: Addison-Wesley, pp. 377–378.

Hirvonen, H., S. Vilhunen, C. Brown, V. Lintunen, and K. N. Laland. 2003. Improving anti-predator responses of hatchery reared salmonids by social learning. In: Fish

Models as Behavior, Conference Proceedings: Fisheries Society of the British Isles Annual Symposium, Norwich, UK, June 30–July 04, 2003. *Journal of Fish Biology* 63:Supplement A. 63(232), 10.1111/j.1095–8649.2003.0216n.x.

Holden, C., and R. Mace. 1997. Phylogenetic analysis of the evolution of lactose digestion in adults. *Human Biology* 69:605–628.

Hoogland, R. D., D. Morris, and N. Tinbergen. 1957. The spines of sticklebacks (*Gasterosteus* and *Pygosteus*) as a means of defence against predators (*Perca* and *Esox*). *Behaviour* 10:205–237.

Hopkins, W. D., and C. Cantalupo. 2004. Handedness in chimpanzees (*Pan troglodytes*) is associated with asymmetries of the primary motor cortex but not with homologous language areas. *Behavioral Neuroscience* 118:1176–1183.

Hoppitt, W.J.E., and K. N. Laland. 2008. Social processes influencing learning in animals: a review of the evidence. *Advances in the Study of Behavior* 38:105–165.

———. 2011. Detecting social learning using networks: a user's guide. *American Journal of Primatology* 73:834–844.

———. 2013. *Social Learning: An Introduction to Mechanisms, Methods, and Models.* Princeton, NJ: Princeton University Press.

Hoppitt, W.J.E., N. J. Boogert, and K. N. Laland. 2010a. Detecting social transmission in networks. *Journal of Theoretical Biology* 263:544–555.

Hoppitt, W.J.E., G. Brown, R. L. Kendal, L. Rendell, A. Thornton, M. Webster, and K. N. Laland. 2008. Lessons from animal teaching. *Trends in Ecology and Evolution* 23:486–493.

Hoppitt, W.J.E., A. Kandler, J. R. Kendal, and K. N. Laland. 2010b. The effect of task structure on diffusion dynamics: implications for diffusion curve and network-based analyses. *Learning and Behavior* 38:243–251.

Horner, V., and A. Whiten. 2005. Causal knowledge and imitation/emulation switching in chimpanzees (*Pan troglodytes*) and children (*Homo sapiens*). *Animal Cognition* 8:164–181.

Horner, V., D. Proctor, K. E. Bonnie, A. Whiten, and F.B.M. de Waal. 2010. Prestige affects cultural learning in chimpanzees. *PLOS ONE* 5:e10625, doi:10.1371/journal.pone.0010625.

Hosey, G. R., M. Jacques, and A. Pitts. 1997. Drinking from tails: social learning of a novel behaviour in a group of ring-tailed lemurs (*Lemur catta*). *Primates* 38:415–422.

Houston, R. G. 1973. Sickle cell anemia and dietary precursors of cyanate. *American Journal of Clinical Nutrition* 26:1261–1264.

Hovers, E. 2012. Invention, reinvention and innovation: makings of Oldowan lithic technology. In: *Origins of Human Innovation and Creativity*, ed. S. Elias. Vol. 16 in *Developments in Quaternary Science*, ed. J.J.M. van der Meer. Maryland Heights, MO: Elsevier, pp. 51–68.

Hrdy, S. 1999. *Mother Nature—Maternal Instincts and How They Shape the Human Species.* New York, NY: Ballantine Books.

Hrubesch, C., S. Preuschoft, and C. van Schaik. 2008. Skill mastery inhibits adoption of observed alternative solutions among chimpanzees (*Pan troglodytes*). *Animal Cognition* 12:209–216.

Huffman, M. A. 1996. Acquisition of innovative cultural behaviors in nonhuman primates: a case study of stone handling, a socially transmitted behavior in Japanese macaques. In: *Social Learning in Animals: The Roots of Culture*, ed. C. M. Heyes and B. G. Galef Jr. Cambridge, MA: Academic, pp. 267–290.

Huffman, M. A., and S. Hirata. 2003. Biological and ecological foundations of primate behavioral tradition. In: *The Biology of Traditions: Models and Evidence*, ed. D. M. Fragaszy and S. Perry. Cambridge, UK: Cambridge University Press, pp. 267–296.

Hugo, V. (1831) 1978. *Notre-Dame of Paris*. London, UK: Penguin Classics.

Humphrey, N. K. 1976. The social function of intellect. In: *Growing Points in Ethology*, ed. P.P.G. Bateson and R. A. Hinde. Cambridge, UK: Cambridge University Press, pp. 303–317.

Hunemeier, T., C.E.G. Amorim, S. Azevedo, V. Contini, V. Acuna-Alonzo, F. Rothhammer, J.-M. Dugoujon, et al. 2012. Evolutionary responses to a constructed niche: ancient Mesoamericans as a model of gene-culture coevolution. *PLOS ONE* 7:e38862, doi:org/10.1371/journal.pone.0038862.

Hunemeier, T., J. Gómez-Valdés, M. Ballesteros-Romero, S. de Azevedo, N. Martínez-Abadías, M. Esparza, T. Sjøvold, et al. 2012. Cultural diversification promotes rapid phenotypic evolution in Xavánte Indians. *Proceedings of the National Academy of Sciences USA* 109(1):73–77.

Hung, C. C., Y. K. Tu, S. H. Chen, and R. C. Chen. 1985. A study of handedness and cerebral speech dominance in right-handed Chinese. *Journal of Neurolinguistics* 1:143–163.

Hunt, G. R., and R. D. Gray. 2003. Diversification and cumulative evolution in New Caledonian crow tool manufacture. *Proceedings of the Royal Society of London B* 270:867–874.

Hurford, J. R. 1999. The evolution of language and of languages. In: *The Evolution of Culture*, ed. R. Dunbar, C. Knight, and C. Power. Edinburgh, UK: Edinburgh University Press, pp. 173–193.

———. 2014. *Origins of Language: A Slim Guide*. Oxford, UK: Oxford Linguistics.

Huttenlocher, J., M. Vasilyeva, E. Cymerman, and S. Levine. 2002. Language input and child syntax. *Cognitive Psychology* 45:337–374.

Iacoboni, M., R. P. Woods, M. Brass, H. Bekkering, J. C. Mazziotta, and G. Rizzolatti. 1999. Cortical mechanisms of human imitation. *Science* 286(5449):2526–2528.

Ihara, Y., K. Aoki, and M. W. Feldman. 2003. Runaway sexual selection with paternal transmission of the male trait and gene-culture determination of the female preference. *Theoretical Population Biology* 63:53–62.

Ingram, J. 1998. *The Barmaid's Brain*. New York, NY: Viking.

Isler, K., and C. P. van Schaik. 2012. How our ancestors broke through the gray ceiling: comparative evidence for cooperative breeding in early *Homo*. *Current Anthropology* 53(6):453–465.

Izagirre, N., I. Garcia, C. Junquera, C. de la Rua, and S. Alonso. 2006. A scan for signatures of positive selection in candidate loci for skin pigmentation in humans. *Molecular Biology and Evolution* 23:1697–1706.

Jablonka, E., and M. J. Lamb. 2005. *Evolution in Four Dimensions*. Cambridge, MA: MIT Press.

Jacobs, L. F., and W. D. Spencer. 1994. Natural space-use patterns and hippocampal size in kangaroo rats. *Brain, Behavior and Evolution* 44:125–132.

James, S. R. 1989. Hominid use of fire in the lower and middle Pleistocene. *Current Anthropology* 30:1–26.

Janik, J. M., and P.J.B. Slater. 1997. Vocal learning in mammals. *Advances in the Study of Behavior* 26:59–99.

Jaradat, A. A. 2007. Biodiversity and sustainable agriculture in the Fertile Crescent. *Yale Forestry & Environmental Science Bulletin* 103:31–57.

Jarvis, E. D. 2004. Learned birdsong and the neurobiology of human language. *Annals of the New York Academy of Sciences* 1016:749–777.

Jensen, K., J. Call, and M. Tomasello. 2007. Chimpanzees are rational maximizers in an ultimatum game. *Science* 318:107–109.

Joffe, T. H., and R.I.M. Dunbar. 1997. Visual and socio-cognitive information processing in primate brain evolution. *Proceedings of the Royal Society of London B* 264:1303–1307.

Johannessen, L. E., T. Slagsvold, and B. T. Hansen. 2006. Effects of social rearing conditions on song structure and repertoire size: experimental evidence from the field. *Animal Behaviour* 72:83–95.

Jolly, A. 1966. Lemur social behavior and primate intelligence. *Science* 153:501–506.

Jones, B. C., L. M. DeBruine, A. C. Little, R. P. Burriss, and D. R. Feinberg. 2007. Social transmission of face preferences among humans. *Proceedings of the Royal Society B* 274:899–903.

Jonxis, J.H.P. 1965. Haemoglobinopathies in West Indian groups of African origin. In: *Abnormal Haemoglobins in Africa*, ed. J.H.P. Jonxis. Oxford, UK: Blackwell, pp. 329–338.

Kambo, J. S., and L.A.M. Galea. 2006. Activational levels of androgens influence risk assessment behaviour but do not influence stress-induced suppression in hippocampal cell proliferation in adult male rats. *Behavioural Brain Research* 175:263–270.

Kameda, T., and D. Nakanishi. 2002. Cost-benefit analysis of social/cultural learning in a nonstationary uncertain environment: an evolutionary simulation and an experiment with human subjects. *Evolution and Human Behavior* 23:373–393.

Kant, M. 2007. *The Cambridge Companion to Ballet*. Cambridge, UK: Cambridge University Press.

Kaplan, H. S., and J. A. Robson. 2002. The emergence of humans: the coevolution of intelligence and longevity with intergenerational transfers. *Proceedings of the National Academy of Sciences USA* 99:10221–10226.

Kaplan, H. S., K. Hill, J. Lancaster, and A. M. Hurtado, 2000. A theory of human life history evolution: diet, intelligence, and longevity. *Evolutionary Anthropology* 9:156–185.

Kappe, S.H.I., A. M. Vaughan, J. A. Boddey, and A. F. Cowman. 2010. That was then but this is now: malaria research in the time of an eradication agenda. *Science* 328:862–866.

Kappeler, P. M., and E. W. Heyman. 1996. Nonconvergence in the evolution of primate life history and socio-ecology. *Biological Journal of the Linnean Society* 59:297–326.

Kappeler, P., and J. Silk. 2009. *Mind the Gap: Tracing the Origins of Human Universals.* New York, NY: Springer.

Kaufman, A. B., and J. C. Kaufman. 2015. *Animal Creativity and Innovation.* San Diego, CA: Academic Press.

Kavaliers, M., E. Choleris, and D. D. Colwell. 2001. Brief exposure to female odors 'emboldens' male mice by reducing predator-induced behavioural and hormonal responses. *Hormones and Behavior* 40:497–509.

Kavaliers, M., D. Colwell, and E. Choleris. 2003. Learning to fear and cope with a natural stressor: individually and socially acquired corticosterone and avoidance responses to biting flies. *Hormones and Behavior* 43:99–107.

Kawai, M. 1965. Newly-acquired pre-cultural behavior of the natural troop of Japanese monkeys on Koshima islet. *Primates* 6:1–30.

Kayser, M., S. Brauer, R. Cordaux, A. Castro, O. Lao, L. Zhivotovsky, C. Moyse-Faurie, et al. 2006. Melanesian and Asian origins of Polynesians: mtDNA and Y chromosome gradients across the Pacific. *Molecular Biology and Evolution* 23:2234–2244.

Keller, E. 2010. *The Mirage of a Space between Nature and Nurture.* Durham, NC: Duke University Press.

Kelley, J. L., and W. J. Swanson. 2008. Dietary change and adaptive evolution of enamelin in humans and among primates. *Genetics* 178:1595–1603.

Kendal, J. R., J. J. Tehrani, and J. Odling-Smee. 2011. Human niche construction in interdisciplinary focus. *Philosophical Transactions of the Royal Society B* 366(1566):785–792.

Kendal, R. L., R. L. Coe, and K. N. Laland. 2005. Age differences in neophilia, exploration, and innovation in family groups of callitrichid monkeys. *American Journal of Primatology* 66:167–188.

Kendal, R. L., I. Coolen, and K. N. Laland. 2004. The role of conformity in foraging when personal and social information conflict. *Behavioral Ecology* 15:269–277.

Kendal, R. L., I. Coolen, Y. van Bergen, and K. N. Laland. 2005. Tradeoffs in the adaptive use of social and asocial learning. *Advances in the Study of Behaviour* 35:333–379.

Kendal, R. L., L. M. Hopper, A. Whiten, S. F. Brosnan, S. P. Lambeth, S. J. Schapiro, and W. Hoppitt. 2015. Chimpanzees copy dominant and knowledgeable individuals: implications for cultural diversity. *Evolution and Human Behavior* 36:65–72.

Kendal, J. R., L. Rendell, T. Pike, and K. N. Laland. 2009. Nine-spined sticklebacks deploy a hill-climbing social learning strategy. *Behavioral Ecology* 20:238–244.

King, B. J. 1986. Extractive foraging and the evolution of primate intelligence. *Human Evolution* 14:361–372.

King, M. C., and A. C. Wilson. 1975. Evolution at two levels in humans and chimpanzees. *Science* 188:107–116.

Kinzler, K. D., K. Shutts, J. DeJesus, and E. S. Spelke. 2009. Accent trumps race in guiding children's social preferences. *Social Cognition* 27(4):623–634.

Kirby, S., H. Cornish, and K. Smith. 2008. Cumulative cultural evolution in the laboratory: an experimental approach to the origins of structure in human language. *Proceedings of the National Academy of Sciences USA* 105:10681–10686.

Kirby, S., M. Dowman, and T. L. Griffiths. 2007. Innateness and culture in the evolution of language. *Proceedings of the National Academy of Sciences USA* 104:5241–5245.

Kirby, S., M. Tamariz, H. Cornish, and K. Smith. 2015. Compression and communication in the cultural evolution of linguistic structure. *Cognition* 141:87–102.

Kirschner, M., and J. Gerhart. 2005. *The Plausibility of Life: Resolving Darwin's Dilemma.* New Haven, CT: Yale University Press.

Kirschner, W. H. 1987. Tradition im Bienenstaat: Kommunikation Zwichen den Imagines und der Brut der Honigbiene Durch Vibrationssignale. PhD diss., Julius-Maximilians Universität.

Klein, R. G. 1999. *The Human Career*, 2nd ed. Chicago, IL: University of Chicago Press.

———. 2000. Archeology and the evolution of human behavior. *Evolutionary Anthropology* 9:17–36.

Kline, M. A., and R. Boyd. 2010. Population size predicts technological complexity in Oceania. *Proceedings of the Royal Society of London B* 277:2559–2564.

Kline, M. A., R. Boyd, and J. Henrich. 2013. Teaching and the life history of cultural transmission in Fijian villages. *Human Nature* 24(4):351–374.

Kohler, W. 1925. The *Mentality of Apes*. Translated from the 2nd revised edition by E. Winter. New York, NY: Harcourt, Brace.

Kokko, H., and M. D. Jennions. 2008. Parental investment, sexual selection and sex ratios. *Journal of Evolutionary Biology* 21:919–948.

Kokko, H., and P. Monaghan. 2001. Predicting the direction of sexual selection. *Ecology Letters* 4:159–165.

Kolata, S., K. Light, and L. D. Matzel. 2008. Domain-specific and domain-general learning factors are expressed in genetically heterogeneous cd-1 mice. *Intelligence* 36:619–629.

Kothbauer-Hellman, R. 1990. On the origin of a tradition: milk bottle opening by titmice. *Zoologischer Anzeiger* 225:353–361.

Kotiaho, J. S. 2001. Costs of sexual traits: a mismatch between theoretical considerations and empirical evidence. *Biological Reviews* 76:365–376.

Koulouriotis, D. E., and A. Xanthopoulos. 2008. Reinforcement learning and evolutionary algorithms for non-stationary multi-armed bandit problems. *Applied Mathematics and Computation* 196:913–922.

Kraak, S.B.M., and F. J. Weissing. 1996. Female preference for nests with many eggs: a cost-benefit analysis of female choice in fish with paternal care. *Behavioral Ecology* 7:353–361.

Kramer, S. N. 1964. *The Sumerians: Their History, Culture, and Character*. Chicago, IL: University of Chicago Press.

Krause, J., and G. D. Ruxton. 2002. *Living in Groups*. Oxford, UK: Oxford University Press.

Krebs, J. R., D. F. Sherry, S. D. Healy, H. Perry, and A. L. Vaccerino. 1989. Hippocampal specialization of food-storing birds. *Proceedings of the National Academy of Sciences USA* 86:1388–1392.

Kruetzen, M., S. Kreicker, C. D. MacLeod, J. Learmonth, A. Kopps, P. Walsham, and S. Allen. 2014. Cultural transmission of tool use by Indo-Pacific bottlenose dolphins (*Tursiops* sp.) provides access to a novel foraging niche. *Proceedings of the Royal Society of London B* 281(1784):20140374, doi:10.1098/rspb.2014.0374.

Krutzen, M., J. Mann, M. R. Heithaus, R. C. Connor, L. Bejder, and W. B. Sherwin. 2005. Cultural transmission of tool use in bottlenose dolphins. *Proceedings of the National Academy of Sciences USA* 102:8939–8943.

Kudo, H., and R.I.M. Dunbar. 2001. Neocortex size and social network size in primates. *Animal Behaviour* 62:711–722.

Kumm, J., K. N. Laland, and M. W. Feldman. 1994. Gene-culture coevolution and sex ratios: the effects of infanticide, sex-selective abortion, and sex-biased parental investment on the evolution of sex ratios. *Theoretical Population Biology* 46:249–278.

Kummer, H., and J. Goodall. 1985. Conditions of innovative behavior in primates. *Philosophical Transactions of the Royal Society of London B* 308:203–214.

Kummer, H., and F. Kurt. 1965. A comparison of social behaviour in captive and wild hamadryas baboons. In: *The Baboon in Medical Research*, ed. H. Vagtbord. Austin, TX: University of Texas Press, pp. 65–80.

Lachlan, R., L. Crooks, and K. N. Laland. 1998. Who follows whom? Shoaling preferences and social learning of foraging information in guppies. *Animal Behaviour* 56:181–190.

Laland, K. N. 1990. A Theoretical Investigation of the Role of Social Transmission in Evolution. PhD diss., University College London.

———. 1994. Sexual selection with a culturally transmitted mating preference. *Theoretical Population Biology* 45:1–15.

———. 2004. Social learning strategies. *Learning & Behavior* 32:4–14.

———. 2016. The origins of language in teaching. *Psychonomic Bulletin and Review*, doi:10.3758/s13423–016–1077–7.

Laland, K. N., and P.P.G. Bateson. 2001. The mechanisms of imitation. *Cybernetics and Systems* 32:195–224.

Laland, K. N., and G. R. Brown. 2006. Niche construction, human behaviour and the adaptive lag hypothesis. *Evolutionary Anthropology* 15:95–104.

———. 2011. *Sense and Nonsense. Evolutionary Perspectives on Human Behaviour*. Oxford, UK: Oxford University Press.

Laland, K. N., and B. G. Galef Jr., eds. 2009. *The Question of Animal Culture*. Cambridge, MA: Harvard University Press.

Laland, K. N., and W.J.E. Hoppitt. 2003. Do animals have culture? *Evolutionary Anthropology* 12:150–159.

Laland, K. N., and V. M. Janik. 2006. The animal cultures debate. *Trends in Ecology and Evolution* 21:542–547.

Laland, K. N., and M. J. O'Brien. 2010. Niche construction theory and archaeology. *Journal of Archaeological Method and Theory* 17:303–322.

Laland, K. N., and H. C. Plotkin. 1991. Excretory deposits surrounding food sites facilitate social learning of food preferences in Norway rats. *Animal Behaviour* 41:997–1005.

———. 1993. Social transmission of food preferences amongst Norway rats by marking of food sites, and by gustatory contact. *Animal Learning and Behavior* 21:35–41.

Laland, K. N., and S. M. Reader. 1999a. Foraging innovation in the guppy. *Animal Behaviour* 57:331–340.

———. 1999b. Foraging innovation is inversely related to competitive ability in male but not in female guppies. *Behavioral Ecology* 10:270–274.

———. 2009. Comparative perspectives on human innovation. In: *Innovation in Cultural Systems*, ed. M. J. O'Brien and S. Shennan. Cambridge, MA: MIT Press, pp. 37–51.

———. 2010. Innovation in Animals. In: *Encyclopaedia of Animal Behaviour*, ed. M. D. Breed and J. Moore. Oxford, UK: Academic, pp. 150–154.

Laland, K. N., and L. R. Rendell. 2013. Cultural memory. *Current Biology* 2317: R736–R740.

Laland, K. N., and K. Williams. 1997. Shoaling generates social learning of foraging information in guppies. *Animal Behaviour* 53:1161–1169.

———. 1998. Social transmission of maladaptive information in the guppy. *Behavioral Ecology* 9:493–499.

Laland, K. N., J. R. Kendal, and R. L. Kendal. 2009. Animal culture: problems and solutions. In: *The Question of Animal Culture*, ed. K. N. Laland and B. G. Galef Jr. Cambridge, MA: Harvard University Press, pp. 174–197.

Laland, K. N., J. Kumm, and M. W. Feldman. 1995. Gene-culture coevolutionary theory: a test case. *Current Anthropology* 36:131–156.

Laland, K. N., J. Kumm, J. D. Van Horn, and M. W. Feldman. 1995. A gene-culture model of handedness. *Behavior Genetics* 25:433–445.

Laland, K. N., F. J. Odling-Smee, and M. W. Feldman. 2000. Niche construction, biological evolution and cultural change. *Behavioral and Brain Sciences* 23:131–146.

Laland, K. N., F. J. Odling-Smee, and S. Myles. 2010. How culture shaped the human genome: bringing genetics and the human sciences together. *Nature Reviews Genetics* 11:137–148.

Laland, K. N., P. J. Richerson, and R. Boyd. 1993. Animal social learning: towards a new theoretical approach. In: *Behavior and Evolution*, ed. P. H. Klopfer, P. P. Bateson, and N. S. Thompson, Vol. 10 of *Perspectives in Ethology*. Berlin, Germany: Plenum, pp. 249–277.

Laland, K. N., K. Sterelny, F. J. Odling-Smee, W.J.E. Hoppitt, and T. Uller. 2011. Cause and effect in biology revisited: Is Mayr's proximate-ultimate dichotomy still useful? *Science* 334:1512–1516.

Laland, K. N., C. Wilkins, and N. S. Clayton. 2016. The evolution of dance. *Current Biology* 26:R1–R21.

Lane, D. 2016. Innovation cascades: artefacts, organization, attributions. *Philosophical Transactions of the Royal Society of London B*, doi:10.1098/rstb.2015.0194.

Langen, T. A. 2000. Prolonged offspring dependence and cooperative breeding in birds. *Behavior Ecology* 11:367–377.

Lao, O., J. M. de Gruijter, K. van Duijn, A. Navarro, and M. Kayser. 2007. Signatures of positive selection in genes associated with human skin pigmentation as revealed from analyses of single nucleotide polymorphisms. *Annals of Human Genetics* 71:354–369.

Largiader, C. R., V. Fries, and T.C.M. Bakker. 2001. Genetic analysis of sneaking and egg-thievery in a natural population of the three-spined stickleback (*Gasterosteus aculeatus*) *Heredity* 48:459–468.

Latane, B. 1981. The psychology of social impact. American Psychologist 36:343–356.

Laubin, R., and G. Laubin. 1977. *Indian Dances of North America*. Norman, OK: University of Oklahoma Press.

Laurent, R., B. Toupance, and R. Chaix. 2012. Non-random mate choice in humans: insights from a genome scan. *Molecular Ecology* 21:587–596.

Lawson, J. 1964. *European Folk Dance*. London, UK: Pitman and Sons.

Leadbeater, E., and L. Chittka. 2007a. The dynamics of social learning in an insect model: the bumblebee (*Bombus terrestris*). *Behavioral Ecology and Sociobiology* 61:1789–1796.

———. 2007b. Social learning in insects—from miniature brains to consensus building. *Current Biology* 17:R703–R713.

Leadbeater, E., N. E. Raine, and L. Chittka. 2006. Social learning: ants and the meaning of teaching. *Current Biology* 16:R323e–R325.

Leca, J. B., N. Gunst, and M. A. Huffman. 2007. Age-related differences in the performance, diffusion, and maintenance of stone handling, a behavioral tradition in Japanese macaques. *Journal of Human Evolution* 53:691–708.

Lee, R. B. 1979. *The !Kung San: Men, Women, and Work in a Foraging Society*. Cambridge, UK: Cambridge University Press.

Lee, R. B., and R. Daly, eds. 1999. *The Cambridge Encyclopedia of Hunters and Gatherers*. Cambridge, UK: Cambridge University Press.

Lefebvre, L. 1995. The opening of milk-bottles by birds: evidence for accelerating learning rates, but against the wave-of-advance model of cultural transmission. *Behavioral Processes* 34:43–53.

———. 2010. Taxonomic counts of cognition in the wild. *Biology Letters* 7:631–633.

Lefebvre, L., S. Ducatez, and J. N. Audet. 2016. Feeding innovations in a nested phylogeny of Neotropical passerines. *Philosophical Transactions of the Royal Society of London B*, doi:10.1098/rstb.2015.0188.

Lefebvre, L., S. M. Reader, and D. Sol. 2004. Brains, innovations and evolution in birds and primates. *Brain, Behavior and Evolution* 63:233–246.

Lefebvre, L., P. Whittle, E. Lascaris, and A. Finkelstein. 1997. Feeding innovations and forebrain size in birds. *Animal Behaviour* 53:549–560.

Lehmann, L., and L. Keller. 2006. The evolution of cooperation and altruism. *Journal of Evolutionary Biology* 19:1365–1376.

Lehmann, L., K. Aoki, and M. W. Feldman. 2011. On the number of independent cultural traits carried by individuals and populations. *Philosophical Transactions of the Royal Society of London B* 366:424–435.

Lepre, C. J., H. Roche, D. V. Kent, S. Harmand, R. L. Quinn, J.-P. Brugal, P.-J. Texier, et al. 2011. An earlier origin for the Acheulian. *Nature* 477:82–85.

Lewens, T. 2015. *Cultural Evolution: Conceptual Challenges.* Oxford, UK: Oxford University Press.

Lewin, R. 1987. *Bones of Contention: Controversies in the Search for Human Origins.* London, UK: Penguin.

Lewin, R., and R. A. Foley. 2004. *Principles of Human Evolution,* 2nd ed. Cambridge, UK: Blackwell.

Lewis, H. M., and K. N. Laland. 2012. Transmission fidelity is the key to the build-up of cumulative culture. *Philosophical Transactions of the Royal Society of London B* 367:2171–2180.

Lewis, J. 2007. Ekila: blood, bodies, and egalitarian societies. *Journal of the Royal Anthropological Institute* 14:297–335.

Lewontin, R. C. 1970. The units of selection. *Annual Review of Ecology, Evolution, and Systematics* 1:1–18.

Li, H., and R. Durbin. 2011. Inference of human population history from individual whole-genome sequences. *Nature* 475:493–496.

Li, H. T., T. T. Zhang, Y. Q. Zhou, Q. H. Huang, and J. Huang. 2006. SLC11A1 (formerly NRAMP1) gene polymorphisms and tuberculosis susceptibility: a meta-analysis. *International Journal of Tuberculosis and Lung Disease* 10:3–12.

Liao, B. Y., and J. Zhang. 2008. Null mutations in human and mouse orthologs frequently result in different phenotypes. *Proceedings of the National Academy of Sciences USA* 105:6987–6992.

Linden, E. 1975. *Apes, Men and Language.* New York, NY: Bookthrift.

Little, A. C., R. P. Burriss, B. C. Jones, L. M. DeBruine, and C. Caldwell. 2008. Social influence in human face preference: men and women are influenced more for long-term than short-term attractiveness decisions. *Evolution and Human Behavior* 29:140–146.

Livingstone, F. B. 1958. Anthropological implications of sickle-cell distribution in west Africa. *American Anthropologist* 60:533–562.

Lloyd Morgan, C. 1912. *Instinct and Experience.* London, UK: Methuen.

Logan, C. J., A. J. Breen, A. H. Taylor, R. D. Gray, and W.J.E. Hoppitt. 2015. How New Caledonian crows solve novel foraging problems and what it means for cumulative culture. *Learning & Behavior* 44:18–28, doi:10.3758/s13420-015-0194-x.

Lonsdorf, E. V. 2006. What is the role of mothers in the acquisition of termite-fishing behaviors in wild chimpanzees (*Pan troglodytes schweinfurthii*)? *Animal Cognition* 9:36–46.

Lonsdorf, E. V., E. A. Pusey, and L. Eberly. 2004. Sex differences in learning in chimpanzees. *Nature* 428:715–716.

López Herráez, D., M. Bauchet, K. Tang, C. Theunert, I. Pugach, J. Li, M. Nandi-neni, et al. 2009. Genetic variation and recent positive selection in worldwide human populations: evidence from nearly 1 million SNPs. *PLOS ONE* 4:e7888, doi:org/10.1371/journal.pone.0007888.

Lorenz, K. 1966. *On Aggression*. London, UK: Routledge.

Lovejoy, P. E. 1989. The impact of the Atlantic slave trade on Africa: a review of the literature. *Journal of African History* 30:365–394.

Lu, H., J. Zhang, K.-B. Liu, N. Wu, Y. Li, K. Zhou, M. Ye, et al. 2009. Earliest domesti-cation of common millet (*Panicum miliaceum*) in East Asia extended to 10,000 years ago. *Proceedings of the National Academy of Sciences USA* 10618:7367–7372.

Lukas, D., and T. Clutton-Brock. 2012. Life histories and the evolution of coop-erative breeding in mammals. *Proceedings of the Royal Society of London B* 279:4065–4070.

Lupfer, G., J. Frieman, and D. L. Coonfield. 2003. Social transmission of flavour pref-erences in social and non-social hamsters. *Journal of Comparative Psychology* 117:449–455.

Lupfer-Johnson, G., and J. Ross. 2007. Dogs acquire food preferences from interacting with recently fed conspecifics. *Behavioural Processes* 10:104–106.

Lynch, M. 1990. The rate of morphological evolution in mammals from the standpoint of the neutral expectation. *American Naturalist* 136(6):727–741.

Lyons, D. E., D. H. Damrosch, J. K. Lin, D. M. Macris, and F. C. Keil. 2011. The scope and limits of over-imitation in the transmission of artefact culture. *Proceedings of the Royal Society B* 366:1158–1167.

Lyons, D. E., A. G. Young, and F. C. Keil. 2007. The hidden structure of overimitation. *Proceedings of the National Academy of Sciences USA* 104(5):19751–19756.

MacLean, E. L., B. Hare, C. L. Nunn, E. Addessi, F. Amici, R. Anderson, F. Aureli, et al. 2014. The evolution of self-control. *Proceedings of the National Academy of Sciences USA* 111:2140–2148.

Macphail, E. M. 1982. *Brain and Intelligence in Vertebrates*. Oxford, UK: Clarendon Press.

Macphail, E. M., and J. J. Bolhuis. 2001. The evolution of intelligence: adaptive special-izations versus general process. *Biological Review of the Cambridge Philosophical Society* 76:341–364.

Madden, J. R., C. Dingle, J. Isden, J. Sarfeld, A. Goldizen, and J. A. Endler. 2012. Male spotted bowerbirds propagate fruit for use in the sexual display. *Current Biology* 22:R264–R265.

Magalon, H., E. Patin, F. Austerlitz, T. Hegay, A. Aldashev, L. Quintana-Murci, and E. Heyer. 2008. Population genetic diversity of the NAT2 gene supports a role of acetylation in human adaptation to farming in Central Asia. *European Journal of Human Genetics* 16:243–251.

Magurran, A. E., and B. H. Seghers. 1994. A cost of sexual harassment in the guppy, *Poecilia reticulata*. *Proceedings of the Royal Society of London B* 258:89–92.

Marino, L. 2006. Absolute brain size: Did we throw the baby out with the bathwater? *Proceedings of the National Academy of Sciences USA* 103:13563–13564.

Marler, P. 1952. Variations in the song of the chaffinch, *Fringilla coelebs*. Ibis 94:458–472.

Marler, P., and S. S. Peters. 1989. *The Comparative Psychology of Audition: Perceiving Complex Sounds*, ed. S. Hulse and R. Dooling. Hillsdale, NJ: Erlbaum, pp. 243–273.

Marler, P., and M. Tamura. 1964. Culturally transmitted patterns of vocal behaviour in sparrows. *Science* 146:1483–1486.

Marlowe, F. 2001. Male contribution to diet and female reproductive success among foragers. *Current Anthropology* 42:755–760.

Marshall-Pescini, S., and A. Whiten. 2008. Chimpanzees (*Pan troglodytes*) and the question of cumulative culture: an experimental approach. *Animal Cognition* 11:449–456.

Martinez del Rio, C. 1993. Do British tits drink milk or just skim the cream? *British Birds* 86:321–322.

Masataka, N. 2003. *The Onset of Language*. Cambridge, UK: Cambridge University Press.

Mason, J. R. 1988. Direct and observational learning by red-winged blackbirds (*Agelaius phoeniceus*): the importance of complex stimuli. In: *Social Learning: Psychological and Biological Perspectives*, ed. B. G. Galef Jr. and T. R. Zentall. Hillsdale, NJ: Erlbaum, pp. 99–117.

Matsuoka, Y., Y. Vigouroux, M. M. Goodman, G. J. Sanchez, E. Buckler, and J. Doebley. 2002. A single domestication for maize shown by multilocus microsatellite genotyping. *Proceedings of the National Academy of Sciences USA* 99:6080–6084.

Matzel, L. D., Y. R. Han, H. S. Grossman, M. S. Karnik, D. Patel, N. Scott, S. M. Specht, and C. C. Gandhi. 2003. Individual differences in the expression of a 'general' learning ability in mice. *Journal of Neuroscience* 23:6423–6433.

Maynard-Smith, J. 1982. *Evolution and the Theory of Games*. Cambridge, UK: Cambridge University Press.

Maynard-Smith, J. 1991. Honest signalling: the Phillip Sidney game. *Animal Behaviour* 42:1034–1035.

MacDonald, K. 2007. Cross-cultural comparison of learning in human hunting. *Human Nature* 18:386–402.

McBrearty, S., and A. S. Brooks. 2000. The revolution that wasn't: a new interpretation of the origin of modern human behaviour. *Journal of Human Evolution* 39:453–563.

McDonagh, D. 1976. *The Complete Guide to Modern Dance*. New York, NY: Doubleday.

McFadyen-Ketchum, S. A., and R. H. Porter. 1989. Transmission of food preferences in spiny mice (*Acomys cahirinus*) via nose–mouth interaction. *Behavioral Ecology and Sociobiology* 24:59–62.

McGrew, W. C. 1992. *Chimpanzee Material Culture: Implications for Human Evolution*. Cambridge, UK: Cambridge University Press.

McGrew, W. C., and C.E.G. Tutin. 1978. Evidence for a social custom in wild chimpanzees? *Man* 13:234–251.

McManus, I. C. 1985. Handedness, language dominance and aphasia. *Psychological Medicine Monograph Supplement* 8:3–40.

McPherron, S. P., Z. Alemseged, C. W. Marean, J. G. Wynn, D. Reed, D. Geraads, R. Bobe, et al. 2010. Evidence for stone-tool-assisted consumption of animal tissues before 3.39 million years ago at Dikika, Ethiopia. *Nature* 466:857–860.

Melis, A., B. Hare, and M. Tomasello. 2006. Engineering cooperation in chimpanzees: tolerance constraints on cooperation. *Animal Behaviour* 72:276–286.

Mellars, P. 1996. *The Neanderthal Legacy*. Princeton, NJ: Princeton University Press.

Menzel, E. W., and C. R. Menzel. 1979. Cognitive, developmental and social aspects of responsiveness to novel objects in a family group of marmosets (*Saguinus fusicollis*). *Behaviour* 70:251–279.

Mery, F., S. Varela, E. Danchin, S. Blanchet, D. Parejo, I. Coolen, and R. Wagner. 2009. Public versus personal information for mate copying in an invertebrate. *Current Biology* 19:730–734.

Mesoudi, A. 2008. An experimental simulation of the 'copy-successful-individuals' cultural learning strategy: adaptive landscapes, producer-scrounger dynamics, and informational access costs. *Evolution and Human Behavior* 29:350–363.

———. 2011. *Cultural Evolution: How Darwinian Theory Can Explain Human Culture and Synthesize the Social Sciences*. Chicago, IL: University of Chicago Press.

Mesoudi, A., and M. J. O'Brien. 2008. The cultural transmission of Great Basin projectile point technology II: an agent-based computer simulation. *American Antiquity* 73(4):627–644.

Mesoudi, A., A. Whiten, and K. N. Laland. 2004. Is human cultural evolution Darwinian? Evidence reviewed from the perspective of *The Origin of Species*. *Evolution* 58:1–11.

———. 2006. Towards a unified science of cultural evolution. *Behavioural and Brain Sciences* 29:329–347.

Michel, J. B., Y. K. Shen, A. P. Aiden, A. Veres, M. K. Gray, The Google Books Team, J. P. Pickett, et al. 2011. Quantitative analysis of culture using millions of digitized books. *Science* 331(6014):176–182.

Milbrath, C. 2013. Socio-cultural selection and the sculpting of the human genome: cultures' directional forces on evolution and development. *New Ideas in Psychology* 31:390–406.

Milinski, M. 1994. Long-term memory for food patches and implications for ideal free distributions in sticklebacks. *Ecology* 75:1150–1156.

Milinski, M., D. Kulling, and R. Kettler. 1990. Tit for tat: sticklebacks (*Gasterosteus aculeatus*) 'trusting' a cooperating partner. *Behavioral Ecology* 1:7–11.

Miller, G. 2001. *The Mating Mind: How Sexual Choice Shaped the Evolution of Human Nature*. New York, NY: Anchor Books.

Mineka, S., and M. Cook. 1988. Social learning and the acquisition of snake fear in monkeys. In: *Social Learning: Psychological and Biological Perspectives*, ed. B. G. Galef Jr. and T. R. Zentall. Hillsdale, NJ: Erlbaum, pp. 51–73.

Mineka, S., and R. Zinbarg. 2006. A contemporary learning theory perspective on the etiology of anxiety disorders: it's not what you thought it was. *American Psychologist* 61:10–26.

Mineka, S., M. Davidson, M. Cook, and R. Keir. 1984. Observational conditioning of snake fear in rhesus monkeys. *Journal of Abnormal Psychology* 93:355–372.

Minsky, M. 1967. *Computation: Finite and Infinite Machines*. Upper Saddle River, NJ: Prentice Hall.

Moglich, M. 1978. Social organization of nest emigration in *Leptothorax* (Hym., Form.) *Insectes Sociaux* 25:205–225.

Moglich, M., and B. Holldobler. 1974. Social carrying behavior and division of labor during nest moving in ants. *Psyche: A Journal of Entomology* 81:219–236.

Montgomery, S. H., I. Capellini, R. A. Barton, and N. I. Mundy. 2010. Reconstructing the ups and downs of primate brain evolution: implications for adaptive hypotheses and *Homo floresiensis*. *BMC Biology* 8:9, doi:10.1186/1741-7007-8-9.

Moore, B. R. 1992. Avian movement imitation and a new form of mimicry: tracing the evolution of a complex form of learning. *Behaviour* 122:231–263.

Moore, B. R. 1996. The Evolution of imitative language. In: *Social Learning in Animals: The Roots of Culture*, ed. C. M. Heyes and B. G. Galef Jr. London, UK: Academic, pp. 245–265.

Morand-Ferron, J., L. Lefebvre, S. M. Reader, D. Sol, and S. Elvin. 2004. Dunking behaviour in Carib grackles. *Animal Behaviour* 68:1267–1274.

Morgan, C. L. 1912. *Instinct and Experience*. London, UK: Methuen.

Morgan, L. H. 1877. *Ancient Society, or Researches in the Lines of Human Progress from Savagery through Barbarism to Civilization*. New York, NY: Holt.

Morgan, M. J., and M. C. Corballis. 1978. The inheritance of laterality. *Behavioral and Brain Science* 2:270–277.

Morgan, T.J.H., and K. N. Laland. 2012. The biological bases of conformity. *Frontiers in Decision Neuroscience*, doi:10.3389/fnins.2012.00087.

Morgan, T.J.H., L. E. Rendell, M. Ehn, W. Hoppitt, and K. N. Laland. 2012. The evolutionary basis of human social learning. *Proceedings of the Royal Society B* 279:653–662.

Morgan, T.J.H., N. Uomini, L. E. Rendell, L. Chouinard-Thuly, S. E. Street, H. M. Lewis, C. P. Cross, et al. 2015. Experimental evidence for the co-evolution of hominin tool-making, teaching and language. *Nature Communications*, doi: 10.1038/ncomms7029.

Morrell, L. J., D. P. Croft, J.R.G. Dyer, B. B. Chapman, J. L. Kelley, K. N. Laland, and J. Krause. 2008. Association patterns and foraging behaviour in natural and artificial guppy shoals. *Animal Behaviour* 76:855–864.

Morris, D. 1967. *The Naked Ape*. London, UK: Vintage.

Movius, H. L., Jr. 1950. A wooden spear of third interglacial age from lower Saxony. *Southwestern Journal of Anthropology* 6:139–142.

Mueller, T., R. O'Hara, S. J. Converse, R. P. Urbanek, and W. F. Fagan. 2013. Social learning of migratory performance. *Science* 341:999–1002.

Müller, G. 2007. Evo-devo: extending the evolutionary synthesis. *Nature Review Genetics* 8:943–950.

Mundinger, P. C. 1980. Animal cultures and a general theory of cultural evolution. *Ethology and Sociobiology* 1:83–223.

Myles, S., E. Hradetzky, J. Engelken, O. Lao, P. Nürnberg, R. J. Trent, X. Wang, et al. 2007. Identification of a candidate genetic variant for the high prevalence of type II diabetes in Polynesians. *European Journal of Human Genetics* 15:584–589.

Myles, S., M. Somel, K. Tang, J. Kelso, and M. Stoneking. 2007. Identifying genes underlying skin pigmentation differences among human populations. *Human Genetics* 120:613–621.

Myles, S., K. Tanq, M. Somel, R. E. Green, J. Kelso, and M. Stoneking. 2008. Identification and analysis of high Fst regions from genome-wide SNP data from three human populations. *Annals of Human Genetics* 72:99–110.

Nasidze, I., D. Quinque, M. Rahmani, S. A. Alemohamad, and M. Stoneking. 2006. Concomitant replacement of language and mtDNA in South Caspian populations of Iran. *Current Biology* 16:668–673.

Navarette, A. F., S. M. Reader, S. E. Street, A. Whalen, and K. N. Laland. 2016. The coevolution of innovation and technical intelligence in primates. *Philosophical Transactions of the Royal Society of London B* 371:20150186, http://dx.doi .org/10.1098/rstb.2015.0186.

Neale, M. C. 1988. Handedness in a sample of volunteer twins. *Behavior Genetics* 18:69–79.

Neel, J. V. 1962. Diabetes mellitus: a "thrifty" genotype rendered detrimental by "progress"? *American Journal of Human Genetics* 14:352–362.

Nettl, B. 2000. An ethnomusicologist contemplates universals in musical sound and musical culture. In: *The Origins of Music*, ed. N. L. Wallin, B. Merker, and S. Brown. Cambridge, MA: MIT Press, pp. 463–472.

Nguyen, D.-Q., C. Webber, and C. P. Ponting. 2006. Bias of selection on human copy-number variants. *PLOS Genetics* 2:e20, doi:10.1371/journal.pgen.0020020.

Nicol, C. J., and S. J. Pope. 1996. The maternal feeding display of domestic hens is sensitive to perceived chick error. *Animal Behaviour* 52:767–774.

Nicolakakis, N., D. Sol, and L. Lefebvre. 2003. Behavioral flexibility predicts species richness in birds, but not extinction risk. *Animal Behaviour* 65:445–452.

Nielsen, R., I. Hellmann, M. Hubisz, C. Bustamante, and A. G. Clark. 2007. Recent and ongoing selection in the human genome. *Nature Reviews Genetics* 8:857–868.

Nightingale, G., N. J. Boogert, K. N. Laland, and W.J.E. Hoppitt. 2015. Quantifying diffusion on social networks: a Bayesian approach. In: *Animal Social Networks: Perspectives and Challenges*, ed. J. Krause, D. Croft, and R. James. Oxford, UK: Oxford University Press, pp. 38–52.

Nihei, Y. 1995. Variations of behaviour of carrion crows (*Corvus corone*) using automobiles as nutcrackers. *Japanese Journal of Ornithology* 44:21–35.

Niño-Mora, J. 2007. Dynamic priority allocation via restless bandit marginal productivity indices. *TOP* 15:161–198.

Noad, M. J., D. H. Cato, M. M. Bryden, M. N. Jenner, and K. C. Jenner. 2000. Cultural revolution in whale songs. *Nature* 408:537.

Nonaka, T., B. Bril, and R. Rein. 2010. How do stone knappers predict and control the outcome of flaking? Implications for understanding early stone tool technology. *Journal of Human Evolution* 59:155–167.

Nowak, M., and R. Highfield. 2011. *Super-cooperators: The Mathematics of Evolution, Altruism and Human Behaviour or Why We Need Each Other to Succeed*. London, UK: Canongate.

Nowak, M. A., and K. Sigmund. 1998. Evolution of indirect reciprocity by image scoring. *Nature* 393(6685):573–577.

Nowicki, S., and W. A. Searcy. 2014. The evolution of vocal learning. *Current Opinion in Neurobiology* 28:48–53.

Nunn, C. L. 2011. *The Comparative Approach in Evolutionary Anthropology and Biology*. Chicago, IL: University of Chicago Press.

Nunn, C. L., and C. P. van Schaik. 2002. Reconstructing the behavioural ecology of extinct primates. In: *Reconstructing Behaviour in the Primate Fossil Record*, ed. J. M. Plavcan, R. F. Kay, W. L. Jungers, and C. P. van Schaik. New York, NY: Plenum, pp. 159–199.

Oberheim, N. A., T. Takano, X. Han, W. He, J. H. Lin, F. Wang, Q. Xu, et al. 2009. Uniquely hominid features of adult human astrocytes. *Journal of Neuroscience* 29:3276–3287.

O'Brien, M. J., and K. N. Laland. 2012. Genes, culture and agriculture: an example of human niche construction. *Current Anthropology* 53:434–470.

Odling-Smee, F. J., and K. N. Laland. 2009. Cultural niche construction: evolution's cradle of language. In: *The Prehistory of Language*, ed. R. Botha and C. Knight. Oxford, UK: Oxford University Press, pp. 99–121.

Odling-Smee, F. J., K. N. Laland, and M. W. Feldman. 1996. Niche construction. *American Naturalist* 147:641–648.

———. 2003. *Niche Construction: The Neglected Process in Evolution*. Princeton, NJ: Princeton University Press.

Ohnuma, K., K. Aoki, and T. Akazawa. 1997. Transmission of tool-making through verbal and non-verbal communication-preliminary experiments in Levallois flake production. *Anthropological Science* 105:159–168.

Oikawa, S., and Y. Itazawa. 1992. Relationship between metabolic rate in vitro and body mass in a marine teleost, porgy (*Pagrus major*). *Fish Physiology and Biochemistry* 10:177–182.

Olalde, I., M. E. Allentoft, F. Sanchez-Quinto, G. Santpere, C. W. Chiang, M. DiGiorgio, J. Prado-Marinez, et al. 2014. Derived immune and ancestral pigmentation alleles in a 7,000-year-old Mesolithic European. *Nature* 507:225–228.

Olsson, A., and E. Phelps. 2007. Social learning of fear. *Nature Neuroscience* 10:1095–1102.

Onishi, K. H., and R. Baillargeon. 2005. Do 15-month-old infants understand false beliefs? *Science* 308:255–258.

Oota, H., W. Settheetham-Ishida, D. Tiwawech, T. Ishida, and M. Stoneking. 2001. Human mtDNA and Y-chromosome variation is correlated with matrilocal versus patrilocal residence. *Nature Genetics* 29:20–21.

Otto, S. P., F. B. Christiansen, and M. W. Feldman. 1995. *Genetic and Cultural Inheritance of Continuous Traits*. Morrison Institute for Population and Resource Studies Paper, no. 0064. Palo Alto, CA: Stanford University Press.

Over, H., and M. Carpenter, M. 2012. Putting the social into social learning: explaining both selectivity and fidelity in children's copying behavior. *Journal of Comparative Psychology* 126(2):182–192.

Oyama, S. 1985. *The Ontogeny of Information: Developmental Systems and Evolution*, 2nd ed. Durham, NC: Duke University Press.

Oyama, S., P. E. Griffiths, and R. D. Gray, eds. 2001. *Cycles of Contingency: Developmental Systems and Evolution*. Cambridge, MA: MIT Press.

Pagel, M. 2012. *Wired for Culture: The Natural History of Human Cooperation*. London, UK: Allen Lang.

Pagel, M., Q. D. Atkinson, and A. Meade. 2007. Frequency of word use predicts rates of lexical evolution throughout Indo-European history. *Nature* 449:717–720.

Papadimitriou, C. H., and J. N. Tsitsiklis. 1999. The complexity of optimal queuing network control. *Mathematics of Operations Research* 24:293–305.

Patel, A. D. 2006. Musical rhythm, linguistic rhythm, and human evolution. *Music Perception* 24:99–104.

Patel, A. D., J. R. Iversen, M. R. Bregman, and I. Schulz. 2009. Experimental evidence for synchronization to a musical beat in a nonhuman animal. *Current Biology* 19:827–830.

Pawlby, S. J. 1977. Imitative interaction. In: *Studies in Mother-Infant Interaction*, ed. H. Scaffer. New York, NY: Academic, pp. 203–224.

Payne, K., and R. Payne. 1985. Large scale changes over 19 years in songs of humpback whales in Bermuda. *Zeitschrift fur Tierpsychologie* 68:89–114.

Pearce-Duvet, J. M. 2006. The origin of human pathogens: evaluating the role of agriculture and domestic animals in the evolution of human disease. *Biological Reviews* 81:369–382.

Pearson, A. T. 1989. *The Teacher: Theory and Practice in Teacher Education*. London, UK: Routledge.

Peck, J. R., and M. W. Feldman. 1986. The evolution of helping behavior in large, randomly mixed populations. *American Naturalist* 127:209–221.

Pedersen, B. H. 1997. The cost of growth in young fish larvae, a review of new hypotheses. *Aquaculture* 155:259–269.

Peng, M. S., J. D. He, C. L. Zhu, S. F. Wu, J. Q. Jin, and Y. P. Zhang, et al. 2012. Lactase persistence may have an independent origin in Tibetan populations from Tibet, China. *Journal of Human Genetics* 57:394–397.

Pennisi, E. 2010. Conquering by copying. *Science* 328:165–167.

Pepperberg, I. M. 1988. The importance of social interaction and observation in the acquisition of communicative competence. In: *Social Learning: Psychological and Biological Perspectives*, ed. T. R. Zentall and B. G. Galef Jr. Hillsdale, NJ: Erlbaum, pp. 279–299.

Perreault, C. 2012. The pace of cultural evolution. *PLOS ONE* 7(9):e45150, doi:10.1371/journal.pone.0045150.

Perry, G. H., N. J. Dominy, K. G. Claw, A. S. Lee, H. Fiegler, R. Redon, J. Werner, et al. 2007. Diet and the evolution of human amylase gene copy number variation. *Nature Genetics* 39:1256–1260.

Perry, S., M. Baker, L. Fedigan, J. Gros-Louis, K. Jack, K. MacKinnon, J. Manson, et al. 2003. Social conventions in wild white-faced capuchin monkeys: evidence for traditions in a neotropical primate. *Current Anthropology* 44:241–268.

Perry, S. 2011. Social traditions and social learning in capuchin monkeys (*Cebus*). *Philosophical Transactions of the Royal Society of London B* 366:988–996.

Petraglia, M., C.B.K. Shipton, and K. Paddayya. 2005. *Hominid Individual Context in Archaeological Investigations of Lower and Middle Palaeolithic Landscapes, Locales and Artefacts*, ed. C. Gamble and M. Porr. London, UK: Routledge.

Petroski, K. 1992. *The Evolution of Useful Things*. New York, NY: Vintage Books.

Piel, F. B., A. P. Patil, R. E. Howes, O. A. Nyangiri, P. W. Gething, T. N. Williams, D. J. Weatherall, and S. I. Hay. 2010. Global distribution of the sickle cell gene and geographical confirmation of the malaria hypothesis. *Nature Communications*, doi:10.1038/ncomms1104.

Pigliucci, M., and G. B. Müller. 2010. *Evolution, The Extended Synthesis*. Cambridge, MA: MIT Press.

Pika, S., K. Liebal, J. Call, and M. Tomasello. 2005. The gestural communication of apes. *Gesture* 5(1–2):41–56.

Pike, T., and K. N. Laland. 2010. Conformist learning in nine-spined sticklebacks' foraging decisions. *Biology Letters* 64:466–468.

Pike, T. W., J. R. Kendal, L. Rendell, and K. N. Laland. 2010. Learning by proportional observation in a species of fish. *Behavioral Ecology* 20:238–244.

Pinker, S. 1995. *The Language Instinct*. New York, NY: Penguin.

———. 2010. The cognitive niche: coevolution of intelligence, sociality, and language. *Proceedings of the National Academy of Sciences USA* 107:8993–8999.

Pinker, S., and R. Jackendoff. 2005. The faculty of language: What's special about it? *Cognition* 95:201–236.

Plath, M., D. Blum, R. Tiedemann, and I. Schlupp. 2008. A visual audience effect in a cavefish. *Behaviour* 145:931–947.

PLOS Biology Synopsis. 2005. Mitochondrial DNA provides a link between Polynesians and Indigenous Taiwanese. *PLOS Biology* 38:e281, doi:10.1371/journal.pbio.0030281.

Plotkin, H. 1994. *Darwin Machines and the Nature of Knowledge*. London, UK: Penguin.

Plotnik, J. M., F.B.M. de Waal, and D. Reiss. 2006. Self-recognition in an Asian elephant. *Proceedings of the National Academy of Sciences USA* 103:17053–17057.

Posadas-Andrews, A., and T. J. Roper. 1983. Social transmission of food preferences in adult rats. *Animal Behaviour* 31:265–271.

Posnansky, M. 1969. Yams and the origins of West African agriculture. *Odu* 1:101–107.

Potts, R. 2013. Hominin evolution in settings of strong environmental variability. *Quaternary Science Reviews* 73:1–13.

Povinelli, D. J., K. E. Nelson, and S. T. Boysen. 1992. Comprehension of role reversal in chimpanzees: Evidence of empathy? *Animal Behaviour* 43:633–640.

Powell, A., S. Shennan, and M. G. Thomas. 2009. Late Pleistocene demography and the appearance of modern human behavior. *Science* 324:1298–1301.

Power, C. 1998. Old wives' tales: the gossip hypothesis and the reliability of cheap signals. In: *Approaches to the Evolution of Language*, ed. J. R. Hurford, M. Studdert-Kennedy, and C. Knight. Cambridge, UK: Cambridge University Press, pp. 111–129.

Prabhakar, S., A. Visel, J. Akiyama, M. Shoukry, K. Lewis, A. Holt, I. Plajzer-Frick, et al. 2008. Human-specific gain of function in a developmental enhancer. *Science* 321:1346–1350.

Pratt, S. C., D. Sumpter, E. Mallon, and N. Franks. 2005. An agent-based model of collective nest choice by the ant (*Temnothorax albipennis*). *Animal Behaviour* 70:1023–1036.

Premack, D. 2007. Human and animal cognition: continuity and discontinuity. *Proceedings of the National Academy of Sciences USA* 104:13861–13867.

Premack, D., and G. Woodruff. 1978. Does the chimpanzee have a theory of mind? *Behavioral and Brain Science* 1:515–526.

Pullium, H. R., and C. Dunford. 1980. *Programmed to Learn*. New York, NY: Columbia University Press.

Purves, D. 1988. *Body and Brain: A Trophic Theory of Neural Connections*. Cambridge, MA: Harvard University Press.

Purvis, A., and A. Rambaut. 1995. Comparative analysis by independent contrasts CAIC: an Apple Macintosh application for analysing comparative data. *Computer Applications in the Biosciences* 11:247–251.

Putt, S. S., A. D. Woods, and R. G. Franciscus. 2014. The role of verbal interaction during experimental bifacial stone tool manufacture. *Lithic Technology* 39:96–112.

Quach, H., L. B. Barreiro, G. Laval, N. Zidane, E. Patin, K. Kidd, J. Kidd, et al. 2009. Signatures of purifying and local positive selection in human miRNAs. *American Journal of Human Genetics* 84:316–327.

Radick, G. 2008. *The Simian Tongue: The Long Debate about Animal Language*. Chicago, IL: University of Chicago Press.

Raihani, N. J., and A. R. Ridley. 2008. Experimental evidence for teaching in wild pied babblers. *Animal Behaviour* 75:3–11.

Rakic, P. 1986. Mechanisms of ocular dominance segregation in the lateral geniculate nucleus: competitive elimination hypothesis. *Trends in Neuroscience* 9:11–15.

Ramsey, G., M. L. Bastian, and C. van Schaik. 2007. Animal innovation defined and operationalized. *Behavioral and Brain Sciences* 30:393–437.

Rapaport, L. M., and G. R. Brown. 2008. Social influences on foraging behaviour in young non-human primates: learning what, where, and how to eat. *Evolutionary Anthropology* 17:189–201.

Ratcliffe, J. M., and H. M. ter Hofstede. 2005. Roosts as information centres: social learning of food preferences in bats. *Biology Letters* 1:72–74.

Reader, S. M. 2000. Social Learning and Innovation: Individual Differences, Diffusion Dynamics and Evolutionary Issues. PhD diss., University of Cambridge.

Reader, S. M., and K. N. Laland. 2000. Diffusion of foraging innovations in the guppy. *Animal Behaviour* 60:175–180.

———. 2001. Primate innovation: sex, age and social rank differences. *International Journal of Primatology* 22:787–805.

———. 2002. Social intelligence, innovation and enhanced brain size in primates. *Proceedings of the National Academy of Sciences USA* 99:4436–4441.

———eds. 2003a. *Animal Innovation.* Oxford, UK: Oxford University Press.

———. 2003b. Animal innovation: an introduction. In: *Animal Innovation*, ed. S. M. Reader and K. N. Laland. Oxford, UK: Oxford University Press, pp. 3–35.

Reader, S. M., and L. Lefebvre. 2001. Social learning and sociality. *Behavioral and Brain Science* 24:353–355.

Reader, S. M., E. Flynn, J. Morand-Ferron, and K. N. Laland. 2016. Innovation in animals and humans: understanding the origins and development of novel and creative behavior. *Philosophical Transactions of the Royal Society B* 371(1690).

Reader, S. M., Y. Hager, and K. N. Laland. 2011. The evolution of primate general and cultural intelligence. *Philosophical Transactions of the Royal Society of London B* 366:1017–1027.

Reader, S. M., J. R. Kendal, and K. N. Laland. 2003. Social learning of foraging sites and escape routes in wild Trinidadian guppies. *Animal Behaviour* 66:729–739.

Remick, A. K., J. Polivy, and P. Pliner. 2009. Internal and external moderators of the effect of variety on food intake. *Psychological Bulletin* 135(3):434–451.

Rendell, L., and H. Whitehead. 2001. Culture in whales and dolphins. *Behavioral and Brain Sciences* 24:309–324.

———. 2015. *The Cultural Lives of Whales and Dolphins.* Chicago, IL: University of Chicago Press.

Rendell, L., R. Boyd, D. Cownden, M. Enquist, K. Eriksson, M. W. Feldman, L. Fogarty, et al. 2010. Why copy others? Insights from the social learning strategies tournament. *Science* 327:208–213.

Rendell, L., R. Boyd, M. Enquist, M. W. Feldman, L. Fogarty, and K. N. Laland. 2011. How copying affects the amount, evenness and persistence of cultural knowledge: insights from the social learning strategies tournament. *Philosophical Transactions of the Royal Society of London B* 366:1118–1128.

Rendell, L., L. Fogarty, W.J.E. Hoppitt, T.J.H. Morgan, M. Webster, and K. N. Laland. 2011. Cognitive culture: theoretical and empirical insights into social learning strategies. *Trends in Cognitive Sciences* 15:68–76.

Rendell, L., L. Fogarty, and K. N. Laland. 2011. Runaway cultural niche construction. *Philosophical Transactions of the Royal Society of London B* 366:823–835.

Reznick, D., and A. P. Yang. 1993. The influence of fluctuating resources on life-history patterns of allocation and plasticity in female guppies. *Ecology* 74:2011–2019.

Richards, M. P., R. J. Schulting, and R.E.M. Hedges. 2003. Archaeology: sharp shift in diet at onset of Neolithic. *Nature* 425:366.

Richerson, P. J. 2013. Rethinking paleoanthropology: a world queerer than we supposed. In: *Evolution of Mind, Brain and Culture*, ed. G. Hatfield and H. Pittman. Philadelphia, PA: University of Pennsylvania Press, pp. 263–302.

Richerson, P. J., and R. Boyd. 1998. The evolution of human ultrasociality. In: *Indoctrinability, Ideology, and Warfare: Evolutionary Perspectives*, ed. I. Eibl-Eibesfeldt and F. K. Salter. New York, NY: Berghahn Books, pp. 71–95.

———. 2005. *Not by Genes Alone: How Culture Transformed Human Evolution*. Chicago, IL: University of Chicago Press.

Richerson, P. J., and J. Henrich. 2012. Tribal social instincts and the cultural evolution of institutions to solve collective action problems. *Cliodynamics* 3:38–80.

Richerson, P. J., R. Baldini, A. Bell, K. Demps, K. Frost, V. Hillis, S. Mathew, et al. 2014. Cultural group selection plays an essential role in explaining human cooperation: a sketch of the evidence. *Behavioural and Brain Science* 28:1–71.

Richerson, P. J., M. Borgerhoff Mulder, and B. J. Vila. 1996. *Principles of Human Ecology*. New York, NY: Simon & Schuster.

Richerson, P., R. Boyd, and R. Bettinger. 2001. Was agriculture impossible during the Pleistocene but mandatory during the Holocene? *American Antiquity* 66: 387–411.

Richerson, P. J., R. Boyd, and J. Henrich. 2010. Gene-culture coevolution in the age of genomics. *Proceedings of the National Academy of Sciences USA* 107:8985–8992.

Riddell, W. I., and K. G. Corl. 1977. Comparative investigation of the relationship between cerebral indices and learning abilities. *Brain, Behavior and Evolution* 14:385–398.

Ridley, M. 2011. *The Rational Optimist*. New York, NY: HarperCollins.

Rilling, J. K., and T. R. Insel. 1999. The primate neocortex in comparative perspective using magnetic resonance imaging. *Journal of Human Evolution* 37:191–223.

Rilling, J. K., M. F. Glasser, T. M. Preuss, X. Ma, T. Zhao, X. Hu, and T.E.J. Behrens. 2008. The evolution of the arcuate fasciculus revealed with comparative DTI. *Nature Neuroscience* 11(4):426–428.

Rizzolatti, G., and L. Craighero. 2004. The mirror-neuron system. Annual Review of *Neuroscience* 27(1):169–192.

Roberts, C. A., and J. E. Buikstra. 2003. *The Bioarchaeology of Tuberculosis: A Global View on a Re-emerging Disease*. Gainesville, FL: University Press of Florida.

Roberts, M. J. 2007. *Integrating the Mind*. New York, NY: Psychology Press.

Roche, H., A. Delagnes, J.-P. Brugal, C. Feibel, M. Kibunjia, V. Mourre, and P.-J. Texier. 1999. Early hominid stone tool production and technical skill 2.34 myr ago in West Turkana, Kenya. *Nature* 399:57–60.

Rogers, A. 1988. Does biology constrain culture? *American Anthropologist* 90:819–813.

Rogers, E. M. 1995. *Diffusion of Innovations*, 4th ed. New York, NY: Free Press.

Rose, M. R., and G. V. Lauder. 1996. *Adaptation*. San Diego, CA: Academic.

Rowe, N. 1996. *The Pictorial Guide to the Living Primates*. New York, NY: Pogonias.

Rowley-Conwy, P., and R. Layton. 2011. Foraging and farming as niche construction: stable and unstable adaptations. *Philosophical Transactions of the Royal Society of London B* 366:849–862.

Russon, A. E. 2003. Innovation and creativity in forest-living rehabilitant orangutans. In: *Animal Innovation*, ed. S. M. Reader and K. N. Laland. Oxford, UK: Oxford University Press, pp. 279–306.

Rutz, C., L. A. Bluff, N. Reed, J. Troscianko, J. Newton, R. Inger, A. Kacelnik, and S. Bearhop. 2010. The ecological significance of tool use in New Caledonian crows. *Science* 329:1523–1526.

Rutz, C., B. C. Klump, L. Komarczyk, R. Leighton, J. Kramer, S. Wischnewski, et al. 2016. Discovery of species-wide tool use in the Hawaiian crow. *Nature* 537:403–407.

Sabbagh, A., P. Darlu, B. Crouau-Roy, and E. S. Poloni. 2011. Arylamine N-acetyltransferase 2 NAT2 genetic diversity and traditional subsistence: a worldwide population survey. *PLOS ONE* 6:e18507, doi:10.1371/journal.pone.0018507.

Sabeti, P. C., S. F. Schaffner, B. Fry, J. Lohmueller, P. Varilly, O. Shamovsky, A. Palma, et al. 2006. Positive natural selection in the human lineage. *Science* 312:1614–1620.

Sabeti, P. C., P. Varilly, B. Fry, J. Lohmueller, E. Hostetter, C. Cotsapas, X. Xie, et al. 2007. Genome-wide detection and characterization of positive selection in human populations. *Nature* 449:913–918.

Sachs, J. L., U. G. Mueller, T. P. Wilcox, and J. J. Bull. 2004. The evolution of cooperation. *Quarterly Review of Biology* 79:135–160.

Saggerson, A. L., D. N. George, and R. C. Honey. 2005. Imitative learning of stimulus response and response-outcome associations in pigeons. *Journal of Experimental Psychology: Animal Behavior Processes* 31:289–300.

Sakai, T., A. Mikami, M. Tomanaga, M. Matsui, J. Suzuki, Y. Hamada, M. Tanaka, et al. 2011. Differential prefrontal white matter development in chimpanzees and humans. *Current Biology* 21:1397–1402.

Sanz, C. M., J. Call, and D. B. Morgan. 2009. Design complexity in termite-fishing tools of chimpanzees (*Pan troglodytes*). *Biology Letters* 5:293–296.

Sargeant, B. L., and J. Mann. 2009. Developmental evidence for foraging traditions in wild bottlenose dolphins. *Animal Behaviour* 78:715–721.

Sargent, R. C., and M. R. Gross. 1993. Williams' principle: an explanation of parental care in teleost fishes. In: *Behavior of Teleost Fishes*, 2nd ed., ed. T. J. Pitcher. London, UK: Chapman and Hall, pp. 333–361.

Schachner, A., T. F. Brady, I. M. Pepperberg, and M. D. Hauser. 2009. Spontaneous motor entrainment to music in multiple vocal mimicking species. *Current Biology* 19:831–836.

Schenker, N. M., W. D. Hopkins, M. A. Spocter, A. R. Garrison, C. D. Stimpson, J. M. Erwin, P. R. Hof, and C. C. Sherwood. 2010. Broca's area homologue in chimpanzees (*Pan troglodytes*): probabilistic mapping, asymmetry and comparison to humans. *Cerebral Cortex* 20:730–742.

Schick, K., and N. Toth. 2006. *Oldowan Case Studies into Earliest Stone Age*, ed. N. Toth and K. Schick. Gosport, UK: Stone Age Institute.

Schlag, K. H. 1998. Why imitate and if so, how? A boundedly rational approach to multi-armed bandits. *Journal of Economic Theory* 78:130–156.

Schoenemann, P. T. 2006. Evolution of the size and functional areas of the human brain. *Annual Review of Anthropology* 35:379–406.

Schönholzer, L. 1958. Beobachtungen über das Trinkverhalten bei Zootieren. *Der Zoologische Garten* (N. F.) 24:345–431.

Schuppli, C., E.J.M. Meulman, S.I.F. Forss, F. Aprilinayati, M. A. van Noordwijk, and C. P. van Schaik. 2016. Observational social learning and socially induced practice of routine skills in immature wild orang-utans. *Animal Behaviour* 119:87–98.

Schuster, S., S. Wohl, M. Griebsch, and I. Klostermeier. 2006. Animal cognition: how archer fish learn to down rapidly moving targets. *Current Biology* 16:378–383.

Sealey, T. D. 2010. *Honeybee Democracy*. Princeton, NJ: Princeton University Press.

Searcy, W. A., and S. Nowicki, S. 2005. *The Evolution of Animal Communication*. Princeton, NJ: Princeton University Press.

Semendeferi, K., K. Teffer, D. P. Buxoeveden, M. S. Park, S. Bludau, K. Amunts, K. Travis, et al. 2011. Spatial organization of neurons in the frontal pole sets humans apart from great apes. *Cerebral Cortex* 21:1485–1497.

Seyfarth, R. M., and D. L. Cheney. 2000. Social awareness in the monkey. *American Zoologist* 40:902–909.

Seyfarth, R. M., D. L. Cheney, and P. Marler. 1980. Vervet monkey alarm calls: semantic communication in a free-ranging primate. *Animal Behaviour* 28:1070–1094.

Shaw, J. 2003. Who built the pyramids? *Harvard Magazine* 7:42–99.

Sherry, D. F., and B. G. Galef Jr. 1984. Cultural transmission without imitation: milk bottle opening by birds. *Animal Behaviour* 32:937–938.

———. 1990. Social learning without imitation: more about milk bottle opening by birds. *Animal Behaviour* 40:987–989.

Sherwood, C., C. D. Stimpson, M. A. Raghanti, D. E. Wildman, M. Uddin, L. Grossman, M. Goodman, et al. 2006. Evolution of increased glia-neuron ratios in the human frontal cortex. *Proceedings of the National Academy of Sciences USA* 103:13606–13611.

Shettleworth, S. J. 2001. Animal cognition and animal behavior. *Animal Behaviour* 61:277–286.

———. 2010. *Cognition, Evolution, and Behavior*, 2nd ed. New York, NY: Oxford University Press.

Shipton, C.B.K., M. Petraglia, and K. Paddayya. 2009. Stone tool experiments and reduction methods at the Acheulean site of Isampur Quarry, India. *Antiquity* 83:769–785.

Shultz, S., and R.I.M. Dunbar. 2006. Both social and ecological factors predict ungulate brain size. *Proceedings of the Royal Society of London B* 273:207–215.

Sieveking, A. 1979. *The Cave Artists*. London, UK: Thames and Hudson.

Silk, J. B. 2002. The form and function of reconciliation in primates. *Annual Review of Anthropology* 31:21–44.

Simonton, D. K. 1995. Exceptional personal influence: an integrative paradigm. *Creativity Research Journal* 8:371–376.

Slagsvold, T., and B. T. Hansen. 2001. Sexual imprinting and the origin of obligate brood parasitism in birds. *American Naturalist* 158:354–367.

Slagsvold, T., and K. L. Wiebe. 2007. Learning the ecological niche. *Proceedings of the Royal Society B* 274:19–23.

———. 2011. Social learning in birds and its role in shaping a foraging niche. *Philosophical Transactions of the Royal Society of London B* 366:969–977.

Slagsvold, T., B. T. Hansen, L. E. Johannessen, and L. T. Lifjeld. 2002. Mate choice and imprinting in birds studied by cross-fostering in the wild. *Proceedings of the Royal Society of London B* 269:1449–1455.

Slagsvold, T., K. Wigdahl Kleiven, A. Eriksen, and L. E. Johannessen. 2013. Vertical and horizontal transmission of nest site preferences in titmice. *Animal Behaviour* 85:323–328.

Slater, P.J.B., and R. F. Lachlan. 2003. Is innovation in bird song adaptive? In: *Animal Innovation*, ed. S. M. Reader and K. N. Laland. Oxford, UK: Oxford University Press, pp. 117–135.

Smit, H. 2014. *The Social Evolution of Human Nature.* Cambridge, UK: Cambridge University Press.

Smith, A. 1776. *The Wealth of Nations.* London, UK: W. Strahan.

Smith, B. D. 1998. *The Emergence of Agriculture.* New York, NY: Freeman.

———. 2001. Low level food production. *Journal of Archaeological Research* 9:1–43.

———. 2007a. The ultimate ecosystem engineers. *Science* 315:1797–1798.

———. 2007b. Niche construction and the behavioral context of plant and animal domestication. *Evolutionary Anthropology* 16:188–199.

Smith, E. A., R. Bliege Bird, and D. W. Bird. 2003. The benefits of costly signaling: Meriam turtle hunters. *Behavioral Ecology* 14:116–126.

Smith, K., and S. Kirby. 2008. Cultural evolution: implications for understanding the human language faculty and its evolution. *Philosophical Transactions of the Royal Society B* 363(1509):3591–3603.

Smith, M. E. 2009. V. Gordon Childe and the urban revolution: a historical perspective on a revolution in urban studies. *Town Planning Review* 80:3–29.

Sol, D. 2003. Behavioral flexibility: A neglected issue in the ecological and evolutionary literature? In: *Animal Innovation*, ed. S. M. Reader and K. N. Laland. Oxford, UK: Oxford University Press, pp. 63–82.

Sol, D., and L. Lefebvre. 2000. Behavioral flexibility predicts invasion success in birds introduced to New Zealand. *Oikos* 90:599–605.

Sol, D., R. P. Duncan, T. M. Blackburn, P. Cassey, and L. Lefebvre. 2005. Big brains, enhanced cognition and response of birds to novel environments. *Proceedings of the National Academy of Sciences USA* 102:5460–5465.

Sol, D., L. Lefebvre, and J. D. Rodríguez-Teijeiro. 2005. Brain size, innovative propensity and migratory behavior in temperate Palaearctic birds. *Proceedings of the Royal Society of London B* 272:1433–1441.

Sol, D., L. Lefebvre, and S. Timmermans. 2002. Behavioral flexibility and invasion success in birds. *Animal Behaviour* 63:495–502.

Sol, D., D. G. Stirling, and L. Lefebvre. 2005. Behavioral drive or behavioral inhibition in evolution: subspecific diversification in Holarctic passerines. *Evolution* 59:2669–2677.

Somel, M., X. Liu, and P. Khaitovich. 2013. Human brain evolution: transcripts, metabolites and their regulators. *Nature Reviews Neuroscience* 14:112–127.

Somel, M., R. Rohlfs, and X. Liu. 2014. Transcriptomic insights into human brain evolution: acceleration, neutrality, heterochrony. *Current Opinion in Genetics and Development* 29:110–119.

Soranzo, N., B. Bufe, P. C. Sabeti, J. F. Wilson, M. E. Weale, R. Marquerie, W. Meyerhof, et al. 2005. Positive selection on a high-sensitivity allele of the human bitter-taste receptor TAS2R16. *Current Biology* 15:1257–1265.

Spencer, H. (1855) 1870. *Principles of Psychology*, 2nd ed. London, UK: Longman.

Spocter, M. A., W. D. Hopkins, S. K. Barks, S. Bianchi, A. E. Hehmeyer, S. M. Anderson, C. D. Stimpson, et al. 2012. Neuropil distribution in the cerebral cortex differs between humans and chimpanzees. *Journal of Comparative Neurology* 520:2917–2929.

———. 1857. Progress: Its law and cause. *Westminster Review* 67:445–485.

Stanford Encyclopedia of Philosophy, http://plato.stanford.edu.

Stanford, C. B., J. Wallis, E. Mpongo, and J. Goodall. 1994. Hunting decisions in wild chimpanzees. *Behaviour* 131(1):1–18.

Stanley, E. L., R. L. Kendal, J. R. Kendal, S. Grounds, and K. N. Laland. 2008. The effects of group size, rate of turnover and disruption to demonstration on the stability of foraging traditions in fish. *Animal Behaviour* 75:565–572.

Stearns, S. C., S. G. Byars, D. R. Govindaraju, and D. Ewbank. 2010. Measuring selection in contemporary human populations. *Nature Reviews Genetics* 11:611–622.

Stedman, H. H., B. W. Kosyak, A. Nelson, D. M. Thesier, L. T. Su, D. W. Low, C. R. Bridges, et al. 2004. Myosin gene mutation correlates with anatomical changes in the human lineage. *Nature* 428:415–418.

Steele, V. 2013. *Dance and Fashion*. New Haven, CT: Yale University Press.

Stefansson, H., A. Helgason, G. Thorliefsson, V. Steinthorsdotti, G. Masson, J. Barnard, A. Baker, et al. 2005. A common inversion under selection in Europeans. *Nature Genetics* 37:129–137.

Steiniger, von, F. 1950. Beitrage zur Sociologie und sonstigen Biologie der Wanderratte. *Zeitschrift fur Tierpsychologie* 7:356–379.

Stel, M., J. Blascovich, C. McCall, J. Mastop, R. B. van Baaren, and R. Vonk. 2010. Mimicking disliked others: effects of *a priori* liking on the mimicry-liking link. *European Journal of Social Psychology* 40:867–880.

Stephan, H., H. Frahm, and G. Baron. 1981. New and revised data on volume of brain structures in insectivores and primates. *Folia Primatologica* 35:1–29.

Stephens, D. 1991. Change, regularity and value in the evolution of learning. *Behavioral Ecology* 2:77–89.

Sterelny, K. 2012a. *The Evolved Apprentice*. Cambridge, MA: MIT Press.

———. 2012b. Language, gesture, skill: the co-evolutionary foundations of language. *Philosophical Transactions of the Royal Society B* 367:2141–2151.

Stoeger, A. S., D. Mietchen, S. Oh, S. de Silva, C. T. Herbst, S. Kwon, and W. T. Fitch. 2012. An Asian elephant imitates human speech. *Current Biology* 22:2144–2148.

Stout, D. 2002. Skill and cognition in stone tool production: an ethnographic case study from Irian Jaya. *Current Anthropology* 43:693–723.

Stout, D., S. Semaw, S., M. J. Rogers, and D. Cauche. 2010. Technological variation in the earliest Oldowan from Gona, Afar, Ethiopia. *Journal of Human Evolution* 58:474–491.

Stout, D., N. Toth, K. Schick, J. Stout, and G. Hutchins. 2000. Stone tool-making and brain activation: position emission tomography PET studies. *Journal of Archaeological Science* 27:1215–1223.

Strandberg-Peshkin, A., D. R. Farine, I. D. Couzin, and M. C. Crofoot. 2015. Shared decision-making drives collective movement in wild baboons. *Science* 348(6241):1358–1361.

Street, S. 2014. Phylogenetic Comparative Investigations of Sexual Selection and Cognitive Evolution in Primates. PhD diss., University of St Andrews.

Striedter, G. F. 2005. *Principles of Brain Evolution*. Sunderland, MA: Sinauer.

Strimling, P., J. Sjostrand, M. Enquist, and K. Eriksson. 2009. Accumulation of independent cultural traits. *Theoretical Population Biology* 76:77–83.

Stringer, C., and P. Andrews. 2005. *The Complete World of Human Evolution*. London, UK: Thames and Hudson.

Su, C. H., P. H. Kuo, C.C.H. Lin, and W. J. Chen. 2005. A school-based twin study of handedness among adolescents in Taiwan. *Behavior Genetics* 35:723–733.

Suddendorf, T. 2013. *The Gap: The Science That Separates Us from Other Animals*. New York, NY: Basic Books.

Suddendorf, T., and M. C. Corballis. 2007. The evolution of foresight: What is mental time travel and is it unique to humans? *Behavioral and Brain Sciences* 30:299–313, and discussion, pp. 313–315.

Swaddle, J. P., M. G. Cathey, M. Correll, and B. P. Hodkinson. 2005. Socially transmitted mate preferences in a monogamous bird: a non-genetic mechanism of sexual selection. *Proceedings of the Royal Society of London B* 272:1053–1058.

Swaney, W., J. R. Kendal, H. Capon, C. Brown, and K. N. Laland. 2001. Familiarity facilitates social learning of foraging behaviour in the guppy. *Animal Behaviour* 62:591–598.

Szamado, S., and E. Szathmary. 2006. Selective scenarios for the emergence of natural language. *Trends in Ecology and Evolution* 21:555–561.

Tang, K., K. R. Thornton, and M. Stoneking. 2007. A new approach for using genome scans to detect recent positive selection in the human genome. *PLOS Biology* 5:e171, doi:10.1371/journal.pbio.0050171.

Tanner, R., R. Ferraro, T. L. Chartrand, J. R. Bettman, and R. van Baaren. 2008. Of chameleons and consumption: the impact of mimicry on choice and preferences. *Journal of Consumer Research* 34:754–766.

Tarr, B., J. Launay, and R.I.M. Dunbar. 2014. Music and social bonding: "self-other" merging and neurohormonal mechanisms. *Frontiers in Psychology* 5:1096, doi:10.3389/fpsyg.2014.01096.

Tattersall, I. 1995. *The Fossil Trail*. Oxford, UK: Oxford University Press.

Taylor, J. 2009. *Not a Chimp: The Hunt to Find the Genes That Make Us Human*. Oxford, UK: Oxford University Press.

Tehrani J. J., and F. Riede. 2008. Towards an archaeology of pedagogy: learning, teaching and the generation of material culture traditions. *World Archaeology* 40:316–331.

Tellier, L. N. 2009. *Urban World History*. Québec City, Québec, Canada: Presses de l'Université du Québec.

Templeton, J. J., and A. Giraldeau. 1996. Vicarious sampling: the use of personal and public information by starlings in a simple patchy environment. *Behavioral Ecology and Sociobiology* 38:105–114.

Teng, E. L., P. Lee, P. C. Yang, and P. C. Chang. 1976. Handedness in a Chinese population: biological, social and pathological factors. *Science* 193:1148–1150.

Tennie, C., J. Call, and M. Tomasello. 2009. Ratcheting up the ratchet: on the evolution of cumulative culture. *Philosophical Transactions of the Royal Society B* 364:2405–2415.

Terrace, H. S. 1979. *How Nim Chimpsky Changed My Mind*. San Francisco, CA: Ziff-Davis.

Texier, P.-J., G. Poraraz, J. E. Parkington, J. P. Rigaud, C. Poggenpoel, C. Miller, C. Tribolo, et al. 2010. A Howiesons Poort tradition of engraving ostrich eggshell containers dated to 60,000 years ago at Diepkloof Rock Shelter, South Africa. *Proceedings of the National Academy of Sciences USA* 107:6180–6185.

Thieme, H. 1997. Lower Palaeolithic hunting spears from Germany. *Nature* 385:807–810.

Thiessen, E. D., E. Hill, and J. R. Saffran. 2005. Infant-directed speech facilitates word segmentation. *Infancy* 7:53–71.

Thorndike, E. L. 1898. Animal intelligence: An experimental study of the associative processes in animals. *The Psychological Review, Series of Monograph Supplements*, Vol. 2, No. 4. New York, NY: Macmillan.

Thornton, A. 2007. Early body condition, time budgets and the acquisition of foraging skills in meerkats. *Animal Behaviour* 75:951–962.

Thornton, A., and K. McAuliffe. 2006. Teaching in wild meerkats. *Science* 313:227–229.

Thornton, A., and N. J. Raihani. 2008. The evolution of teaching. *Animal Behaviour* 75:1823–1836.

Thornton, A., J. Samson, and T. Clutton-Brock. 2010. Multi-generational persistence of traditions in neighbouring meerkat groups. *Proceedings of the Royal Society of London B* 277:3623–3629.

Thorpe, W. H. 1956. *Learning and Instinct in Animals*. London: Methuen.

Tilman, D. 1982. *Resource Competition and Community Structure*. Princeton, NJ: Princeton University Press.

Tomasello, M. 1994. The question of chimpanzee culture. In *Chimpanzee Cultures,* ed. R. Wrangham, W. McGrew, F. de Waal, and P. Heltne. Cambridge, MA: Harvard University Press, pp. 301–317.

———. 1999. *The Cultural Origins of Human Cognition.* Cambridge, MA: Harvard University Press.

———. 2008. *Origins of Human Communication.* Cambridge, MA: MIT Press.

———. 2009. *Why We Cooperate.* Cambridge, MA: MIT Press.

———. 2010. Human culture in evolutionary perspective. In: *Advances in Culture and Psychology*, ed. M. J. Gelfand, C. Chui, and Y. Hong. Oxford, UK: Oxford University Press, pp. 5–51.

Tomasello, M., and J. Call. 1997. *Primate Cognition.* New York, NY: Oxford University Press.

Tomasello, M., M. Carpenter, J. Call, T. Behne, and H. Moll. 2005. Understanding and sharing intentions: the origins of cultural cognition. *Behavioral and Brain Sciences* 28:675–735.

Tomasello, M., B. Hare, H. Lehmann, and J. Call. 2007. Reliance on head versus eyes in the gaze following of great apes and human infants: the cooperative eye hypothesis. *Journal of Human Evolution* 52:314–320.

Toth, N. 1985. Archaeological evidence for preferential right handedness in the Lower and Middle Pleistocene, and its possible implications. *Journal of Human Evolution* 14:607–614.

———. 1987. Behavioral inferences from early stone artifact assemblages: an experimental model. *Journal of Human Evolution* 16:763–787.

Trejaut, J. A., T. Kivisild, J. H. Loo, C. L. Lee, C. L. He, C. J. Hsu, Z. Y. Li, et al. 2005. Traces of archaic mitochondrial lineages persist in Austronesian-speaking Formosan populations. *PLOS Biology* 38:e247, doi:10.1371/journal.pbio.0030247.

Trivers, R. L. 1971. The evolution of reciprocal altruism. *Quarterly Review of Biology* 46:35–57.

———. 1972. Parent investment and sexual selection. In: *Sexual Selection and the Descent of Man: 1871–1971*, ed. B. Campbell. Chicago, IL: Aldine, pp. 136–179.

———. 1974. Parent-offspring conflict. *American Zoologist* 14:249–264.

Truswell, R. 2015. Dendrophobia in bonobo comprehension of spoken English. Paper presented at the 11th International Conference for The Evolution of Language, New Orleans, LA, March 2016, http://evolang.org/neworleans/papers/87.html.

Turing, A. M. 1937. On computable numbers, with an application to the Entscheidungsproblem. *Proceedings of the London Mathematical Society*, 2nd ser., 42:230–265.

Twigg, G. 1975. *The Brown Rat.* New Pomfret, VT: David and Charles.

Tylor, E. B. 1871. *Primitive Culture: Researches into the Development of Mythology, Philosophy, Religion, Art, and Custom,* 2 vols. London, UK: John Murray.

Uddin, M., D. E. Wildman, G. Liu, W. Xu, R. M. Johnson, P. R. Hof, G. Kapatos, et al. 2004. Sister grouping of chimpanzees and humans as revealed by genome-wide phylogenetic analysis of brain gene expression profiles. *Proceedings of the National Academy of Sciences USA* 101:2957–2962.

Uller, T. 2012. Parental effects in development and evolution. In: *Evolution of Parental Care*, ed. N. J. Royle, P. Smiseth, and M. Kölliker. Oxford, UK: Oxford University Press.

Uomini, N. T. 2009. The prehistory of handedness: archaeological data and comparative ethology. *Journal of Human Evolution* 57:411–419.

Uomini, N. T. 2011. Handedness in Neanderthals. In: *Neanderthal Lifeways, Subsistence and Technology*, ed. N. J. Conard and J. Richter. Heidelberg, Germany: Springer, pp. 139–154.

Uomini, N. T., and G. F. Meyer. 2013. Shared brain lateralization patterns in language and Acheulean stone tool production: a functional transcranial Doppler ultrasound study. *PLOS ONE* 8(8):e72693, doi:10.1371/journal.pone.0072693.

Valsecchi, P., and B. G. Galef. 1989. Social influences on the food preferences of house mice (*Mus musculus*). *International Journal of Comparative Psychology* 2: 245–256.

van Baaren, R. B., R. W. Holland, K. Kawakami, and A. Van Knippenberg. 2004. Mimicry and prosocial behavior. *Psychological Science* 15:71–74.

van Baaren, R. B., L. Janssen, T. L. Chartrand, A. Dijksterhuis, et al. 2009. Where is the love? The social aspects of mimicry. *Philosophical Transactions of the Royal Society of London B* 364:2381–2389.

van Bergen, Y. 2004. An Investigation into the Adaptive Use of Social and Asocial Information. PhD diss., University of Cambridge.

van Bergen, Y., I. Coolen, and K. N. Laland. 2004. Ninespined sticklebacks exploit the most reliable source when public and private information conflict. *Proceedings of the Royal Society of London B* 271:957–962.

van der Maas, H.L.J., C. V. Dolan, R.P.P.P. Grasman, J. M. Wicherts, H. M. Huizenga, and M.E.J. Raijmakers. 2006. A dynamical model of general intelligence: the positive manifold of intelligence by mutualism. *Psychological Review* 113: 842–861.

van der Post, D. J., and P. Hogeweg. 2009. Cultural inheritance and diversification of diet in variable environments. *Animal Behaviour* 78:155–166.

van Schaik, C. P., R. O. Deaner, and M. Y. Merrill. 1999. The conditions for tool use in primates: implications for the evolution of material culture. *Journal of Human Evolution* 36:719–741.

van Schaik, C. P. 2009. Geographic variation in the behavior of wild great apes: Is it really Cultural? In: *The Question of Animal Culture*, ed. K. N. Laland and B. G. Galef. Cambridge, UK: Cambridge University Press, pp. 70–98.

van Schaik, C. P., and J. M. Burkart. 2011. Social learning and evolution: the cultural intelligence hypothesis. *Philosophical Transactions of the Royal Society of London B* 366:1008–1016.

van Schaik, C. P., M. A. van Noordwijk, and S. A. Wich. 2003. Innovation in wild Bornean orangutans (*Pongo pygmaeus wurmbii*). *Behaviour* 143:839–876.

Van Swol, L. M. 2003. The effects of nonverbal mirroring on perceived persuasiveness, agreement with an imitator, and reciprocity in a group discussion. *Communication Research* 304:461–480.

REFERENCES **437**

Vigilant, L., M. Stoneking, H. Harpending, K. Hawkes, and A. C. Wilson. 1991. African populations and the evolution of human mitochondrial DNA. *Science* 253:1503–1507.

Vitousek, P. M., H. A. Mooney, J. Lubchenko, and J. M. Mellilo. 1997. Human domination of earth's ecosystems. *Science* 277:494–499.

Voight, B. F., S. Kudaravalli, X. Wen, and J. K. Pritchard. 2006. A map of recent positive selection in the human genome. *PLOS Biology* 4:e72, doi:10.1371/journal.pbio.0040072.

von Frisch, K. 1967. *The Dance Language and Orientation of Bees*. Cambridge, MA: Harvard University Press.

Vygotsky, L. (1934) 1986. *Thought and Language*. Cambridge, MA: MIT Press.

Wakano, J. Y., K. Aoki, and M. W. Feldman. 2004. Evolution of social learning: a mathematical analysis. *Theoretical Population Biology* 66:249–258.

Walden, T. A., and T. A. Ogan. 1988. The development of social referencing. *Child Development* 59(5):1230–1240.

Wallace, A. R. 1869. Geological climates and the origin of species. *Quarterly Review* 126:359–394.

Wallman, J. 1992. *Aping Language*. New York, NY: Cambridge University Press.

Wang, E. T., G. Kodama, P. Baldi, and R. K. Moyzis. 2006. Global landscape of recent inferred Darwinian selection for *Homo sapiens*. *Proceedings of the National Academy of Sciences USA* 103:135–140.

Warner, R. R. 1988. Traditionality of mating-site preferences in a coral reef fish. *Nature* 335:719–721.

Warner, R. R. 1990. Male versus female influences on mating-site determination in a coral-reef fish. *Animal Behaviour* 39:540–548.

Warren, D. M., M. Stern, R. Duggirala, T. D. Dyer, and L. Almasy. 2006. Heritability and linkage analysis of hand, foot, and eye preference in Mexican Americans. *Laterality* 11:508–524.

Washburn, S. L., and C. Lancaster. 1968. The evolution of hunting. In: *Man the Hunter*, ed. R. B. Lee and I. DeVore. Venice, Italy: Aldine, pp. 293–303.

Waters, C. N., J. Zalasiewicz, C. Summerhayes, A. D. Barnosky, C. Poirier, A. Gałuszka, A. Cearreta, et al. 2016. The Anthropocene is functionally and stratigraphically distinct from the Holocene. *Science* 351(6269):aad2622, doi:10.1126/science.aad2622.

Waterson, N. 1978. *The Development of Communication*. Chichester, UK: Wiley.

Watson, S. K., S. W. Townsend, A. M. Schel, C. Wilke, E. K. Wallace, L. Cheng, L. West, and K. E. Slocombe. 2014. Vocal learning in the functionally referential food grunts of chimpanzees. *Current Biology* 25:495–499, doi:org/10.1016/j.cub.2014.12.032.

Weatherall, D., O. Akinyanju, S. Fucharoen, N. Olivieri, and P. Musgrove. 2006. In: *Disease Control Priorities in Developing Countries*, ed. D. T. Jamison, J. G. Breman, A. R. Measham, G. Alleyne, M. Claeson, D. B. Evans, P. Jha, et al. Oxford, UK: Oxford University Press, pp. 663–680.

Webster, M. M., and K. N. Laland. 2008. Social learning strategies and predation risk: minnows copy only when using private information would be costly. *Proceedings. of the Royal Society of London B* 275:2869–2876.

————. 2010. Reproductive state affects reliance on public information in stickle-backs. *Proceedings of the Royal Society Series B* 278:619–627, doi:10.1098/rspb .2010.1562.

————. 2012. Social information, conformity and the opportunity costs paid by forag-ing fish. *Behavioral Ecology and Sociobiology* 66:797–809, doi:10.1007/s00265 –012–1328–1.

————. 2013. The learning mechanism underlying public information use in ninespine sticklebacks (*Pungitius pungitius*). *Journal of Comparative Psychology* 127:154–165.

Webster, M. M., E. L. Adams, E. L., and K. N. Laland. 2008. Diet-specific chemical cues influence association preferences and patch use in a shoaling fish. *Animal Behaviour* 76:17–23.

Webster, M. M., N. Atton, W. Hoppitt, and K. N. Laland. 2013. Environmental com-plexity influences association network structure and network-based diffusion of foraging information in fish shoals. *American Naturalist* 181:235–244.

Webster, M. M., A.J.W. Ward, and P.J.B. Hart. 2008. Shoal and prey patch choice by co-occurring fish and prawns: inter-taxa use of socially transmitted cues. *Proceedings of the Royal Society of London B* 275:203–208.

Wen, N., P. A. Herrman, and C. H. Legare. 2016. Ritual increases children's affiliation with in-group members. *Evolution and Human Behavior* 37:54–60.

West, S. A., A. S. Griffin, and A. Gardner. 2007. Social semantics: altruism, cooper-ation, mutualism, strong reciprocity and group selection. *Journal of Evolutionary Biology* 20:415–432.

West, S. A., C. El Mouden, and A. Gardner. 2011. Sixteen common misconceptions about the evolution of cooperation in humans. *Evolution and Human Behavior* 32:231–262.

West-Eberhard, M. J. 2003. *Developmental Plasticity and Evolution*. Oxford, UK: Oxford University Press.

Whalen, A., D. Cownden, and K. N. Laland. 2015. The learning of action sequences through social transmission. *Animal Cognition* 18:1093–1103. doi:10.1007/s10071 –015–0877–x.

Wheeler, B. C., and J. Fischer. 2012. Functionally referential signals: a promising par-adigm whose time has passed. *Evolutionary Anthropology* 21:195–205.

————. 2015. The blurred boundaries of functional reference: a response to Scaran-tion & Clay. *Animal Behaviour* 100:e9–e13, doi:10.1016/j.anbehav.2014.11.007.

White, D. J. 2004. Influences of social learning on mate-choice decisions. *Learning and Behavior* 32:105–113.

White, D. J., and B. G. Galef. 2000. 'Culture' in quail: social influences on mate choices of female *Coturnix japonica*. *Animal Behaviour* 59:975–979.

Whitehead, H. 1998. Cultural selection and genetic diversity in matrilineal whales. *Science* 282:1708–1711.

Whitehead, H., and L. Rendell. 2015. *The Cultural Lives of Whales and Dolphins*. Chi-cago, IL: University of Chicago Press.

Whiten, A. 1998. Imitation of the sequential structure of actions by chimpanzees (*Pan troglodytes*). *Journal of Comparative Psychology* 112:270–281.

————. 2011. The scope of culture in chimpanzees, humans and ancestral apes. *Philosophical Transactions of the Royal Society of London B* 366:997–1007.

Whiten, A., and R. W. Byrne. 1997. *Machiavellian Intelligence II. Extensions and Evaluations.* Cambridge, UK: Cambridge University Press.

Whiten, A., and D. Custance. 1996. Studies of imitation in chimpanzees and children. In: *Social Learning in Animals: The Roots of Culture*, ed. C. M. Heyes and B. G. Galef Jr. San Diego, CA: Academic, pp. 291–318.

Whiten, A., and D. Erdal. 2012. The human socio-cognitive niche and its evolutionary origins. *Philosophical Transactions of the Royal Society of London B* 367:2119–2129.

Whiten, A., and C. P. van Schaik. 2007. The evolution of animal 'cultures' and social intelligence. *Philosophical Transactions of the Royal Society of London B* 363:603–620.

Whiten, A., J. Goodall, W. C. McGrew, T. Nishida, V. Reynolds, Y. Sugiyama, C.E.G. Tutin, et al. 1999. Cultures in chimpanzees. *Nature* 399:682–685.

Whiten, A., J. Goodall, W. C. McGrew, T. Nishida, V. Reynolds, Y. Sugiyama, C.E.G. Tutin, et al. 2001. Charting cultural variation in chimpanzees. *Behaviour* 138:1481–1516.

Whiten, A., R. Hinde, K. N. Laland, and C. Stringer. 2011. Introduction. Discussion Meeting issue 'Culture Evolves,' ed. A. Whiten, R. A. Hinde, C. B. Stringer, and K. N. Laland. *Philosophical Transactions of the Royal Society of London B* 366:938–948.

Whiten, A., V. Horne, and S. Marchall-Pescini. 2003. Cultural panthropology. *Evolutionary Anthropology* 12:92–105.

Whiten, A., N. McGuigan, S. Marshall-Pescini, and L. M. Hopper. 2009. Emulation, imitation, over-imitation and the scope of culture for child and chimpanzee. *Philosophical Transactions of the Royal Society B* 364:2417–2428.

Whiten, A., A. Spiteri, V. Horner, K. E. Bonnie, S. P. Lambeth, S. J. Schapiro, and F.B.M. de Waal. 2007. Transmission of multiple traditions within and between chimpanzee groups. *Current Biology* 17:1038–1043.

Williamson, S. H., M. J. Hubisz, A. G. Clark, B. A. Payseur, C. D. Bustamante, and R. Nielsen. 2007. Localizing recent adaptive evolution in the human genome. *PLOS Genetics* 3:e90, doi:10.1371/journal.pgen.0030090.

Wilson, A. C. 1985. The molecular basis of evolution. *Scientific American* 253:148–157.

————. 1991. From molecular evolution to body and brain evolution. In: *Perspectives on Cellular Regulation: From Bacteria to Cancer*, ed. J. Campisi and A. B. Pardee. New York, NY: John Wiley/A. R. Liss, pp. 331–340.

Wilson, E. O. 1975. *Sociobiology: The New Synthesis.* Cambridge, MA: Harvard University Press.

————. 1978. *On Human Nature.* Cambridge, MA: Harvard University Press.

Wingfield, J. C., S. E. Lynn, and K. K. Soma. 2001. Avoiding the 'costs' of testosterone: ecological bases of hormone-behaviour interactions. *Brain Behavior and Evolution* 57:239–251.

Winterhalder, B., and D. Kennett. 2006. Behavioral ecology and the transition from hunting and gathering to agriculture. In: *Behavioral Ecology and the Transition to Agriculture*, ed. D. Kennett and B. Winterhalder. Berkeley, CA: University of California Press, pp. 1–21.

Winterhalder, B., and E. A. Smith. 2000. Analysing adaptive strategies: human behavioral ecology at twenty-five. *Evolutionary Anthropology* 9:51–72.

Wisenden, B. D., D. P. Chivers, and R.J.F. Smith. 1997. Learned recognition of predation risk by damselfly larvae on the basis of chemical cues. *Journal of Chemical Ecology* 23:137–151.

Witte, K., and R. Massmann. 2003. Female sailfin mollies, *Poecilia latipinna*, remember males and copy the choice of others after 1 day. *Animal Behaviour* 65:1151–1159.

Witte, K., and M. J. Ryan. 2002. Mate choice in the sailfin molly, *Poecilia latipinna*, in the wild. *Animal Behaviour* 63:94–949.

Wollstonecroft, M. 2011. Investigating the role of food processing in human evolution: a niche construction approach. *Archaeological and Anthropological Sciences* 3:141–150.

Wolpert, D. M., R. C. Miall, and M. Kawato. 1998. Internal models in the cerebellum. *Trends in Cognitive Sciences* 2:338–347.

Wood, D., J. S. Bruner, and G. Ross. 1976. The role of tutoring in problem solving. *Child Psychology and Psychiatry* 17:89–100.

Wood, L., R. L. Kendal, and E. Flynn. 2012. Context dependent model-based biases in cultural transmission: children's imitation is affected by model age over model knowledge state. *Evolution and Human Behavior* 104:367–381.

Wright, H. E., Jr. 1977. Environmental change and the origin of agriculture in the Old and New Worlds. In: *Origins of Agriculture*, ed. C. A. Reed. The Hague, Netherlands: Mouton, pp. 281–318.

Wyles, J. S., J. G. Kunkel, and A. C. Wilson. 1983. Birds, behavior, and anatomical evolution. *Proceedings of the National Academy of Sciences USA* 80:4394–4397.

Wynn, T., A. Hernandez-Aguilar, L. F. Marchant, and W. C. McGrew. 2011. "An ape's view of the Oldowan" revisited. *Evolutionary Anthropology* 20:181–197.

Yabar, Y., L. Johnston, L. Miles, and V. Peace. 2006. Implicit behavioral mimicry: investigating the impact of group membership. *Journal of Nonverbal Behavior* 30:97–113.

Yang, C. 2013. Ontogeny and phylogeny of language. *Proceedings of the National Academy of Sciences USA* 110:6323–6327.

Yeakel, J. D., M. M. Pires, L. Rudolf, N. J. Dominy, P. L. Koch, P. R. Guimarães Jr., and T. Gross. 2014. Collapse of an ecological network in ancient Egypt. *Proceedings of the National Academy of Sciences USA* 110:14472–14477.

Young, H. G. 1987. Herring gull preying on rabbits. *British Birds* 80:630.

Zaidel, D. W. 2013. Cognition and art: the current interdisciplinary approach. *WIREs Cognitive Science* 4:431–439.

Zeder, M. A. 2012. The broad spectrum revolution at 40: resource diversity, intensification and an alternative to optimal foraging explanations. *Journal of Anthropological Archaeology* 31(3):241–264.

Zeder, M. A. 2016. Domestication as a model system for niche construction theory. *Evolutionary Ecology* 30:325–348.

Zeder, M. A., D. G. Bradley, E. Emshwiller, and B. D. Smith, eds. 2006. *Documenting Domestication: New Genetic and Archaeological Paradigms*. Berkeley, CA: University of California Press.

Zentall, T. R., and B. G. Galef, eds. 1988. *Social Learning: Psychological and Biological Perspectives*. London, UK: Erlbaum.

Ziman, J. 2000. *Technological Evolution as an Evolutionary Process*. Cambridge, UK: Cambridge University Press.

Zuberbuhler, K. 2005. The phylogenetic roots of language—evidence from primate communication and cognition. *Current Directions in Psychological Science* 14: 126–130.

INDEX